Undergraduate Texts in Mathematics

Undergraduate Texts in Mathematics

Abbott: Understanding Analysis.
Anglin: Mathematics: A Concise History and Philosophy.
Readings in Mathematics.
Anglin/Lambek: The Heritage of Thales.
Readings in Mathematics.
Apostol: Introduction to Analytic Number Theory. Second edition.
Armstrong: Basic Topology.
Armstrong: Groups and Symmetry.
Axler: Linear Algebra Done Right. Second edition.
Beardon: Limits: A New Approach to Real Analysis.
Bak/Newman: Complex Analysis. Second edition.
Banchoff/Wermer: Linear Algebra Through Geometry. Second edition.
Beck/Robins: Computing the Continuous Discretely
Berberian: A First Course in Real Analysis.
Bix: Conics and Cubics: A Concrete Introduction to Algebraic Curves.
Brèmaud: An Introduction to Probabilistic Modeling.
Bressoud: Factorization and Primality Testing.
Bressoud: Second Year Calculus.
Readings in Mathematics.
Brickman: Mathematical Introduction to Linear Programming and Game Theory.
Browder: Mathematical Analysis: An Introduction.
Buchmann: Introduction to Cryptography. Second Edition.
Buskes/van Rooij: Topological Spaces: From Distance to Neighborhood.
Callahan: The Geometry of Spacetime: An Introduction to Special and General Relavitity.
Carter/van Brunt: The Lebesgue– Stieltjes Integral: A Practical Introduction.
Cederberg: A Course in Modern Geometries. Second edition.
Chamber-Loir: A Field Guide to Algebra
Childs: A Concrete Introduction to Higher Algebra. Second edition.
Chung/AitSahlia: Elementary Probability Theory: With Stochastic Processes and an Introduction to Mathematical Finance. Fourth edition.
Cox/Little/O'Shea: Ideals, Varieties, and Algorithms. Second edition.
Croom: Basic Concepts of Algebraic Topology.
Cull/Flahive/Robson: Difference Equations. From Rabbits to Chaos
Curtis: Linear Algebra: An Introductory Approach. Fourth edition.
Daepp/Gorkin: Reading, Writing, and Proving: A Closer Look at Mathematics.
Devlin: The Joy of Sets: Fundamentals of Contemporary Set Theory. Second edition.
Dixmier: General Topology.
Driver: Why Math?
Ebbinghaus/Flum/Thomas: Mathematical Logic. Second edition.
Edgar: Measure, Topology, and Fractal Geometry.
Elaydi: An Introduction to Difference Equations. Third edition.
Erdős/Surányi: Topics in the Theory of Numbers.
Estep: Practical Analysis on One Variable.
Exner: An Accompaniment to Higher Mathematics.
Exner: Inside Calculus.
Fine/Rosenberger: The Fundamental Theory of Algebra.
Fischer: Intermediate Real Analysis.
Flanigan/Kazdan: Calculus Two: Linear and Nonlinear Functions. Second edition.
Fleming: Functions of Several Variables. Second edition.
Foulds: Combinatorial Optimization for Undergraduates.
Foulds: Optimization Techniques: An Introduction.
Franklin: Methods of Mathematical Economics.
Frazier: An Introduction to Wavelets Through Linear Algebra.
Gamelin: Complex Analysis.
Ghorpade/Limaye: A Course in Calculus and Real Analysis
Gordon: Discrete Probability.
Hairer/Wanner: Analysis by Its History.
Readings in Mathematics.
Halmos: Finite-Dimensional Vector Spaces. Second edition.
Halmos: Naive Set Theory.
Hämmerlin/Hoffmann: Numerical Mathematics.
Readings in Mathematics.
Harris/Hirst/Mossinghoff: Combinatorics and Graph Theory.
Hartshorne: Geometry: Euclid and Beyond.
Hijab: Introduction to Calculus and Classical Analysis.
Hilton/Holton/Pedersen: Mathematical Reflections: In a Room with Many Mirrors.
Hilton/Holton/Pedersen: Mathematical Vistas: From a Room with Many Windows.
Iooss/Joseph: Elementary Stability and Bifurcation Theory. Second Edition.

(*continued after index*)

Matthias Beck   Sinai Robins

Computing the Continuous Discretely

Integer-Point Enumeration in Polyhedra

Matthias Beck
Mathematics Department
San Francisco State University
San Francisco, CA 94132
USA
beck@math.sfsu.edu

Sinai Robins
Department of Mathematics
Temple University
Philadelphia, PA 19122
USA
srobins@math.temple.edu

ISBN-13: 978-0-387-29139-0 eISBN-13: 978-0-387-46112-0

Library of Congress Control Number: 2006931787

Mathematics Subject Classifications (2000): 05Axx 11Dxx 11Pxx 11Hxx 52Bxx 52Cxx 68Rxx

Printed on acid-free paper.

9 8 7 6 5 4 3 2

springer.com

To Tendai

To my mom, Michal Robins

with all our love.

Preface

The world is continuous, but the mind is discrete.

David Mumford

We seek to bridge some critical gaps between various fields of mathematics by studying the interplay between the continuous volume and the discrete volume of polytopes. Examples of polytopes in three dimensions include crystals, boxes, tetrahedra, and any convex object whose faces are all flat. It is amusing to see how many problems in combinatorics, number theory, and many other mathematical areas can be recast in the language of polytopes that exist in some Euclidean space. Conversely, the versatile structure of polytopes gives us number-theoretic and combinatorial information that flows naturally from their geometry.

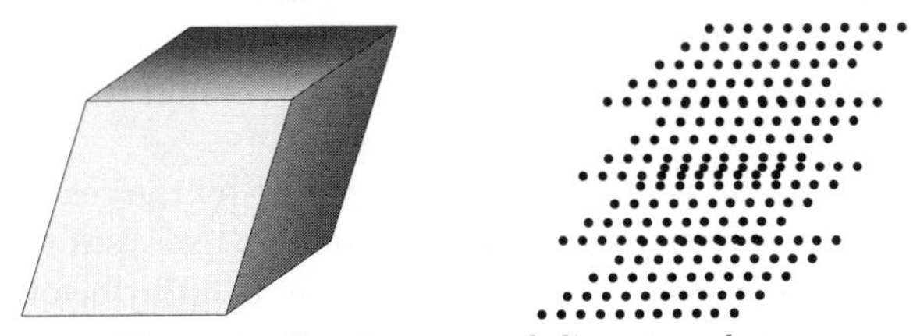

Fig. 0.1. Continuous and discrete volume.

The *discrete volume* of a body $\mathcal{P}$ can be described intuitively as the number of grid points that lie inside $\mathcal{P}$, given a fixed grid in Euclidean space. The *continuous volume* of $\mathcal{P}$ has the usual intuitive meaning of volume that we attach to everyday objects we see in the real world.

Indeed, the difference between the two realizations of volume can be thought of in physical terms as follows. On the one hand, the quantum-level grid imposed by the molecular structure of reality gives us a discrete notion of space and hence discrete volume. On the other hand, the Newtonian notion of continuous space gives us the continuous volume. We see things continuously at the Newtonian level, but in practice we often compute things discretely at the quantum level. Mathematically, the grid we impose in space—corresponding to the grid formed by the atoms that make up an object—helps us compute the usual continuous volume in very surprising and charming ways, as we shall discover.

In order to see the continuous/discrete interplay come to life among the three fields of combinatorics, number theory, and geometry, we begin our focus with the simple-to-state *coin-exchange problem* of Frobenius. The beauty of this concrete problem is that it is easy to grasp, it provides a useful computational tool, and yet it has most of the ingredients of the deeper theories that are developed here.

In the first chapter, we give detailed formulas that arise naturally from the Frobenius coin-exchange problem in order to demonstrate the interconnections between the three fields mentioned above. The coin-exchange problem provides a scaffold for identifying the connections between these fields. In the ensuing chapters we shed this scaffolding and focus on the interconnections themselves:

(1) Enumeration of integer points in polyhedra—combinatorics,
(2) Dedekind sums and finite Fourier series—number theory,
(3) Polygons and polytopes—geometry.

We place a strong emphasis on computational techniques, and on computing volumes by counting integer points using various old and new ideas. Thus, the formulas we get should not only be pretty (which they are!) but should also allow us to efficiently compute volumes by using some nice functions. In the very rare instances of mathematical exposition when we have a formulation that is both "easy to write" and "quickly computable," we have found a mathematical nugget. We have endeavored to fill this book with such mathematical nuggets.

Much of the material in this book is developed by the reader in the more than 200 exercises. Most chapters contain warm-up exercises that do not depend on the material in the chapter and can be assigned before the chapter is read. Some exercises are central, in the sense that current or later material depends on them. Those exercises are marked with ♣, and we give detailed hints for them at the end of the book. Most chapters also contain lists of open research problems.

It turns out that even a fifth grader can write an interesting paper on integer-point enumeration [145], while the subject lends itself to deep investigations that attract the current efforts of leading researchers. Thus, it is an area of mathematics that attracts our innocent childhood questions as well as our refined insight and deeper curiosity. The level of study is highly appropriate for a junior/senior undergraduate course in mathematics. In fact, this book is ideally suited to be used for a *capstone course*. Because the three topics outlined above lend themselves to more sophisticated exploration, our book has also been used effectively for an introductory graduate course.

To help the reader fully appreciate the scope of the connections between the continuous volume and the discrete volume, we begin the discourse in two dimensions, where we can easily draw pictures and quickly experiment. We gently introduce the functions we need in higher dimensions (Dedekind sums) by looking at the coin-exchange problem geometrically as the discrete volume of a generalized triangle, called a simplex.

The initial techniques are quite simple, essentially nothing more than expanding rational functions into partial fractions. Thus, the book is easily accessible to a student who has completed a standard college calculus and linear algebra curriculum. It would be useful to have a basic understanding of partial fraction expansions, infinite series, open and closed sets in $\mathbb{R}^d$, complex numbers (in particular, roots of unity), and modular arithmetic.

An important computational tool that is harnessed throughout the text is the *generating function* $f(x) = \sum_{m=0}^{\infty} a(m)\, x^m$, where the $a(m)$'s form any sequence of numbers that we are interested in analyzing. When the infinite sequence of numbers $a(m), m = 0, 1, 2, \dots$, is embedded into a single generating function $f(x)$, it is often true that for hitherto unforeseen reasons, we can rewrite the whole sum $f(x)$ in a surprisingly compact form. It is the rewriting of these generating functions that allows us to understand the combinatorics of the relevant sequence $a(m)$. For us, the sequence of numbers might be the number of ways to partition an integer into given coin denominations, or the number of points in an increasingly large body, and so on. Here we find yet another example of the interplay between the discrete and the continuous: we are given a *discrete* set of numbers $a(m)$, and we then carry out analysis on the generating function $f(x)$ in the *continuous* variable x.

What Is the Discrete Volume?

The physically intuitive description of the discrete volume given above rests on a sound mathematical footing as soon as we introduce the notion of a lattice. The grid is captured mathematically as the collection of all integer points in Euclidean space, namely $\mathbb{Z}^d = \{(x_1, \dots, x_d) : \text{ all } x_k \in \mathbb{Z}\}$. This discrete collection of equally spaced points is called a *lattice*. If we are given a geometric body $\mathcal{P}$, its discrete volume is simply defined as the number of lattice points inside $\mathcal{P}$, that is, the number of elements in the set $\mathbb{Z}^d \cap \mathcal{P}$.

Intuitively, if we shrink the lattice by a factor k and count the number of newly shrunken lattice points inside $\mathcal{P}$, we obtain a better approximation for the volume of $\mathcal{P}$, relative to the volume of a single cell of the shrunken lattice. It turns out that after the lattice is shrunk by an integer factor k, the number $\#\left(\mathcal{P} \cap \frac{1}{k}\mathbb{Z}^d\right)$ of shrunken lattice points inside an *integral polytope* $\mathcal{P}$

is magically a polynomial in k. This counting function $\#\left(\mathcal{P} \cap \frac{1}{k}\mathbb{Z}^d\right)$ is known as the *Ehrhart polynomial* of $\mathcal{P}$. If we kept shrinking the lattice by taking a limit, we would of course end up with the continuous volume that is given by the usual Riemannian integral definition of calculus:

$$\operatorname{vol} \mathcal{P} = \lim_{k \to \infty} \# \left(\mathcal{P} \cap \frac{1}{k}\mathbb{Z}^d \right) \frac{1}{k^d} \, .$$

However, pausing at fixed dilations of the lattice gives surprising flexibility for the computation of the volume of $\mathcal{P}$ and for the number of lattice points that are contained in $\mathcal{P}$.

Thus, when the body $\mathcal{P}$ is an integral polytope, the error terms that measure the discrepancy between the discrete volume and the usual continuous volume are quite nice; they are given by Ehrhart polynomials, and these enumeration polynomials are the content of Chapter 3.

The Fourier–Dedekind Sums Are the Building Blocks: Number Theory

Every polytope has a discrete volume that is expressible in terms of certain finite sums that are known as *Dedekind sums*. Before giving their definition, we first motivate these sums with some examples that illustrate their building-block behavior for lattice-point enumeration. To be concrete, consider for example a 1-dimensional polytope given by an interval $\mathcal{P} = [0, a]$, where a is any positive real number. It is clear that we need the greatest integer function $\lfloor x \rfloor$ to help us enumerate the lattice points in $\mathcal{P}$, and indeed the answer is $\lfloor a \rfloor + 1$.

Next, consider a 1-dimensional line segment that is sitting in the 2-dimensional plane. Let's pick our segment $\mathcal{P}$ so that it begins at the origin and ends at the lattice point (c, d). As becomes apparent after a moment's thought, the number of lattice points on this finite line segment involves an old friend, namely the greatest common divisor of c and d. The exact number of lattice points on the line segment is $\gcd(c, d) + 1$.

To unify both of these examples, consider a triangle $\mathcal{P}$ in the plane whose vertices have rational coordinates. It turns out that a certain finite sum is completely natural because it simultaneously extends both the greatest integer function and the greatest common divisor, although the latter is less obvious. An example of a Dedekind sum in two dimensions that arises naturally in the formula for the discrete volume of the rational triangle $\mathcal{P}$ is the following:

$$s(a, b) = \sum_{m=1}^{b-1} \left(\frac{m}{b} - \frac{1}{2} \right) \left(\frac{ma}{b} - \left\lfloor \frac{ma}{b} \right\rfloor - \frac{1}{2} \right) .$$

The definition makes use of the greatest integer function. Why do these sums also resemble the greatest common divisor? Luckily, the Dedekind sums satisfy a remarkable reciprocity law, quite similar to the Euclidean algorithm that computes the gcd. This reciprocity law allows the Dedekind sums to be computed in roughly $\log(b)$ steps rather than the b steps that are implied by the definition above. The reciprocity law for $s(a,b)$ lies at the heart of some amazing number theory that we treat in an elementary fashion, but that also comes from the deeper subject of modular forms and other modern tools.

We find ourselves in the fortunate position of viewing an important tip of an enormous mountain of ideas, submerged by the waters of geometry. As we delve more deeply into these waters, more and more hidden beauty unfolds for us, and the Dedekind sums are an indispensable tool that allow us to see further as the waters get deeper.

The Relevant Solids Are Polytopes: Geometry

The examples we have used, namely line segments and polygons in the plane, are special cases of polytopes in all dimensions. One way to define a polytope is to consider the *convex hull* of a finite collection of points in Euclidean space $\mathbb{R}^d$. That is, suppose someone gives us a set of points $\mathbf{v}_1, \dots, \mathbf{v}_n$ in $\mathbb{R}^d$. The polytope determined by the given points $\mathbf{v}_j$ is defined by all linear combinations $c_1\mathbf{v}_1 + c_2\mathbf{v}_2 + \cdots + c_n\mathbf{v}_n$, where the coefficients c_j are nonnegative real numbers that satisfy the relation $c_1 + c_2 + \cdots + c_n = 1$. This construction is called the *vertex description* of the polytope.

There is another equivalent definition, called the *hyperplane description* of the polytope. Namely, if someone hands us the linear inequalities that define a finite collection of half-spaces in $\mathbb{R}^d$, we can define the associated polytope as the simultaneous intersection of the half-spaces defined by the given inequalities.

There are some "obvious" facts about polytopes that are intuitively clear to most students but are, in fact, subtle and often nontrivial to prove from first principles. Two of these facts, namely that every polytope has both a vertex and a hyperplane description, and that every polytope can be triangulated, form a crucial basis to the material we will develop in this book. We carefully prove both facts in the appendices. The two main statements in the appendices are intuitively clear, so that novices can skip over their proofs without any detriment to their ability to compute continuous and discrete volumes of polytopes. All theorems in the text (including those in the appendices) are proved from first principles, with the exception of the last chapter, where we assume basic notions from complex analysis.

The text naturally flows into two parts, which we now explicate.

Part I

We have taken great care in making the content of the chapters flow seamlessly from one to the next, over the span of the first six chapters.

- Chapters 1 and 2 introduce some basic notions of generating functions, in the visually compelling context of discrete geometry, with an abundance of detailed motivating examples.

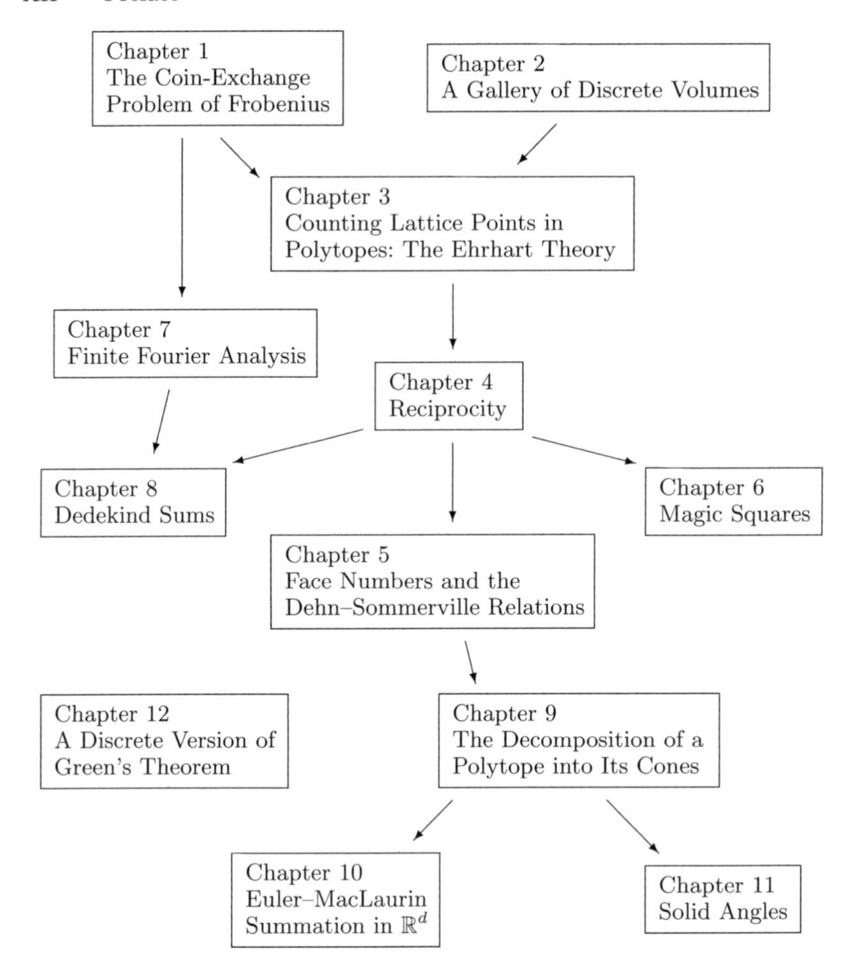

Fig. 0.2. The partially ordered set of chapter dependencies.

- Chapters 3, 4, and 5 develop the full Ehrhart theory of discrete volumes of rational polytopes.
- Chapter 6 is a "dessert" chapter, in that it enables us to use the theory developed to treat the enumeration of *magic squares*, an ancient topic that enjoys active current research.

Part II

We now begin anew.

- Having attained experience with numerous examples and results about integer polytopes, we are ready to learn about the *Dedekind sums* of Chapter 8, which form the atomic pieces of the discrete volume polynomials. On the other hand, to fully understand Dedekind sums, we need to understand *finite Fourier analysis*, which we therefore develop from first principles in Chapter 7, using only partial fractions.

- Chapter 9 answers a simple yet tricky question: how does the finite geometric series in one dimension extend to higher-dimensional polytopes? *Brion's theorem* give the elegant and decisive answer to this question.
- Chapter 10 extends the interplay between the continuous volume and the discrete volume of a polytope (already studied in detail in Part I) by introducing *Euler–Maclaurin summation* formulas in all dimensions. These formulas compare the continuous Fourier transform of a polytope to its discrete Fourier transform, yet the material is completely self-contained.
- Chapter 11 develops an exciting extension of Ehrhart theory that defines and studies the *solid angles* of a polytope; these are the natural extensions of 2-dimensional angles to higher dimensions.
- Finally, we end with another "dessert" chapter that uses complex analytic methods to find an integral formula for the discrepancy between the discrete and continuous areas enclosed by a closed curve in the plane.

Because polytopes are both theoretically useful (in triangulated manifolds, for example) and practically essential (in computer graphics, for example) we use them to link results in number theory and combinatorics. There are many research papers being written on these interconnections, even as we speak, and it is impossible to capture them all here; however, we hope that these modest beginnings will give the reader who is unfamiliar with these fields a good sense of their beauty, inexorable connectedness, and utility. We have written a gentle invitation to what we consider a gorgeous world of counting and of links between the fields of combinatorics, number theory, and geometry for the general mathematical reader.

There are a number of excellent books that have a nontrivial intersection with ours and contain material that complements the topics discussed here. We heartily recommend the monographs of Barvinok [12] (on general convexity topics), Ehrhart [81] (the historic introduction to Ehrhart theory), Ewald [82] (on connections to algebraic geometry), Hibi [96] (on the interplay of algebraic combinatorics with polytopes), Miller–Sturmfels [132] (on computational commutative algebra), and Stanley [172] (on general enumerative problems in combinatorics).

Acknowledgments

We have had the good fortune of receiving help from many gracious people in the process of writing this book. First and foremost, we thank the students of the classes in which we could try out this material, at Binghamton University (SUNY), San Francisco State University, and Temple University. We are indebted to our MSRI/Banff 2005 graduate summer school students. We give special thanks to Kristin Camenga and Kevin Woods, who ran the problem sessions for this summer school, detected numerous typos, and provided us with many interesting suggestions for this book. We are grateful for the generous support for the summer school from the Mathematical Sciences Research Institute, the Pacific Institute of Mathematics, and the Banff International Research Station.

Many colleagues supported this endeavor, and we are particularly grateful to everyone notifying us of mistakes, typos, and good suggestions: Daniel Antonetti, Alexander Barvinok, Nathanael Berglund, Andrew Beyer, Tristram Bogart, Garry Bowlin, Benjamin Braun, Robin Chapman, Yitwah Cheung, Jessica Cuomo, Dimitros Dais, Aaron Dall, Jesus De Loera, David Desario, Mike Develin, Ricardo Diaz, Michael Dobbins, Jeff Doker, Han Minh Duong, Richard Ehrenborg, David Einstein, Joseph Gubeladze, Christian Haase, Mary Halloran, Friedrich Hirzebruch, Brian Hopkins, Serkan Hoşten, Benjamin Howard, Piotr Maciak, Evgeny Materov, Asia Matthews, Peter McMullen, Martín Mereb, Ezra Miller, Mel Nathanson, Julian Pfeifle, Peter Pleasants, Jorge Ramírez Alfonsín, Bruce Reznick, Adrian Riskin, Steven Sam, Junro Sato, Kim Seashore, Melissa Simmons, Richard Stanley, Bernd Sturmfels, Thorsten Theobald, Read Vanderbilt, Andrew Van Herick, Sven Verdoolaege, Michèle Vergne, Julie Von Bergen, Neil Weickel, Carl Woll, Zhiqiang Xu, Jon Yaggie, Ruriko Yoshida, Thomas Zaslavsky, Günter Ziegler, and two anonymous referees. We will collect corrections, updates, etc., at the Internet website

`math.sfsu.edu/beck/ccd.html`.

We are indebted to the Springer editorial staff, first and foremost to Mark Spencer, for facilitating the publication process in an always friendly and supportive spirit. We thank David Kramer for the impeccable copyediting, Frank Ganz for sharing his LaTeX expertise, and Felix Portnoy for the seamless production process.

Matthias Beck would like to express his deepest gratitude to Tendai Chitewere, for her patience, support, and unconditional love. He thanks his family for always being there for him. Sinai Robins would like to thank Michal Robins, Shani Robins, and Gabriel Robins for their relentless support and understanding during the completion of this project. We both thank all the cafés we have inhabited over the past five years for enabling us to turn their coffee into theorems.

San Francisco
Philadelphia
May 2007

Matthias Beck
Sinai Robins

Contents

Part I The Essentials of Discrete Volume Computations

Part II Beyond the Basics

Part I

The Essentials of Discrete Volume Computations

1

The Coin-Exchange Problem of Frobenius

The full beauty of the subject of generating functions emerges only from tuning in on both channels: the discrete and the continuous.

Herbert Wilf [187]

Suppose we're interested in an infinite sequence of numbers $(a_k)_{k=0}^{\infty}$ that arises geometrically or recursively. Is there a "good formula" for a_k as a function of k? Are there identities involving various a_k's? Embedding this sequence into the **generating function**

$$F(z) = \sum_{k \geq 0} a_k \, z^k$$

allows us to retrieve answers to the questions above in a surprisingly quick and elegant way. We can think of $F(z)$ as lifting our sequence a_k from its discrete setting into the continuous world of functions.

1.1 Why Use Generating Functions?

To illustrate these concepts, we warm up with the classic example of the **Fibonacci sequence** f_k, named after Leonardo Pisano Fibonacci (1170–1250?)[1] and defined by the recursion

$$f_0 = 0, \ f_1 = 1, \ \text{ and } \ f_{k+2} = f_{k+1} + f_k \ \text{ for } \ k \geq 0 \,.$$

This gives the sequence $(f_k)_{k=0}^{\infty} = (0, 1, 1, 2, 3, 5, 8, 13, 21, 34, \dots)$ (see also [165, Sequence A000045]). Now let's see what generating functions can do for us. Let

[1] For more information about Fibonacci, see
`http://www-groups.dcs.st-and.ac.uk/~history/Biographies/Fibonacci.html`.

$$F(z) = \sum_{k \geq 0} f_k \, z^k .$$

We embed both sides of the recursion identity into their generating functions:

$$\sum_{k \geq 0} f_{k+2} \, z^k = \sum_{k \geq 0} \left(f_{k+1} + f_k \right) z^k = \sum_{k \geq 0} f_{k+1} \, z^k + \sum_{k \geq 0} f_k \, z^k . \tag{1.1}$$

The left–hand side of (1.1) is

$$\sum_{k \geq 0} f_{k+2} \, z^k = \frac{1}{z^2} \sum_{k \geq 0} f_{k+2} \, z^{k+2} = \frac{1}{z^2} \sum_{k \geq 2} f_k \, z^k = \frac{1}{z^2} \left(F(z) - z \right),$$

while the right–hand side of (1.1) is

$$\sum_{k \geq 0} f_{k+1} \, z^k + \sum_{k \geq 0} f_k \, z^k = \frac{1}{z} F(z) + F(z) .$$

So (1.1) can be restated as

$$\frac{1}{z^2} \left(F(z) - z \right) = \frac{1}{z} F(z) + F(z) ,$$

or

$$F(z) = \frac{z}{1 - z - z^2} .$$

It's fun to check (e.g., with a computer) that when we expand the function F into a power series, we indeed obtain the Fibonacci numbers as coefficients:

$$\frac{z}{1 - z - z^2} = z + z^2 + 2\, z^3 + 3\, z^4 + 5\, z^5 + 8\, z^6 + 13\, z^7 + 21\, z^8 + 34\, z^9 + \cdots .$$

Now we use our favorite method of handling rational functions: the partial fraction expansion. In our case, the denominator factors as $1 - z - z^2 = \left(1 - \frac{1+\sqrt{5}}{2} \, z \right) \left(1 - \frac{1-\sqrt{5}}{2} \, z \right)$, and the partial fraction expansion is (see Exercise 1.1)

$$F(z) = \frac{z}{1 - z - z^2} = \frac{1/\sqrt{5}}{1 - \frac{1+\sqrt{5}}{2} \, z} - \frac{1/\sqrt{5}}{1 - \frac{1-\sqrt{5}}{2} \, z} . \tag{1.2}$$

The two terms suggest the use of the **geometric series**

$$\sum_{k \geq 0} x^k = \frac{1}{1 - x} \tag{1.3}$$

(see Exercise 1.2) with $x = \frac{1+\sqrt{5}}{2} \, z$ and $x = \frac{1-\sqrt{5}}{2} \, z$, respectively:

$$\begin{aligned} F(z) = \frac{z}{1 - z - z^2} &= \frac{1}{\sqrt{5}} \sum_{k \geq 0} \left(\frac{1 + \sqrt{5}}{2} \, z \right)^k - \frac{1}{\sqrt{5}} \sum_{k \geq 0} \left(\frac{1 - \sqrt{5}}{2} \, z \right)^k \\ &= \sum_{k \geq 0} \frac{1}{\sqrt{5}} \left(\left(\frac{1 + \sqrt{5}}{2} \right)^k - \left(\frac{1 - \sqrt{5}}{2} \right)^k \right) z^k . \end{aligned}$$

Comparing the coefficients of z^k in the definition of $F(z) = \sum_{k\geq 0} f_k\, z^k$ and the new expression above for $F(z)$, we discover the closed form expression for the Fibonacci sequence

$$f_k = \frac{1}{\sqrt{5}}\left(\frac{1+\sqrt{5}}{2}\right)^k - \frac{1}{\sqrt{5}}\left(\frac{1-\sqrt{5}}{2}\right)^k .$$

This method of decomposing a rational generating function into partial fractions is one of our key tools. Because we will use partial fractions time and again throughout this book, we record the result on which this method is based.

Theorem 1.1 (Partial fraction expansion). *Given any rational function*

$$F(z) := \frac{p(z)}{\prod_{k=1}^{m} (z-a_k)^{e_k}} ,$$

where p is a polynomial of degree less than $e_1 + e_2 + \cdots + e_m$ and the a_k's are distinct, there exists a decomposition

$$F(z) = \sum_{k=1}^{m} \left(\frac{c_{k,1}}{z-a_k} + \frac{c_{k,2}}{(z-a_k)^2} + \cdots + \frac{c_{k,e_k}}{(z-a_k)^{e_k}} \right),$$

where $c_{k,j} \in \mathbb{C}$ are unique.

One possible proof of this theorem is based on the fact that the polynomials form a Euclidean domain. For readers who are acquainted with this notion, we outline this proof in Exercise 1.35.

1.2 Two Coins

Let's imagine that we introduce a new coin system. Instead of using pennies, nickels, dimes, and quarters, let's say we agree on using 4-cent, 7-cent, 9-cent, and 34-cent coins. The reader might point out the following flaw of this new system: certain amounts cannot be changed (that is, created with the available coins), for example, 2 or 5 cents. On the other hand, this deficiency makes our new coin system more interesting than the old one, because we can ask the question, "which amounts can be changed?" In fact, we will prove in Exercise 1.20 that there are only finitely many integer amounts that *cannot* be changed using our new coin system. A natural question, first tackled by Ferdinand Georg Frobenius (1849–1917),[2] and James Joseph Sylvester (1814–1897)[3] is,

[2] For more information about Frobenius, see
`http://www-groups.dcs.st-and.ac.uk/~history/Biographies/Frobenius.html`.

[3] For more information about Sylvester, see
`http://www-groups.dcs.st-and.ac.uk/~history/Biographies/Sylvester.html`.

"what is the *largest* amount that cannot be changed?" As mathematicians, we like to keep questions as general as possible, and so we ask, given coins of denominations $a_1, a_2, \dots, a_d$, which are positive integers without any common factor, can you give a formula for the largest amount that cannot be changed using the coins $a_1, a_2, \dots, a_d$? This problem is known as the *Frobenius coin-exchange problem.*

To be precise, suppose we're given a set of positive integers

$$A = \{a_1, a_2, \dots, a_d\}$$

with $\gcd(a_1, a_2, \dots, a_d) = 1$ and we call an integer n **representable** if there exist nonnegative integers $m_1, m_2, \dots, m_d$ such that

$$n = m_1 a_1 + \cdots + m_d a_d .$$

In the language of coins, this means that we can change the amount n using the coins $a_1, a_2, \dots, a_d$. The Frobenius problem (often called the *linear Diophantine problem of Frobenius*) asks us to find the largest integer that is not representable. We call this largest integer the **Frobenius number** and denote it by $g(a_1, \dots, a_d)$. The following theorem gives us a pretty formula for $d = 2$.

Theorem 1.2. *If a_1 and a_2 are relatively prime positive integers, then*

$$g(a_1, a_2) = a_1 a_2 - a_1 - a_2 .$$

This simple-looking formula for g inspired a great deal of research into formulas for $g(a_1, a_2, \dots, a_d)$ with only limited success; see the notes at the end of this chapter. For $d = 2$, Sylvester gave the following result.

Theorem 1.3 (Sylvester's theorem). *Let a_1 and a_2 be relatively prime positive integers. Exactly half of the integers between 1 and $(a_1 - 1)(a_2 - 1)$ are representable.*

Our goal in this chapter is to prove these two theorems (and a little more) using the machinery of partial fractions. We approach the Frobenius problem through the study of the **restricted partition function**

$$p_A(n) := \#\left\{(m_1, \dots, m_d) \in \mathbb{Z}^d : \text{ all } m_j \geq 0,\ m_1 a_1 + \cdots + m_d a_d = n\right\},$$

the number of partitions of n using only the elements of A as parts.[4] In view of this partition function, $g(a_1, \dots, a_d)$ is the largest integer n for which $p_A(n) = 0$.

There is a beautiful geometric interpretation of the restricted partition function. The geometric description begins with the set

[4] A **partition** of a positive integer n is a multiset (i.e., a set in which we allow repetition) $\{n_1, n_2, \dots, n_k\}$ of positive integers such that $n = n_1 + n_2 + \cdots + n_k$. The numbers $n_1, n_2, \dots, n_k$ are called the **parts** of the partition.

$$\mathcal{P} = \left\{ (x_1, \dots, x_d) \in \mathbb{R}^d : \text{ all } x_j \geq 0, \ x_1 a_1 + \cdots + x_d a_d = 1 \right\}. \tag{1.4}$$

The nth **dilate** of any set $S \subseteq \mathbb{R}^d$ is

$$\left\{ (nx_1, nx_2, \dots, nx_d) : (x_1, \dots, x_d) \in S \right\}.$$

The function $p_A(n)$ counts precisely those integer points that lie in the nth integer dilate of the body $\mathcal{P}$. The dilation process in this context is tantamount to replacing $x_1a_1 + \cdots + x_da_d = 1$ in the definition of $\mathcal{P}$ by $x_1a_1 + \cdots + x_da_d = n$. The set $\mathcal{P}$ turns out to be a *polytope*. We can easily picture $\mathcal{P}$ and its dilates for dimension $d \leq 3$; Figure 1.1 shows the three-dimensional case.

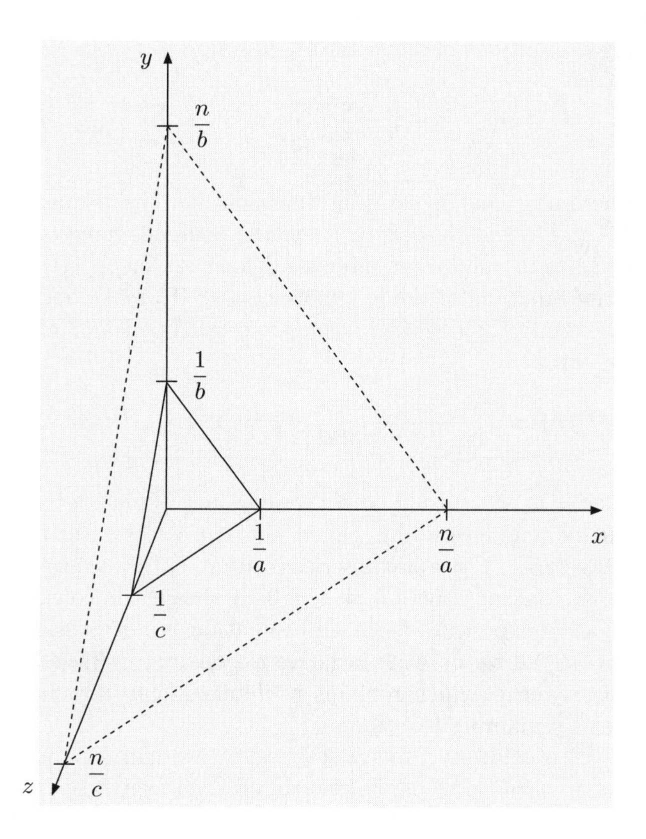

Fig. 1.1. $d = 3$.

1.3 Partial Fractions and a Surprising Formula

We first concentrate on the case $d = 2$ and study

$$p_{\{a,b\}}(n) = \#\left\{(k,l) \in \mathbb{Z}^2 : \, k,l \geq 0, \; ak + bl = n\right\}.$$

Recall that we require a and b to be relatively prime. To begin our discussion, we start playing with generating functions. Consider the product of the following two geometric series:

$$\left(\frac{1}{1-z^a}\right)\left(\frac{1}{1-z^b}\right) = \left(1 + z^a + z^{2a} + \cdots\right)\left(1 + z^b + z^{2b} + \cdots\right)$$

(see Exercise 1.2). If we multiply out all the terms we'll get a power series all of whose exponents are linear combinations of a and b. In fact, the coefficient of z^n in this power series counts the number of ways that n can be written as a nonnegative linear combination of a and b. In other words, these coefficients are precisely evaluations of our counting function $p_{\{a,b\}}$:

$$\left(\frac{1}{1-z^a}\right)\left(\frac{1}{1-z^b}\right) = \sum_{k\geq 0}\sum_{l\geq 0} z^{ak} z^{bl} = \sum_{n\geq 0} p_{\{a,b\}}(n)\, z^n.$$

So this function is the generating function for the sequence of integers $\left(p_{\{a,b\}}(n)\right)_{n=0}^{\infty}$. The idea is now to study the compact function on the left.

We would like to uncover an interesting formula for $p_{\{a,b\}}(n)$ by looking at the generating function on the left more closely. To make our computational life easier, we study the *constant term* of a related series; namely, $p_{\{a,b\}}(n)$ is the constant term of

$$f(z) := \frac{1}{(1-z^a)(1-z^b)\, z^n} = \sum_{k\geq 0} p_{\{a,b\}}(k)\, z^{k-n}.$$

The latter series is not quite a power series, since it includes terms with negative exponents. These series are called *Laurent series*, after Pierre Alphonse Laurent (1813–1854). For a power series (centered at 0), we could simply evaluate the corresponding function at $z = 0$ to obtain the constant term; once we have negative exponents, such an evaluation is not possible. However, if we first subtract all terms with negative exponents, we'll get a power series whose constant term (which remains unchanged) can now be computed by evaluating this remaining function at $z = 0$.

To be able to compute this constant term, we will expand f into partial fractions. As a warm-up to partial fraction decompositions, we first work out a one-dimensional example. Let's denote the first a^{th} root of unity by

$$\xi_a := e^{2\pi i/a} = \cos\frac{2\pi}{a} + i\sin\frac{2\pi}{a};$$

then all the ath roots of unity are $1, \xi_a, \xi_a^2, \xi_a^3, \dots, \xi_a^{a-1}$.

Example 1.4. Let's find the partial fraction expansion of $\frac{1}{1-z^a}$. The poles of this function are located at all a^{th} roots of unity ξ_a^k for $k = 0, 1, \dots, a-1$. So we expand

$$\frac{1}{1-z^a} = \sum_{k=0}^{a-1} \frac{C_k}{z - \xi_a^k}.$$

How do we find the coefficients C_k? Well,

$$C_k = \lim_{z \to \xi_a^k} \left(z - \xi_a^k\right) \left(\frac{1}{1-z^a}\right) = \lim_{z \to \xi_a^k} \frac{1}{-a\, z^{a-1}} = -\frac{\xi_a^k}{a},$$

where we have used L'Hôpital's rule in the penultimate equality. Therefore, we arrive at the expansion

$$\frac{1}{1-z^a} = -\frac{1}{a} \sum_{k=0}^{a-1} \frac{\xi_a^k}{z - \xi_a^k}.$$ □

Returning to restricted partitions, the poles of f are located at $z = 0$ with multiplicity n, at $z = 1$ with multiplicity 2, and at all the other a^{th} and b^{th} roots of unity with multiplicity 1 because a and b are relatively prime. Hence our partial fraction expansion looks like

$$f(z) = \frac{A_1}{z} + \frac{A_2}{z^2} + \cdots + \frac{A_n}{z^n} + \frac{B_1}{z-1} + \frac{B_2}{(z-1)^2} + \sum_{k=1}^{a-1} \frac{C_k}{z - \xi_a^k} + \sum_{j=1}^{b-1} \frac{D_j}{z - \xi_b^j}. \tag{1.5}$$

We invite the reader to compute the coefficients (Exercise 1.21)

$$\begin{aligned} C_k &= -\frac{1}{a\,(1 - \xi_a^{kb})\, \xi_a^{k(n-1)}}, \\ D_j &= -\frac{1}{b\left(1 - \xi_b^{ja}\right) \xi_b^{j(n-1)}}. \end{aligned} \tag{1.6}$$

To compute B_2, we multiply both sides of (1.5) by $(z-1)^2$ and take the limit as $z \to 1$ to obtain

$$B_2 = \lim_{z \to 1} \frac{(z-1)^2}{(1-z^a)\,(1-z^b)\, z^n} = \frac{1}{ab},$$

by applying L'Hôpital's rule twice, for example. For the more interesting constant B_1, we compute

$$B_1 = \lim_{z \to 1} (z-1) \left(\frac{1}{(1-z^a)\,(1-z^b)\, z^n} - \frac{\frac{1}{ab}}{(z-1)^2} \right) = \frac{1}{ab} - \frac{1}{2a} - \frac{1}{2b} - \frac{n}{ab},$$

again by applying L'Hôpital's rule.

We don't need to compute the coefficients $A_1, \dots, A_n$, since they contribute only to the terms with negative exponents, which we can safely neglect; these terms do not contribute to the constant term of f. Once we have the other coefficients, the constant term of the Laurent series of f is—as we said above—the following function evaluated at 0:

$$p_{\{a,b\}}(n) = \left(\frac{B_1}{z-1} + \frac{B_2}{(z-1)^2} + \sum_{k=1}^{a-1} \frac{C_k}{z - \xi_a^k} + \sum_{j=1}^{b-1} \frac{D_j}{z - \xi_b^j} \right) \Bigg|_{z=0}$$
$$= -B_1 + B_2 - \sum_{k=1}^{a-1} \frac{C_k}{\xi_a^k} - \sum_{j=1}^{b-1} \frac{D_j}{\xi_b^j} \, .$$

With (1.6) in hand, this simplifies to

$$p_{\{a,b\}}(n) = \frac{1}{2a} + \frac{1}{2b} + \frac{n}{ab} + \frac{1}{a} \sum_{k=1}^{a-1} \frac{1}{(1 - \xi_a^{kb}) \xi_a^{kn}} + \frac{1}{b} \sum_{j=1}^{b-1} \frac{1}{(1 - \xi_b^{ja}) \xi_b^{jn}} \, . \tag{1.7}$$

Encouraged by this initial success, we now proceed to analyze each sum in (1.7) with the hope of recognizing them as more familiar objects.

For the next step we need to define the **greatest-integer function** $\lfloor x \rfloor$, which denotes the greatest integer less than or equal to x. A close sibling to this function is the **fractional-part function** $\{x\} = x - \lfloor x \rfloor$. To readers not familiar with the functions $\lfloor x \rfloor$ and $\{x\}$ we recommend working through Exercises 1.3–1.5.

What we'll do next is study a special case, namely $b = 1$. This is appealing because $p_{\{a,1\}}(n)$ simply counts integer points in an interval:

$$\begin{aligned} p_{\{a,1\}}(n) &= \# \left\{ (k,l) \in \mathbb{Z}^2 : \ k, l \geq 0, \ ak + l = n \right\} \\ &= \# \left\{ k \in \mathbb{Z} : \ k \geq 0, \ ak \leq n \right\} \\ &= \# \left\{ k \in \mathbb{Z} : \ 0 \leq k \leq \frac{n}{a} \right\} \\ &= \left\lfloor \frac{n}{a} \right\rfloor + 1 \, . \end{aligned}$$

(See Exercise 1.3.) On the other hand, in (1.7) we just computed a different expression for this function, so that

$$\frac{1}{2a} + \frac{1}{2} + \frac{n}{a} + \frac{1}{a} \sum_{k=1}^{a-1} \frac{1}{(1 - \xi_a^k) \, \xi_a^{kn}} \ = \ p_{\{a,1\}}(n) \ = \ \left\lfloor \frac{n}{a} \right\rfloor + 1 \, .$$

With the help of the fractional-part function $\{x\} = x - \lfloor x \rfloor$, we have derived a formula for the following sum over a^{th} roots of unity:

$$\frac{1}{a} \sum_{k=1}^{a-1} \frac{1}{(1 - \xi_a^k) \, \xi_a^{kn}} = - \left\{ \frac{n}{a} \right\} + \frac{1}{2} - \frac{1}{2a} \, . \tag{1.8}$$

We're almost there: we invite the reader (Exercise 1.22) to show that

$$\frac{1}{a}\sum_{k=1}^{a-1}\frac{1}{\left(1-\xi_a^{bk}\right)\xi_a^{kn}}=\frac{1}{a}\sum_{k=1}^{a-1}\frac{1}{\left(1-\xi_a^{k}\right)\xi_a^{b^{-1}kn}}, \tag{1.9}$$

where b^{-1} is an integer such that $b^{-1}b \equiv 1 \bmod a$, and to conclude that

$$\frac{1}{a}\sum_{k=1}^{a-1}\frac{1}{\left(1-\xi_a^{bk}\right)\xi_a^{kn}}=-\left\{\frac{b^{-1}n}{a}\right\}+\frac{1}{2}-\frac{1}{2a}. \tag{1.10}$$

Now all that's left to do is to substitute this expression back into (1.7), which yields the following beautiful formula due to Tiberiu Popoviciu (1906–1975).

Theorem 1.5 (Popoviciu's theorem). *If a and b are relatively prime, then*

$$p_{\{a,b\}}(n)=\frac{n}{ab}-\left\{\frac{b^{-1}n}{a}\right\}-\left\{\frac{a^{-1}n}{b}\right\}+1,$$

where $b^{-1}b \equiv 1 \bmod a$ and $a^{-1}a \equiv 1 \bmod b$. □

1.4 Sylvester's Result

Before we apply Theorem 1.5 to obtain the classical Theorems 1.2 and 1.3, we return for a moment to the geometry behind the restricted partition function $p_{\{a,b\}}(n)$. In the two-dimensional case (which is the setting of Theorem 1.5), we are counting integer points $(x,y) \in \mathbb{Z}^2$ on the line segments defined by the constraints

$$ax+by=n, \qquad x,y \geq 0.$$

As n increases, the line segment gets dilated. It is not too far-fetched (although Exercise 1.13 teaches us to be careful with such statements) to expect that the likelihood for an integer point to lie on the line segment increases with n. In fact, one might even guess that the number of points on the line segment increases linearly with n, since the line segment is a one-dimensional object. Theorem 1.5 quantifies the previous statement in a very precise form: $p_{\{a,b\}}(n)$ has the "leading term" n/ab, and the remaining terms are bounded as functions in n. Figure 1.2 shows the geometry behind the counting function $p_{\{4,7\}}(n)$ for the first few values of n. Note that the thick line segment for $n = 17 = 4 \cdot 7 - 4 - 7$ is the last one that does not contain any integer point.

Lemma 1.6. *If a and b are relatively prime positive integers and $n \in [1, ab-1]$ is not a multiple of a or b, then*

$$p_{\{a,b\}}(n)+p_{\{a,b\}}(ab-n)=1.$$

In other words, for n between 1 and $ab-1$ and not divisible by a or b, exactly one of the two integers n and $ab-n$ is representable in terms of a and b.

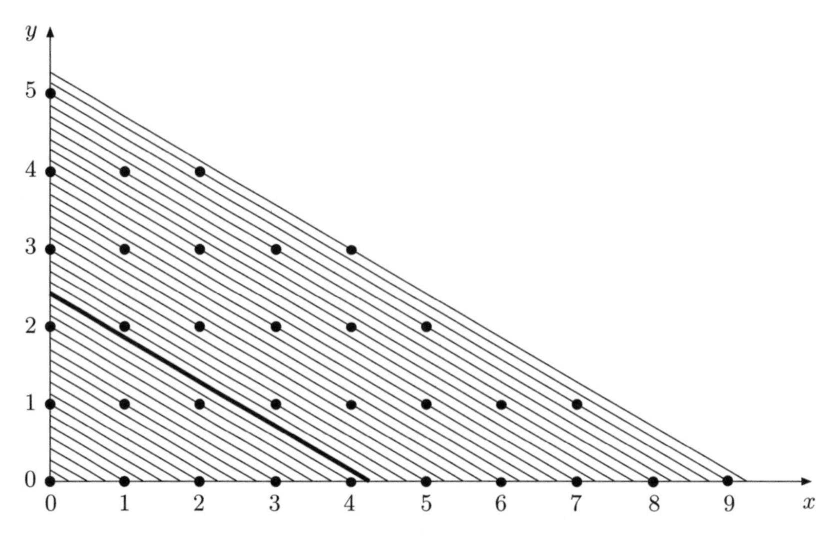

Fig. 1.2. $4x + 7y = n, \ n = 1, 2, \ldots$

Proof. This identity follows directly from Theorem 1.5:

$$\begin{aligned}
p_{\{a,b\}}(ab - n) &= \frac{ab - n}{ab} - \left\{ \frac{b^{-1}(ab - n)}{a} \right\} - \left\{ \frac{a^{-1}(ab - n)}{b} \right\} + 1 \\
&= 2 - \frac{n}{ab} - \left\{ \frac{-b^{-1}n}{a} \right\} - \left\{ \frac{-a^{-1}n}{b} \right\} \\
&\overset{(\star)}{=} -\frac{n}{ab} + \left\{ \frac{b^{-1}n}{a} \right\} + \left\{ \frac{a^{-1}n}{b} \right\} \\
&= 1 - p_{\{a,b\}}(n) \, .
\end{aligned}$$

Here, $(\star)$ follows from the fact that $\{-x\} = 1 - \{x\}$ if $x \notin \mathbb{Z}$ (see Exercise 1.5). □

Proof of Theorem 1.2. We have to show that $p_{\{a,b\}}(ab - a - b) = 0$ and that $p_{\{a,b\}}(n) > 0$ for every $n > ab - a - b$. The first assertion follows with Exercise 1.24, which states that $p_{\{a,b\}}(a + b) = 1$, and Lemma 1.6. To prove the second assertion, we note that for any integer m, $\left\{ \frac{m}{a} \right\} \leq 1 - \frac{1}{a}$. Hence for any positive integer n,

$$p_{\{a,b\}}(ab - a - b + n) \geq \frac{ab - a - b + n}{ab} - \left(1 - \frac{1}{a} \right) - \left(1 - \frac{1}{b} \right) + 1 = \frac{n}{ab} > 0 \, .$$

□

Proof of Theorem 1.3. Recall that Lemma 1.6 states that for n between 1 and $ab - 1$ and not divisible by a or b, exactly one of n and $ab - n$ is representable. There are

$$ab - a - b + 1 = (a - 1)(b - 1)$$

integers between 1 and $ab-1$ that are not divisible by a or b. Finally, we note that $p_{\{a,b\}}(n) > 0$ if n is a multiple of a or b, by the very definition of $p_{\{a,b\}}(n)$. Hence the number of nonrepresentable integers is $\frac{1}{2}(a-1)(b-1)$. □

Note that we have proved even more. Essentially by Lemma 1.6, every positive integer less than ab has at most one representation. Hence, the representable integers less than ab are *uniquely* representable (see also Exercise 1.25).

1.5 Three and More Coins

What happens to the complexity of the Frobenius problem if we have more than two coins? Let's go back to our restricted partition function

$$p_A(n) = \#\left\{(m_1, \dots, m_d) \in \mathbb{Z}^d : \text{ all } m_j \geq 0, \ m_1 a_1 + \cdots + m_d a_d = n\right\},$$

where $A = \{a_1, \dots, a_d\}$. By the very same reasoning as in Section 1.3, we can easily write down the generating function for $p_A(n)$:

$$\sum_{n \geq 0} p_A(n)\, z^n = \left(\frac{1}{1-z^{a_1}}\right)\left(\frac{1}{1-z^{a_2}}\right)\cdots\left(\frac{1}{1-z^{a_d}}\right).$$

We use the same methods that were exploited in Section 1.3 to recover our function $p_A(n)$ as the constant term of a useful generating function. Namely,

$$p_A(n) = \operatorname{const}\left(\frac{1}{(1-z^{a_1})(1-z^{a_2})\cdots(1-z^{a_d})\,z^n}\right).$$

We now expand the function on the right into partial fractions. For reasons of simplicity we assume in the following that $a_1, \dots, a_d$ are *pairwise* relatively prime; that is, no two of the integers $a_1, a_2, \dots, a_d$ have a common factor. Then our partial fraction expansion looks like

$$\begin{aligned} f(z) &= \frac{1}{(1-z^{a_1})\cdots(1-z^{a_d})\,z^n} \\ &= \frac{A_1}{z} + \frac{A_2}{z^2} + \cdots + \frac{A_n}{z^n} + \frac{B_1}{z-1} + \frac{B_2}{(z-1)^2} + \cdots + \frac{B_d}{(z-1)^d} \\ &\quad + \sum_{k=1}^{a_1-1} \frac{C_{1k}}{z-\xi_{a_1}^k} + \sum_{k=1}^{a_2-1} \frac{C_{2k}}{z-\xi_{a_2}^k} + \cdots + \sum_{k=1}^{a_d-1} \frac{C_{dk}}{z-\xi_{a_d}^k}. \end{aligned} \tag{1.11}$$

By now we're experienced in computing partial fraction coefficients, so that the reader will easily verify that (Exercise 1.29)

$$C_{1k} = -\frac{1}{a_1\left(1-\xi_{a_1}^{ka_2}\right)\left(1-\xi_{a_1}^{ka_3}\right)\cdots\left(1-\xi_{a_1}^{ka_d}\right)\xi_{a_1}^{k(n-1)}}. \tag{1.12}$$

As before, we don't have to compute the coefficients $A_1, \dots, A_n$, because they don't contribute to the constant term of f. For the computation of $B_1, \dots, B_d$, we may use a symbolic manipulation program such as `Maple` or `Mathematica`. Again, once we have calculated these coefficients, we can compute the constant term of f by dropping all negative exponents and evaluating the remaining function at 0:

$$\begin{aligned} p_A(n) &= \left(\frac{B_1}{z-1} + \dots + \frac{B_d}{(z-1)^d} + \sum_{k=1}^{a_1-1} \frac{C_{1k}}{z-\xi_{a_1}^k} + \dots + \sum_{k=1}^{a_d-1} \frac{C_{dk}}{z-\xi_{a_d}^k} \right) \Bigg|_{z=0} \\ &= -B_1 + B_2 - \dots + (-1)^d B_d - \sum_{k=1}^{a_1-1} \frac{C_{1k}}{\xi_{a_1}^k} - \sum_{k=1}^{a_2-1} \frac{C_{2k}}{\xi_{a_2}^k} - \dots - \sum_{k=1}^{a_d-1} \frac{C_{dk}}{\xi_{a_d}^k} . \end{aligned}$$

Substituting the expression we found for C_{1k} into the latter sum over the nontrivial a_1^{th} roots of unity, for example, gives rise to

$$\frac{1}{a_1} \sum_{k=1}^{a_1-1} \frac{1}{\left(1-\xi_{a_1}^{ka_2}\right)\left(1-\xi_{a_1}^{ka_3}\right)\cdots\left(1-\xi_{a_1}^{ka_d}\right)\xi_{a_1}^{kn}} .$$

This motivates the definition of the **Fourier–Dedekind sum**

$$s_n\left(a_1, a_2, \dots, a_m; b\right) := \frac{1}{b} \sum_{k=1}^{b-1} \frac{\xi_b^{kn}}{\left(1-\xi_b^{ka_1}\right)\left(1-\xi_b^{ka_2}\right)\cdots\left(1-\xi_b^{ka_m}\right)} . \tag{1.13}$$

We will study these sums in detail in Chapter 8. With this definition, we have arrived at the following result.

Theorem 1.7. *The restricted partition function for $A = \{a_1, a_2, \dots, a_d\}$, where the a_k's are pairwise relatively prime, can be computed as*

$$\begin{aligned} p_A(n) = -B_1 + B_2 - \dots + (-1)^d B_d + s_{-n}\left(a_2, a_3, \dots, a_d; a_1\right) \\ + s_{-n}\left(a_1, a_3, a_4, \dots, a_d; a_2\right) + \dots + s_{-n}\left(a_1, a_2, \dots, a_{d-1}; a_d\right) . \end{aligned}$$

Here $B_1, B_2, \dots, B_d$ are the partial fraction coefficients in the expansion (1.11). □

Example 1.8. We give the restricted partition functions for $d = 3$ and 4. These closed-form formulas have proven useful in the refined analysis of the periodicity that is inherent in the restricted partition function $p_A(n)$. For example, one can visualize the graph of $p_{\{a,b,c\}}(n)$ as a "wavy parabola," as its formula plainly shows.

$$p_{\{a,b,c\}}(n) = \frac{n^2}{2abc} + \frac{n}{2}\left(\frac{1}{ab} + \frac{1}{ac} + \frac{1}{bc}\right) + \frac{1}{12}\left(\frac{3}{a} + \frac{3}{b} + \frac{3}{c} + \frac{a}{bc} + \frac{b}{ac} + \frac{c}{ab}\right)$$
$$+ \frac{1}{a}\sum_{k=1}^{a-1} \frac{1}{\left(1-\xi_a^{kb}\right)\left(1-\xi_a^{kc}\right)\xi_a^{kn}} + \frac{1}{b}\sum_{k=1}^{b-1} \frac{1}{\left(1-\xi_b^{kc}\right)\left(1-\xi_b^{ka}\right)\xi_b^{kn}}$$
$$+ \frac{1}{c}\sum_{k=1}^{c-1} \frac{1}{\left(1-\xi_c^{ka}\right)\left(1-\xi_c^{kb}\right)\xi_c^{kn}},$$

$$p_{\{a,b,c,d\}}(n) = \frac{n^3}{6abcd} + \frac{n^2}{4}\left(\frac{1}{abc} + \frac{1}{abd} + \frac{1}{acd} + \frac{1}{bcd}\right)$$
$$+ \frac{n}{12}\left(\frac{3}{ab} + \frac{3}{ac} + \frac{3}{ad} + \frac{3}{bc} + \frac{3}{bd} + \frac{3}{cd} + \frac{a}{bcd} + \frac{b}{acd} + \frac{c}{abd} + \frac{d}{abc}\right)$$
$$+ \frac{1}{24}\left(\frac{a}{bc} + \frac{a}{bd} + \frac{a}{cd} + \frac{b}{ad} + \frac{b}{ac} + \frac{b}{cd} + \frac{c}{ab} + \frac{c}{ad} + \frac{c}{bd}\right.$$
$$\left. + \frac{d}{ab} + \frac{d}{ac} + \frac{d}{bc}\right) - \frac{1}{8}\left(\frac{1}{a} + \frac{1}{b} + \frac{1}{c} + \frac{1}{d}\right)$$
$$+ \frac{1}{a}\sum_{k=1}^{a-1} \frac{1}{\left(1-\xi_a^{kb}\right)\left(1-\xi_a^{kc}\right)\left(1-\xi_a^{kd}\right)\xi_a^{kn}}$$
$$+ \frac{1}{b}\sum_{k=1}^{b-1} \frac{1}{\left(1-\xi_b^{kc}\right)\left(1-\xi_b^{kd}\right)\left(1-\xi_b^{ka}\right)\xi_b^{kn}}$$
$$+ \frac{1}{c}\sum_{k=1}^{c-1} \frac{1}{\left(1-\xi_c^{kd}\right)\left(1-\xi_c^{ka}\right)\left(1-\xi_c^{kb}\right)\xi_c^{kn}}$$
$$+ \frac{1}{d}\sum_{k=1}^{d-1} \frac{1}{\left(1-\xi_d^{ka}\right)\left(1-\xi_d^{kb}\right)\left(1-\xi_d^{kc}\right)\xi_d^{kn}}. \qquad \square$$

Notes

1. The theory of generating functions has a long and powerful tradition. We only touch on its utility. For those readers who would like to dig a little deeper into the vast generating-function garden, we strongly recommend Herb Wilf's *generatingfunctionology* [187] and László Lovász's *Combinatorial Problems and Exercises* [122]. The reader might wonder why we do not stress convergence aspects of the generating functions we play with. Almost all of our series are geometric series and have trivial convergence properties. In the spirit of not muddying the waters of lucid mathematical exposition, we omit such convergence details.

2. The Frobenius problem is named after Georg Frobenius, who apparently liked to raise this problem in his lectures [41]. Theorem 1.2 is one of the famous folklore results and might be one of the most misquoted theorems in all of mathematics. People usually cite James J. Sylvester's problem in [177], but his paper contains Theorem 1.3 rather than 1.2. In fact, Sylvester's problem had previously appeared as a theorem in [176]. It is not known who first discovered or proved Theorem 1.2. It is very conceivable that Sylvester knew about it when he came up with Theorem 1.3.

3. The linear Diophantine problem of Frobenius should not be confused with the *postage-stamp problem*. The latter problem asks for a similar determination, but adds an additional independent bound on the size of the integer solutions to the linear equation.

4. Theorem 1.5 has an interesting history. The earliest appearance of this result that we are aware of is in a paper by Tiberiu Popoviciu [148]. Popoviciu's formula has since been resurrected at least twice [161, 183].

5. Fourier–Dedekind sums first surfaced implicitly in Sylvester's work (see, e.g., [175]) and explicitly in connection with restricted partition functions in [104]. They were rediscovered in [25], in connection with the Frobenius problem. The papers [157, 83] contain interesting connections to Bernoulli and Euler polynomials. We will resume the study of the Fourier–Dedekind sums in Chapter 8.

6. As we already mentioned above, the Frobenius problem for $d \geq 3$ is much harder than the case $d = 2$ that we have discussed. Certainly beyond $d = 3$, the Frobenius problem is wide open, though much effort has been put into its study. The literature on the Frobenius problem is vast, and there is still much room for improvement. The interested reader might consult the comprehensive monograph [153], which surveys the references to almost all articles dealing with the Frobenius problem and gives about 40 open problems and conjectures related to the Frobenius problem. To give a flavor, we mention two landmark results that go beyond $d = 2$.

The first one concerns the generating function $r(z) := \sum_{k \in R} z^k$, where R is the set of all integers representable by a given set of relatively prime positive integers $a_1, a_2, \dots, a_d$. It is not hard to see (Exercise 1.34) that $r(z) = p(z) / \left(1 - z^{a_1}\right)\left(1 - z^{a_2}\right) \cdots \left(1 - z^{a_d}\right)$ for some polynomial p. This rational generating function contains all the information about the Frobenius problem; for example, the Frobenius number is the total degree of the function $\frac{1}{1-z} - r(z)$. Hence the Frobenius problem reduces to finding the polynomial p, the numerator of r. Marcel Morales [134, 135] and Graham Denham [73] discovered the remarkable fact that for $d = 3$, the polynomial p has either 4 or 6 terms. Moreover, they gave semi-explicit formulas for p. The Morales–Denham theorem implies that the Frobenius number in the case $d = 3$ is quickly computable, a result that is originally due, in various disguises, to Jürgen Herzog [95], Harold Greenberg [89], and J. Leslie Davison [65]. As much as there seems to be a well-defined border between the cases $d = 2$ and $d = 3$, there also seems to be such a border between the cases $d = 3$ and $d = 4$: Henrik Bresinsky [43] proved that for $d \geq 4$, there is no absolute bound for the number of terms in the numerator p, in sharp contrast to the Morales–Denham theorem.

On the other hand, Alexander Barvinok and Kevin Woods [14] proved that for fixed d, the rational generating function $r(z)$ can be written as a "short" sum of rational functions; in particular, r can be efficiently computed when d is fixed. A corollary of this fact is that the Frobenius number can be efficiently computed when d is fixed; this theorem is due to Ravi Kannan [105]. On the other hand, Jorge Ramírez-Alfonsín [152] proved that trying to efficiently compute the Frobenius number is hopeless if d is left as a variable.

While the above results settle the theoretical complexity of the computation of the Frobenius number, practical algorithms are a completely different matter. Both Kannan's and Barvinok–Woods's ideas seem complex enough that nobody has yet tried to implement them. Currently, the fastest algorithm is presented in [32].

Exercises

1.1. ♣ Check the partial fraction expansion (1.2):

$$\frac{z}{1-z-z^2} = \frac{1/\sqrt{5}}{1-\frac{1+\sqrt{5}}{2}\,z} - \frac{1/\sqrt{5}}{1-\frac{1-\sqrt{5}}{2}\,z}.$$

1.2. ♣ Suppose z is a complex number, and n is a positive integer. Show that

$$(1-z)\left(1+z+z^2+\cdots+z^n\right) = 1 - z^{n+1},$$

and use this to prove that if $|z| < 1$,

$$\sum_{k \geq 0} z^k = \frac{1}{1-z}.$$

1.3. ♣ Find a formula for the number of lattice points in $[a, b]$ for arbitrary real numbers a and b.

1.4. Prove the following. Unless stated differently, $n \in \mathbb{Z}$ and $x, y \in \mathbb{R}$.

(a) $\lfloor x + n \rfloor = \lfloor x \rfloor + n$.
(b) $\lfloor x \rfloor + \lfloor y \rfloor \leq \lfloor x + y \rfloor \leq \lfloor x \rfloor + \lfloor y \rfloor + 1$.
(c) $\lfloor x \rfloor + \lfloor -x \rfloor = \begin{cases} 0 & \text{if } x \in \mathbb{Z}, \\ -1 & \text{otherwise.} \end{cases}$
(d) For $n \in \mathbb{Z}_{>0}$, $\left\lfloor \frac{\lfloor x \rfloor}{n} \right\rfloor = \left\lfloor \frac{x}{n} \right\rfloor$.
(e) $-\lfloor -x \rfloor$ is the least integer greater than or equal to x, denoted by $\lceil x \rceil$.

(f) $\lfloor x + 1/2 \rfloor$ is the nearest integer to x (and if two integers are equally near to x, it is the larger of the two).
(g) $\lfloor x \rfloor + \lfloor x + 1/2 \rfloor = \lfloor 2x \rfloor$.
(h) If m and n are positive integers, $\left\lfloor \frac{m}{n} \right\rfloor$ is the number of integers among $1, \dots, m$ that are divisible by n.
(i) ♣ If $m \in \mathbb{Z}_{>0}, n \in \mathbb{Z}$, then $\left\lfloor \frac{n-1}{m} \right\rfloor = - \left\lfloor \frac{-n}{m} \right\rfloor - 1$.
(j) ♣ If $m \in \mathbb{Z}_{>0}, n \in \mathbb{Z}$, then $\left\lfloor \frac{n-1}{m} \right\rfloor + 1$ is the least integer greater than or equal to n/m.

1.5. Rewrite in terms of the fractional-part function as many of the above identities as you can make sense of.

1.6. Suppose m and n are relatively prime positive integers. Prove that

$$\sum_{k=0}^{m-1} \left\lfloor \frac{kn}{m} \right\rfloor = \sum_{j=0}^{n-1} \left\lfloor \frac{jm}{n} \right\rfloor = \frac{1}{2}(m-1)(n-1) .$$

1.7. Prove the following identities. They will become handy at least twice: when we study partial fractions, and when we discuss finite Fourier series. For $\phi, \psi \in \mathbb{R}, n \in \mathbb{Z}_{>0}, m \in \mathbb{Z}$,

(a) $e^{i0} = 1$,
(b) $e^{i\phi}\, e^{i\psi} = e^{i(\phi+\psi)}$,
(c) $1/e^{i\phi} = e^{-i\phi}$,
(d) $e^{i(\phi+2\pi)} = e^{i\phi}$,
(e) $e^{2\pi i} = 1$,
(f) $\left|e^{i\phi}\right| = 1$,
(g) $\frac{d}{d\phi}\, e^{i\phi} = i\, e^{i\phi}$,
(h) $\sum_{k=0}^{n-1} e^{2\pi i k m/n} = \begin{cases} n & \text{if } n|m, \\ 0 & \text{otherwise,} \end{cases}$
(i) $\sum_{k=1}^{n-1} k\, e^{2\pi i k/n} = \frac{n}{e^{2\pi i/n} - 1}$.

1.8. Suppose $m, n \in \mathbb{Z}$ and $n > 0$. Find a closed form for $\sum_{k=0}^{n-1} \left\{ \frac{k}{n} \right\} e^{2\pi i k m/n}$ (as a function of m and n).

1.9. ♣ Suppose m and n are relatively prime integers, and n is positive. Show that

$$\left\{ e^{2\pi i m k/n} : \; 0 \le k < n \right\} = \left\{ e^{2\pi i j/n} : \; 0 \le j < n \right\}$$

and

$$\left\{ e^{2\pi i m k/n} : \; 0 < k < n \right\} = \left\{ e^{2\pi i j/n} : \; 0 < j < n \right\} .$$

Conclude that if f is any complex-valued function, then

$$\sum_{k=0}^{n-1} f\left(e^{2\pi i m k/n}\right) = \sum_{j=0}^{n-1} f\left(e^{2\pi i j/n}\right)$$

and

$$\sum_{k=1}^{n-1} f\left(e^{2\pi imk/n}\right) = \sum_{j=1}^{n-1} f\left(e^{2\pi ij/n}\right).$$

1.10. Suppose n is a positive integer. If you know what a *group* is, prove that the set $\{e^{2\pi ik/n} : 0 \le k < n\}$ forms a cyclic group of order n (under multiplication in $\mathbb{C}$).

1.11. Fix $n \in \mathbb{Z}_{>0}$. For an integer m, let $(m \bmod n)$ denote the least nonnegative integer in $G_1 := \mathbb{Z}_n$ to which m is congruent. Let's denote by $\star$ addition modulo n, and by $\circ$ the following composition:

$$\left\{\frac{m_1}{n}\right\} \circ \left\{\frac{m_2}{n}\right\} = \left\{\frac{m_1 + m_2}{n}\right\},$$

defined on the set $G_2 := \left\{\left\{\frac{m}{n}\right\} : m \in \mathbb{Z}\right\}$. Define the following functions:

$$\phi\left((m \bmod n)\right) = e^{2\pi im/n},$$
$$\psi\left(e^{2\pi im/n}\right) = \left\{\frac{m}{n}\right\},$$
$$\chi\left(\left\{\frac{m}{n}\right\}\right) = (m \bmod n).$$

Prove the following:

$$\phi\left((m_1 \bmod n) \star (m_2 \bmod n)\right) = \phi\left((m_1 \bmod n)\right) \phi\left((m_2 \bmod n)\right),$$
$$\psi\left(e^{2\pi im_1/n} e^{2\pi im_2/n}\right) = \psi\left(e^{2\pi im_1/n}\right) \circ \psi\left(e^{2\pi im_2/n}\right),$$
$$\chi\left(\left\{\frac{m_1}{n}\right\} \circ \left\{\frac{m_2}{n}\right\}\right) = \chi\left(\left\{\frac{m_1}{n}\right\}\right) \star \chi\left(\left\{\frac{m_2}{n}\right\}\right).$$

Prove that the three maps defined above, namely ϕ, ψ, and χ, are one-to-one. Again, for the reader who is familiar with the notion of a *group*, let G_3 be the group of nth roots of unity. What we have shown is that the three groups G_1, G_2, and G_3 are all isomorphic. It is very useful to cycle among these three isomorphic groups.

1.12. ♣ Given integers a, b, c, d, form the line segment in $\mathbb{R}^2$ joining the point (a, b) to (c, d). Show that the number of integer points on this line segment is $\gcd(a - c, b - d) + 1$.

1.13. Give an example of a line with

(a) no lattice point;
(b) one lattice point;
(c) an infinite number of lattice points.

In each case, state—if appropriate—necessary conditions about the (ir)rationality of the slope.

1.14. Suppose a line $y = mx + b$ passes through the lattice points (p_1, q_1) and (p_2, q_2). Prove that it also passes through the lattice points

$$\left(p_1 + k(p_2 - p_1), q_1 + k(q_2 - q_1)\right), \quad k \in \mathbb{Z}.$$

1.15. Given positive irrational numbers p and q with $\frac{1}{p} + \frac{1}{q} = 1$, show that $\mathbb{Z}_{>0}$ is the disjoint union of the two integer sequences $\{\lfloor pn \rfloor : n \in \mathbb{Z}_{>0}\}$ and $\{\lfloor qn \rfloor : n \in \mathbb{Z}_{>0}\}$. This theorem from 1894 is due to Lord Rayleigh and was rediscovered in 1926 by Sam Beatty. Sequences of the form $\{\lfloor pn \rfloor : n \in \mathbb{Z}_{>0}\}$ are often called *Beatty sequences.*

1.16. Let $a, b, c, d \in \mathbb{Z}$. We say that $\{(a, b), (c, d)\}$ is a *lattice basis* of $\mathbb{Z}^2$ if any lattice point $(m, n) \in \mathbb{Z}^2$ can be written as

$$(m, n) = p\,(a, b) + q\,(c, d)$$

for some $p, q \in \mathbb{Z}$. Prove that if $\{(a, b), (c, d)\}$ and $\{(e, f), (g, h)\}$ are lattice bases of $\mathbb{Z}^2$ then there exists an integer matrix M with determinant ± 1 such that

$$\begin{pmatrix} a & b \\ c & d \end{pmatrix} = M \begin{pmatrix} e & f \\ g & h \end{pmatrix}.$$

Conclude that the determinant of $\begin{pmatrix} a & b \\ c & d \end{pmatrix}$ is ± 1.

1.17. ♣ Prove that a triangle with vertices on the integer lattice has no other interior/boundary lattice points if and only if it has area $\frac{1}{2}$. (*Hint:* You may begin by "doubling" the triangle to form a parallelogram.)

1.18. Let's define a *northeast lattice path* as a path through lattice points that uses only the steps $(1, 0)$ and $(0, 1)$. Let L_n be the line defined by $x + 2y = n$. Prove that the number of northeast lattice paths from the origin to a lattice point on L_n is the $(n+1)^{\text{th}}$ Fibonacci number f_{n+1}.

1.19. Compute the coefficients of the Taylor series of $1/(1-z)^2$ expanded at $z = 0$

(a) by a counting argument,
(b) by differentiating the geometric series.

Generalize.

1.20. ♣ Prove that if $a_1, a_2, \ldots, a_d \in \mathbb{Z}_{>0}$ do not have a common factor then the Frobenius number $g(a_1, \ldots, a_d)$ is well defined.

1.21. ♣ Compute the partial fraction coefficients (1.6).

1.22. ♣ Prove (1.9): For relatively prime positive integers a and b,

$$\frac{1}{a}\sum_{k=1}^{a-1}\frac{1}{\left(1-\xi_a^{bk}\right)\xi_a^{kn}}=\frac{1}{a}\sum_{k=1}^{a-1}\frac{1}{\left(1-\xi_a^{k}\right)\xi_a^{b^{-1}kn}},$$

where $b^{-1}b \equiv 1 \bmod a$, and deduce from this (1.10), namely,

$$\frac{1}{a}\sum_{k=1}^{a-1}\frac{1}{\left(1-\xi_a^{bk}\right)\xi_a^{kn}}=-\left\{\frac{b^{-1}n}{a}\right\}+\frac{1}{2}-\frac{1}{2a}.$$

(*Hint:* Use Exercise 1.9.)

1.23. Prove that for relatively prime positive integers a and b,

$$p_{\{a,b\}}(n+ab)=p_{\{a,b\}}(n)+1.$$

1.24. ♣ Show that if a and b are relatively prime positive integers, then

$$p_{\{a,b\}}(a+b)=1.$$

1.25. To extend the Frobenius problem, let us call an integer n *k-representable* if $p_A(n)=k$; that is, n can be represented in exactly k ways using the integers in the set A. Define $g_k = g_k(a_1,\dots,a_d)$ to be the largest k-representable integer. Prove:

(a) Let $d=2$. For any $k\in\mathbb{Z}_{\geq 0}$ there is an N such that all integers larger than N have at least k representations (and hence $g_k(a,b)$ is well defined).
(b) $g_k(a,b)=(k+1)ab-a-b$.
(c) Given $k\geq 2$, the smallest k-representable integer is $ab(k-1)$.
(d) The smallest interval containing all uniquely representable integers is $[\min(a,b), g_1(a,b)]$.
(e) Given $k\geq 2$, the smallest interval containing all k-representable integers is $[g_{k-2}(a,b)+a+b, g_k(a,b)]$.
(f) There are exactly $ab-1$ integers that are uniquely representable. Given $k\geq 2$, there are exactly ab k-representable integers.
(g) Extend all of this to $d\geq 3$ (see open problems).

1.26. Find a formula for $p_{\{a\}}(n)$.

1.27. Prove the following recursion formula:

$$p_{\{a_1,\dots,a_d\}}(n)=\sum_{m\geq 0}p_{\{a_1,\dots,a_{d-1}\}}(n-ma_d).$$

(Here we use the convention that $p_A(n)=0$ if $n<0$.) Use it in the case $d=2$ to give an alternative proof of Theorem 1.2.

1.28. Prove the following extension of Theorem 1.5: Suppose $\gcd(a,b) = d$. Then

$$p_{\{a,b\}}(n) = \begin{cases} \frac{nd}{ab} - \left\{ \frac{\beta n}{a} \right\} - \left\{ \frac{\alpha n}{b} \right\} + 1 & \text{if } d|n, \\ 0 & \text{otherwise,} \end{cases}$$

where $\beta \frac{b}{d} \equiv 1 \bmod \frac{a}{d}$, and $\alpha \frac{a}{d} \equiv 1 \bmod \frac{b}{d}$.

1.29. ♣ Compute the partial fraction coefficient (1.12).

1.30. Find a formula for $p_{\{a,b,c\}}(n)$ for the case $\gcd(a,b,c) \neq 1$.

1.31. ♣ With $A = \{a_1, a_2, \dots, a_d\} \subset \mathbb{Z}_{>0}$, let

$$p_A^\circ(n) := \# \left\{ (m_1, \dots, m_d) \in \mathbb{Z}^d : \text{ all } m_j > 0, \; m_1 a_1 + \dots + m_d a_d = n \right\};$$

that is, $p_A^\circ(n)$ counts the number of partitions of n using only the elements of A as parts, *where each part is used at least once.* Find formulas for p_A° for $A = \{a\}, A = \{a,b\}, A = \{a,b,c\}, A = \{a,b,c,d\}$, where a, b, c, d are pairwise relatively prime positive integers. Observe that in all examples, the counting functions p_A and p_A° satisfy the algebraic relation

$$p_A^\circ(-n) = (-1)^{d-1} p_A(n).$$

1.32. Prove that $p_A^\circ(n) = p_A(n - a_1 - a_2 - \dots - a_d)$. (Here, as usual, $A = \{a_1, a_2, \dots, a_d\}$.) Conclude that in the examples of Exercise 1.31 the algebraic relation

$$p_A(-t) = (-1)^{d-1} p_A(t - a_1 - a_2 - \dots - a_d)$$

holds.

1.33. For relatively prime positive integers a, b, let

$$R := \{am + bn : \; m, n \in \mathbb{Z}_{\geq 0}\},$$

the set of all integers representable by a and b. Prove that

$$\sum_{k \in R} z^k = \frac{1 - z^{ab}}{(1 - z^a)(1 - z^b)}.$$

Use this rational generating function to give alternative proofs of Theorems 1.2 and 1.3.

1.34. For relatively prime positive integers $a_1, a_2, \dots, a_d$, let

$$R := \{m_1 a_1 + m_2 a_2 + \dots + m_d a_d : \; m_1, m_2, \dots, m_d \in \mathbb{Z}_{\geq 0}\},$$

the set of all integers representable by $a_1, a_2, \dots, a_d$. Prove that

$$r(z) := \sum_{k \in R} z^k = \frac{p(z)}{(1 - z^{a_1})(1 - z^{a_2}) \cdots (1 - z^{a_d})}$$

for some polynomial p.

1.35. Prove Theorem 1.1: Given any rational function $\frac{p(z)}{\prod_{k=1}^{m}(z-a_k)^{e_k}}$, where p is a polynomial of degree less than $e_1+e_2+\cdots+e_m$ and the a_k's are distinct, there exists a decomposition

$$\sum_{k=1}^{m}\left(\frac{c_{k,1}}{z-a_k}+\frac{c_{k,2}}{(z-a_k)^2}+\cdots+\frac{c_{k,e_k}}{(z-a_k)^{e_k}}\right),$$

where the $c_{k,j}\in\mathbb{C}$ are unique.

Here is an outline of one possible proof. Recall that the set of polynomials (over $\mathbb{R}$ or $\mathbb{C}$) forms a *Euclidean domain*, that is, given any two polynomials $a(z), b(z)$, there exist polynomials $q(z), r(z)$ with $\deg(r)<\deg(b)$, such that

$$a(z)=b(z)q(z)+r(z).$$

Applying this procedure repeatedly (the *Euclidean algorithm*) gives the greatest common divisor of $a(z)$ and $b(z)$ as a linear combination of them, that is, there exist polynomials $c(z)$ and $d(z)$ such that $a(z)c(z)+b(z)d(z)=\gcd\left(a(z),b(z)\right)$.

Step 1: Apply the Euclidean algorithm to show that there exist polynomials u_1, u_2 such that

$$u_1(z)\left(z-a_1\right)^{e_1}+u_2(z)\left(z-a_2\right)^{e_2}=1.$$

Step 2: Deduce that there exist polynomials v_1, v_2 with $\deg\left(v_k\right)<e_k$ such that

$$\frac{p(z)}{\left(z-a_1\right)^{e_1}\left(z-a_2\right)^{e_2}}=\frac{v_1(z)}{\left(z-a_1\right)^{e_1}}+\frac{v_2(z)}{\left(z-a_2\right)^{e_2}}.$$

(*Hint:* Long division.)

Step 3: Repeat this procedure to obtain a partial fraction decomposition for

$$\frac{p(z)}{\left(z-a_1\right)^{e_1}\left(z-a_2\right)^{e_2}\left(z-a_3\right)^{e_3}}.$$

Open Problems

1.36. Come up with a new approach or a new algorithm for the Frobenius problem in the $d=4$ case.

1.37. There are a very good lower [65] and several upper bounds [153, Chapter 3] for the Frobenius number. Come up with improved upper bounds.

1.38. Solve Vladimir I. Arnold's Problems 1999-8 through 1999-11 [7]. To give a flavor, we mention two of the problems explicitly:

(a) Explore the statistics of $g(a_1, a_2, \ldots, a_d)$ for typical large $a_1, a_2, \ldots, a_d$. It is conjectured that $g(a_1, a_2, \ldots, a_d)$ grows asymptotically like a constant times $\sqrt[d-1]{a_1 a_2 \cdots a_d}$.
(b) Determine what fraction of the integers in the interval $[0, g(a_1, a_2, \ldots, a_d)]$ is representable, for typical large $a_1, a_2, \ldots, a_d$. It is conjectured that this fraction is asymptotically equal to $\frac{1}{d}$. (Theorem 1.3 implies that this conjecture is true in the case $d = 2$.)

1.39. Study vector generalizations of the Frobenius problem [155, 164].

1.40. There are several special cases of $A = \{a_1, a_2, \ldots, a_d\}$ for which the Frobenius problem is solved, for example, arithmetic sequences [153, Chapter 3]. Study these special cases in light of the generating function $r(x)$, defined in the Notes and in Exercise 1.34.

1.41. Study the generalized Frobenius number g_k (defined in Exercise 1.25), e.g., in light of the Morales–Denham theorem mentioned in the Notes. Derive formulas for special cases, e.g., arithmetic sequences.

1.42. For which $0 \le n \le b-1$ is $s_n(a_1, a_2, \ldots, a_d; b) = 0$?

2

A Gallery of Discrete Volumes

Few things are harder to put up with than a good example.

Mark Twain (1835–1910)

A unifying theme of this book is the study of the number of integer points in polytopes, where the polytopes lives in a real Euclidean space $\mathbb{R}^d$. The integer points $\mathbb{Z}^d$ form a lattice in $\mathbb{R}^d$, and we often call the integer points *lattice points.* This chapter carries us through concrete instances of lattice-point enumeration in various integral and rational polytopes. There is a tremendous amount of research taking place along these lines, even as the reader is looking at these pages.

2.1 The Language of Polytopes

A polytope in dimension 1 is a closed interval; the number of integer points in $\left[\frac{a}{b}, \frac{c}{d}\right]$ is easily seen to be $\left\lfloor \frac{c}{d} \right\rfloor - \left\lfloor \frac{a-1}{b} \right\rfloor$ (Exercise 2.1). A 2-dimensional convex polytope is a **convex polygon**: a compact convex subset of $\mathbb{R}^2$ bounded by a simple, closed curve that is made up of finitely many line segments.

In general dimension d, a **convex polytope** is the convex hull of finitely many points in $\mathbb{R}^d$. To be precise, given any finite point set $\{\mathbf{v}_1, \mathbf{v}_2, \dots, \mathbf{v}_n\} \subset \mathbb{R}^d$, the polytope $\mathcal{P}$ is the smallest convex set containing those points; that is,

$$\mathcal{P} = \{\lambda_1 \mathbf{v}_1 + \lambda_2 \mathbf{v}_2 + \cdots + \lambda_n \mathbf{v}_n : \text{ all } \lambda_k \geq 0 \text{ and } \lambda_1 + \lambda_2 + \cdots + \lambda_n = 1\}.$$

This definition is called the **vertex description** of $\mathcal{P}$, and we use the notation

$$\mathcal{P} = \operatorname{conv}\{\mathbf{v}_1, \mathbf{v}_2, \dots, \mathbf{v}_n\},$$

the convex hull of $\mathbf{v}_1, \mathbf{v}_2, \dots, \mathbf{v}_n$. In particular, a polytope is a *closed* subset of $\mathbb{R}^d$. Many polytopes we will study, however, are not defined this way, but rather as bounded intersections of finitely many half-spaces and hyperplanes. One example is the polytope $\mathcal{P}$ defined by (1.4) in Chapter 1. This **hyperplane description** of a polytope is, in fact, equivalent to the vertex description. The fact that every polytope has both a vertex and a hyperplane description is highly nontrivial, both algorithmically and conceptually. We carefully work out a proof in Appendix A.

The **dimension** of a polytope $\mathcal{P}$ is the dimension of the affine space

$$\operatorname{span} \mathcal{P} := \{\mathbf{x} + \lambda(\mathbf{y} - \mathbf{x}) : \mathbf{x}, \mathbf{y} \in \mathcal{P}, \lambda \in \mathbb{R}\}$$

spanned by $\mathcal{P}$. If $\mathcal{P}$ has dimension d, we use the notation $\dim \mathcal{P} = d$ and call $\mathcal{P}$ a d-polytope. Note that $\mathcal{P} \subset \mathbb{R}^d$ does not necessarily have dimension d. For example, the polytope $\mathcal{P}$ defined by (1.4) has dimension $d - 1$.

Given a convex polytope $\mathcal{P} \subset \mathbb{R}^d$, we say that the hyperplane $H = \left\{\mathbf{x} \in \mathbb{R}^d : \mathbf{a} \cdot \mathbf{x} = b\right\}$ is a **supporting hyperplane** of $\mathcal{P}$ if $\mathcal{P}$ lies entirely on one side of H, that is, $\mathcal{P} \subset \left\{\mathbf{x} \in \mathbb{R}^d : \mathbf{a} \cdot \mathbf{x} \leq b\right\}$ or $\mathcal{P} \subset \left\{\mathbf{x} \in \mathbb{R}^d : \mathbf{a} \cdot \mathbf{x} \geq b\right\}$. A **face** of $\mathcal{P}$ is a set of the form $\mathcal{P} \cap H$, where H is a supporting hyperplane of $\mathcal{P}$. Note that $\mathcal{P}$ itself is a face of $\mathcal{P}$, corresponding to the *degenerate hyperplane* $\mathbb{R}^d$,[1] and the empty set $\varnothing$ is a face of $\mathcal{P}$, corresponding to a hyperplane that does not meet $\mathcal{P}$. The $(d-1)$-dimensional faces are called **facets**, the 1-dimensional faces **edges**, and the 0-dimensional faces **vertices** of $\mathcal{P}$. Vertices are the "extreme points" of a polytope.

A convex d-polytope has at least $d+1$ vertices. A convex d-polytope with exactly $d+1$ vertices is called a d-**simplex**. Every 1-dimensional convex polytope is a 1-simplex, namely, a line segment. The 2-dimensional simplices are the triangles, the 3-dimensional simplices the tetrahedra.

A convex polytope $\mathcal{P}$ is called **integral** if all of its vertices have integer coordinates, and $\mathcal{P}$ is called **rational** if all of its vertices have rational coordinates.

2.2 The Unit Cube

As a warm-up example, we begin with the **unit d-cube** $\square := [0,1]^d$, which simultaneously offers simple geometry and an endless fountain of research questions. The vertex description of $\square$ is given by the set of 2^d vertices $\{(x_1, x_2, \ldots, x_d) :$ all $x_k = 0$ or $1\}$. The hyperplane description is

$$\square = \left\{(x_1, x_2, \ldots, x_d) \in \mathbb{R}^d : 0 \leq x_k \leq 1 \text{ for all } k = 1, 2, \ldots, d\right\}.$$

Thus, there are the $2d$ bounding hyperplanes $x_1 = 0$, $x_1 = 1$, $x_2 = 0$, $x_2 = 1$, ..., $x_d = 0$, $x_d = 1$.

[1] In the remainder of the book, we will reserve the term *hyperplane* for nondegenerate hyperplanes, i.e., sets of the form $\left\{\mathbf{x} \in \mathbb{R}^d : \mathbf{a} \cdot \mathbf{x} = b\right\}$, where not all of the entries of $\mathbf{a}$ are zero.

We now compute the discrete volume of any integer dilate of $\square$. That is, we seek the number of integer points $t\,\square \cap \mathbb{Z}^d$ for all $t \in \mathbb{Z}_{>0}$. Here $t\mathcal{P}$ denotes the dilated polytope

$$\{(tx_1, tx_2, \dots, tx_d) : (x_1, x_2, \dots, x_d) \in \mathcal{P}\},$$

for any polytope $\mathcal{P}$. What is the discrete volume of $\square$? We dilate by the positive integer t, as depicted in Figure 2.1, and count:

$$\#\left(t\,\square \cap \mathbb{Z}^d\right) = \#\left([0,t]^d \cap \mathbb{Z}^d\right) = (t+1)^d.$$

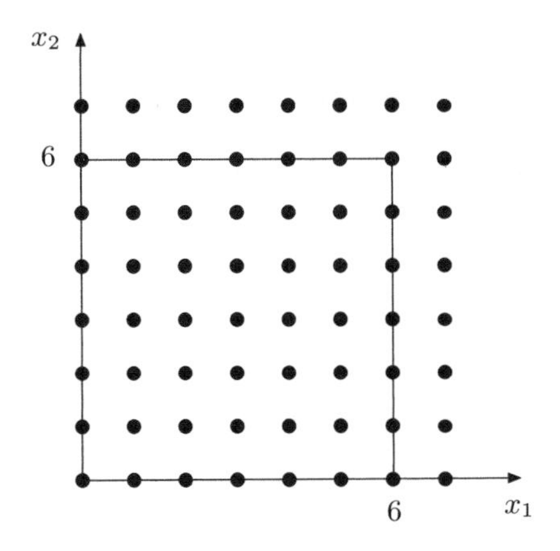

Fig. 2.1. The sixth dilate of $\square$ in dimension 2.

We generally denote the **lattice-point enumerator** for the tth dilates of $\mathcal{P} \subset \mathbb{R}^d$ by

$$L_{\mathcal{P}}(t) := \#\left(t\mathcal{P} \cap \mathbb{Z}^d\right),$$

a useful object that we also call the **discrete volume** of $\mathcal{P}$. We may also think of leaving $\mathcal{P}$ fixed and shrinking the integer lattice:

$$L_{\mathcal{P}}(t) = \#\left(\mathcal{P} \cap \frac{1}{t}\,\mathbb{Z}^d\right).$$

With this convention, $L_{\square}(t) = (t+1)^d$, a polynomial in the integer variable t. Notice that the coefficients of this polynomial are the **binomial coefficients** $\binom{d}{k}$, defined through

$$\binom{m}{n} := \frac{m(m-1)(m-2)\cdots(m-n+1)}{n!} \tag{2.1}$$

for $m \in \mathbb{C}, n \in \mathbb{Z}_{>0}$.

What about the *interior* $\Box^\circ$ of the cube? The number of interior integer points in $t\,\Box^\circ$ is

$$L_{\Box^\circ}(t) = \#\left(t\,\Box^\circ \cap \mathbb{Z}^d\right) = \#\left((0,t)^d \cap \mathbb{Z}^d\right) = (t-1)^d .$$

Notice that this polynomial equals $(-1)^d L_\Box(-t)$, the evaluation of the polynomial $L_\Box(t)$ at negative integers, up to a sign.

We now introduce another important tool for analyzing any polytope $\mathcal{P}$, namely the generating function of $L_\mathcal{P}$:

$$\mathrm{Ehr}_\mathcal{P}(z) := 1 + \sum_{t\ge 1} L_\mathcal{P}(t)\, z^t .$$

This generating function is also called the **Ehrhart series** of $\mathcal{P}$.

In our case, the Ehrhart series of $\mathcal{P} = \Box$ takes on a special form. To illustrate, we define the **Eulerian number** $A(d,k)$ through

$$\sum_{j\ge 0} j^d\, z^j = \frac{\sum_{k=0}^{d} A(d,k)\, z^k}{(1-z)^{d+1}} . \tag{2.2}$$

It is not hard to prove that the polynomial $\sum_{k=1}^{d} A(d,k)\, z^k$ is the numerator of the rational function

$$\left(z\frac{d}{dz}\right)^d \left(\frac{1}{1-z}\right) = \underbrace{z\frac{d}{dz}\cdots z\frac{d}{dz}}_{d \text{ times}} \left(\frac{1}{1-z}\right) .$$

The Eulerian numbers have many fascinating properties, including

$$\begin{aligned} A(d,k) &= A(d, d+1-k)\,, \\ A(d,k) &= (d-k+1)\, A(d-1,k-1) + k\, A(d-1,k)\,, \\ \sum_{k=0}^{d} A(d,k) &= d!\,, \\ A(d,k) &= \sum_{j=0}^{k} (-1)^j \binom{d+1}{j} (k-j)^d , \end{aligned} \tag{2.3}$$

for all $1 \le k \le d$. The first few Eulerian numbers $A(d,k)$ for $0 \le k \le d$ are

$d = 0$: 1
$d = 1$: 0 1
$d = 2$: 0 1 1
$d = 3$: 0 1 4 1
$d = 4$: 0 1 11 11 1
$d = 5$: 0 1 26 66 26 1
$d = 6$: 0 1 57 302 302 57 1 .

(See also [165, Sequence A008292].)

With this definition, we can now express the Ehrhart series of $\square$ in terms of Eulerian numbers:

$$\begin{aligned}\mathrm{Ehr}_{\square}(z) &= 1 + \sum_{t \geq 1} (t+1)^d \, z^t = \sum_{t \geq 0} (t+1)^d \, z^t = \frac{1}{z} \sum_{t \geq 1} t^d \, z^t \\ &= \frac{\sum_{k=1}^{d} A(d,k) \, z^{k-1}}{(1-z)^{d+1}} \, .\end{aligned}$$

To summarize, we have proved the following theorem.

Theorem 2.1. *Let $\square$ be the unit d-cube.*

(a) *The lattice-point enumerator of $\square$ is the polynomial*

$$L_{\square}(t) = (t+1)^d = \sum_{k=0}^{d} \binom{d}{k} t^k .$$

(b) *Its evaluation at negative integers yields the relation*

$$(-1)^d L_{\square}(-t) = L_{\square^{\circ}}(t) \, .$$

(c) *The Ehrhart series of $\square$ is* $\mathrm{Ehr}_{\square}(z) = \frac{\sum_{k=1}^{d} A(d,k) z^{k-1}}{(1-z)^{d+1}}$. $\square$

2.3 The Standard Simplex

The **standard simplex** Δ in dimension d is the convex hull of the $d+1$ points $\mathbf{e}_1, \mathbf{e}_2, \ldots, \mathbf{e}_d$ and the origin; here $\mathbf{e}_j$ is the unit vector $(0, \ldots, 1, \ldots, 0)$, with a 1 in the jth position. Figure 2.2 shows Δ for $d = 3$. On the other hand, Δ can also be realized by its hyperplane description, namely

$$\Delta = \left\{ (x_1, x_2 \ldots, x_d) \in \mathbb{R}^d : \, x_1 + x_2 + \cdots + x_d \leq 1 \text{ and all } x_k \geq 0 \right\} .$$

In the case of the standard simplex, the dilate $t\Delta$ is now given by

$$t\Delta = \left\{ (x_1, x_2, \ldots, x_d) \in \mathbb{R}^d : \, x_1 + x_2 + \cdots + x_d \leq t \text{ and all } x_k \geq 0 \right\} .$$

To compute the discrete volume of Δ, we would like to use the methods developed in Chapter 1, but there's an extra twist. The counting functions in Chapter 1 were defined by equalities, whereas the standard simplex is defined by an *inequality.* We are trying to count all integer solutions $(m_1, m_2, \ldots, m_d) \in \mathbb{Z}_{\geq 0}^d$ to

$$m_1 + m_2 + \cdots + m_d \leq t \, . \tag{2.4}$$

To translate this inequality in d variables into an equality in $d+1$ variables, we introduce a *slack variable* $m_{d+1} \in \mathbb{Z}_{\geq 0}$, which picks up the difference between the right-hand and left-hand sides of (2.4). So the number

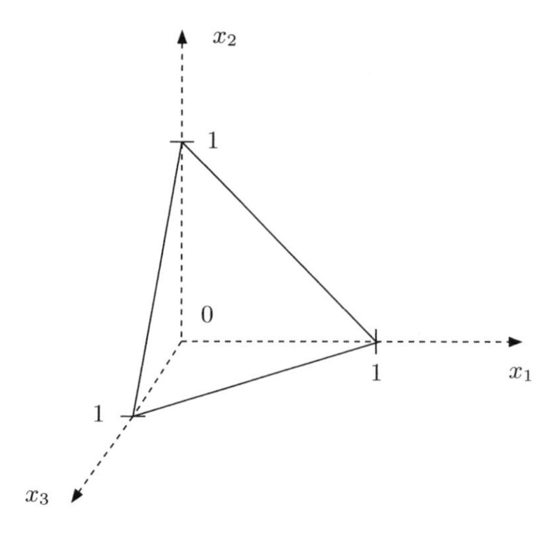

Fig. 2.2. The standard simplex Δ in dimension 3.

of solutions $(m_1, m_2, \ldots, m_d) \in \mathbb{Z}^d_{\geq 0}$ to (2.4) equals the number of solutions $(m_1, m_2, \ldots, m_{d+1}) \in \mathbb{Z}^{d+1}_{\geq 0}$ to

$$m_1 + m_2 + \cdots + m_{d+1} = t\,.$$

Now the methods of Chapter 1 apply:

$$\begin{aligned}
&\# \left(t\Delta \cap \mathbb{Z}^d \right) \\
&\quad = \operatorname{const} \left(\left(\sum_{m_1 \geq 0} z^{m_1} \right) \left(\sum_{m_2 \geq 0} z^{m_2} \right) \cdots \left(\sum_{m_{d+1} \geq 0} z^{m_{d+1}} \right) z^{-t} \right) \\
&\quad = \operatorname{const} \left(\frac{1}{(1-z)^{d+1} z^t} \right) . \qquad (2.5)
\end{aligned}$$

In contrast with Chapter 1, we do not require a partial fraction expansion but simply use the **binomial series**

$$\frac{1}{(1-z)^{d+1}} = \sum_{k \geq 0} \binom{d+k}{d} z^k \qquad (2.6)$$

for $d \geq 0$. The constant-term identity (2.5) requires us to find the coefficient of z^t in the binomial series (2.6), which is $\binom{d+t}{d}$. Hence the discrete volume of Δ is given by $L_\Delta(t) = \binom{d+t}{d}$, a polynomial in the integer variable t of degree d. Incidentally, the coefficients of this polynomial function in t have an alternative life in traditional combinatorics:

$$L_\Delta(t) = \frac{1}{d!} \sum_{k=0}^{d} (-1)^{d-k} \operatorname{stirl}(d+1, k+1)\, t^k,$$

where $\operatorname{stirl}(n, j)$ is the *Stirling number of the first kind* (see Exercise 2.11). We also notice that (2.6) is, by definition, the Ehrhart series of Δ.

Let us repeat this computation for the *interior* Δ° of the standard d-simplex. Now we introduce a slack variable $m_{d+1} > 0$, so that strict inequality is forced:

$$\begin{aligned} L_{\Delta^\circ}(t) &= \#\left\{(m_1, m_2, \ldots, m_d) \in \mathbb{Z}^d_{>0} : m_1 + m_2 + \cdots + m_d < t\right\} \\ &= \#\left\{(m_1, m_2, \ldots, m_{d+1}) \in \mathbb{Z}^{d+1}_{>0} : m_1 + m_2 + \cdots + m_{d+1} = t\right\}. \end{aligned}$$

Now

$$\begin{aligned} L_{\Delta^\circ}(t) &= \operatorname{const}\left(\left(\sum_{m_1>0} z^{m_1}\right)\left(\sum_{m_2>0} z^{m_2}\right)\cdots\left(\sum_{m_{d+1}>0} z^{m_{d+1}}\right) z^{-t}\right) \\ &= \operatorname{const}\left(\left(\frac{z}{1-z}\right)^{d+1} z^{-t}\right) \\ &= \operatorname{const}\left(z^{d+1-t}\sum_{k\ge 0}\binom{d+k}{d} z^k\right) \\ &= \binom{t-1}{d}. \end{aligned}$$

It is a fun exercise to prove that

$$(-1)^d\binom{d-t}{d} = \binom{t-1}{d} \tag{2.7}$$

(see Exercise 2.10). We have arrived at our destination:

Theorem 2.2. *Let Δ be the standard d-simplex.*

(a) *The lattice-point enumerator of Δ is the polynomial $L_\Delta(t) = \binom{d+t}{d}$.*
(b) *Its evaluation at negative integers yields $(-1)^d L_\Delta(-t) = L_{\Delta^\circ}(t)$.*
(c) *The Ehrhart series of Δ is $\operatorname{Ehr}_\Delta(z) = \frac{1}{(1-z)^{d+1}}$.* □

2.4 The Bernoulli Polynomials as Lattice-Point Enumerators of Pyramids

There is a fascinating connection between the Bernoulli polynomials and certain pyramids over unit cubes. The **Bernoulli polynomials** $B_k(x)$ are defined through the generating function

$$\frac{z\, e^{xz}}{e^z - 1} = \sum_{k\ge 0} \frac{B_k(x)}{k!}\, z^k \tag{2.8}$$

and are ubiquitous in the study of the Riemann zeta function, among other objects; they are named after Jacob Bernoulli (1654–1705).[2] The Bernoulli polynomials will play a prominent role in Chapter 10 in the context of Euler–Maclaurin summation. The first few Bernoulli polynomials are

$$\begin{aligned}
B_0(x) &= 1\,, \\
B_1(x) &= x - \frac{1}{2}\,, \\
B_2(x) &= x^2 - x + \frac{1}{6}\,, \\
B_3(x) &= x^3 - \frac{3}{2}x^2 + \frac{1}{2}x\,, \\
B_4(x) &= x^4 - 2x^3 + x^2 - \frac{1}{30}\,, \\
B_5(x) &= x^5 - \frac{5}{2}x^4 + \frac{5}{3}x^3 - \frac{1}{6}x\,, \\
B_6(x) &= x^6 - 3x^5 + \frac{5}{2}x^4 + \frac{1}{2}x^2 + \frac{1}{42}\,, \\
B_7(x) &= x^7 - \frac{7}{2}x^6 + \frac{7}{2}x^5 + \frac{7}{6}x^3 + \frac{1}{6}x\,.
\end{aligned}$$

The **Bernoulli numbers** are $B_k := B_k(0)$ (see also [165, Sequences A000367 & A002445]) and have the generating function

$$\frac{z}{e^z - 1} = \sum_{k \geq 0} \frac{B_k}{k!}\, z^k.$$

Lemma 2.3. *For integers $d \geq 1$ and $n \geq 2$,*

$$\sum_{k=0}^{n-1} k^{d-1} = \frac{1}{d}\left(B_d(n) - B_d\right).$$

Proof. We play with the generating function of $\frac{B_d(n) - B_d}{d!}$:

$$\begin{aligned}
\sum_{d \geq 0} \frac{B_d(n) - B_d}{d!}\, z^d &= z\, \frac{e^{nz} - 1}{e^z - 1} = z \sum_{k=0}^{n-1} e^{kz} = z \sum_{k=0}^{n-1} \sum_{j \geq 0} \frac{(kz)^j}{j!} \\
&= \sum_{j \geq 0} \left(\sum_{k=0}^{n-1} k^j \right) \frac{z^{j+1}}{j!} = \sum_{j \geq 1} \left(\sum_{k=0}^{n-1} k^{j-1} \right) \frac{z^j}{(j-1)!}\,.
\end{aligned}$$

Now compare coefficients on both sides. □

---

[2] For more information about Bernoulli, see
http://www-groups.dcs.st-and.ac.uk/~history/Biographies/Bernoulli_Jacob.html.

Consider a $(d-1)$-dimensional unit cube embedded into $\mathbb{R}^d$ and form a d-dimensional pyramid by adjoining one more vertex at $(0, 0, \ldots, 0, 1)$, as depicted in Figure 2.3. More precisely, this geometric object has the following hyperplane description:

$$\mathcal{P} = \left\{ (x_1, x_2, \ldots, x_d) \in \mathbb{R}^d : \, 0 \le x_1, x_2, \ldots, x_{d-1} \le 1 - x_d \le 1 \right\}. \tag{2.9}$$

By definition, $\mathcal{P}$ is contained in the unit d-cube; in fact, its vertices are a subset of the vertices of the d-cube.

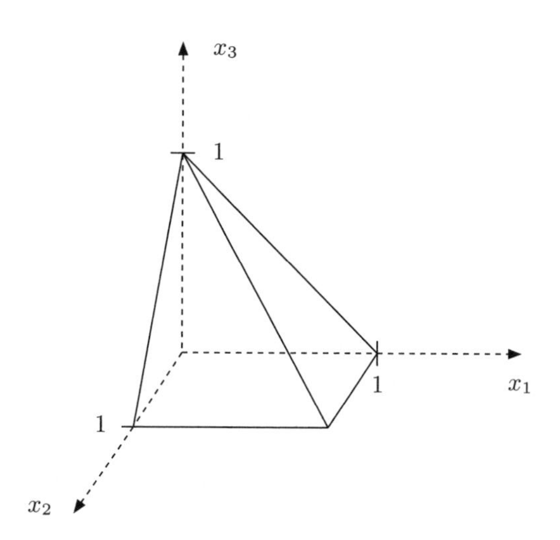

Fig. 2.3. The pyramid $\mathcal{P}$ in dimension 3.

We now count lattice points in integer dilates of $\mathcal{P}$. This number equals

$$\# \left\{ (m_1, m_2, \ldots, m_d) \in \mathbb{Z}^d : \, 0 \le m_k \le t - m_d \le t \text{ for all } k = 1, 2, \ldots, d-1 \right\}.$$

In this case we just count the solutions to $0 \le m_k \le t - m_d \le t$ directly: once we pick the integer m_d (between 0 and t), we have $t - m_d + 1$ independent choices for each of the integers $m_1, m_2, \ldots, m_{d-1}$. Hence

$$L_{\mathcal{P}}(t) = \sum_{m_d=0}^{t} (t - m_d + 1)^{d-1} = \sum_{k=1}^{t+1} k^{d-1} = \frac{1}{d} \left(B_d(t+2) - B_d \right), \tag{2.10}$$

by Lemma 2.3. This is, naturally, a polynomial in t.

We now turn our attention to the number of *interior* lattice points in $\mathcal{P}$:

$$L_{\mathcal{P}^\circ}(t) = \# \left\{ (m_1, m_2, \ldots, m_d) \in \mathbb{Z}^d : \begin{array}{c} 0 < m_k < t - m_d < t \\ \text{for all } k = 1, 2, \ldots, d-1 \end{array} \right\}.$$

By a similar counting argument,

$$L_{\mathcal{P}^\circ}(t) = \sum_{m_d=1}^{t-1} (t - m_d - 1)^{d-1} = \sum_{k=0}^{t-2} k^{d-1} = \frac{1}{d} \left(B_d(t-1) - B_d \right).$$

Incidentally, the Bernoulli polynomials are known (Exercise 2.15) to have the symmetry

$$B_d(1-x) = (-1)^d B_d(x). \tag{2.11}$$

This identity coupled with the fact (Exercise 2.16) that

$$B_d = 0 \text{ for all odd } d \geq 3 \tag{2.12}$$

gives the relation

$$\begin{aligned} L_{\mathcal{P}}(-t) &= \frac{1}{d} \left(B_d(-t+2) - B_d \right) = \frac{1}{d} \left(B_d\left(1 - (t-1)\right) - B_d \right) \\ &= (-1)^d \frac{1}{d} \left(B_d(t-1) - B_d \right) = (-1)^d L_{\mathcal{P}^\circ}(t). \end{aligned}$$

Next we compute the Ehrhart series of $\mathcal{P}$. We can actually do this in somewhat greater generality. Namely, for a $(d-1)$-polytope $\mathcal{Q}$ with vertices $\mathbf{v}_1, \mathbf{v}_2, \dots, \mathbf{v}_m$, define $\operatorname{Pyr}(\mathcal{Q})$, the **pyramid over** $\mathcal{Q}$, as the convex hull of $(\mathbf{v}_1, 0), (\mathbf{v}_2, 0), \dots, (\mathbf{v}_m, 0), (0, \dots, 0, 1)$. In our example above, the d-polytope $\mathcal{P}$ is equal to $\operatorname{Pyr}(\square)$ for the unit $(d-1)$-cube $\square$. The number of integer points in $t \operatorname{Pyr}(\mathcal{Q})$ is, by construction,

$$L_{\operatorname{Pyr}(\mathcal{Q})}(t) = 1 + L_{\mathcal{Q}}(1) + L_{\mathcal{Q}}(2) + \cdots + L_{\mathcal{Q}}(t) = 1 + \sum_{j=1}^{t} L_{\mathcal{Q}}(j),$$

because in $t \operatorname{Pyr}(\mathcal{Q})$, there is one lattice point with x_d-coordinate t, $L_{\mathcal{Q}}(1)$ lattice points with x_d-coordinate $t-1$, $L_{\mathcal{Q}}(2)$ lattice points with x_d-coordinate $t-2$, etc., up to $L_{\mathcal{Q}}(t)$ lattice points with $x_d = 0$. Figure 2.4 shows the instance $t = 3$ for a pyramid over a square.

This identity for $L_{\operatorname{Pyr}(\mathcal{Q})}(t)$ allows us to compute the Ehrhart series of $\operatorname{Pyr}(\mathcal{Q})$ from the Ehrhart series of $\mathcal{Q}$:

Theorem 2.4. $\operatorname{Ehr}_{\operatorname{Pyr}(\mathcal{Q})}(z) = \dfrac{\operatorname{Ehr}_{\mathcal{Q}}(z)}{1-z}.$

Proof.

$$\begin{aligned} \operatorname{Ehr}_{\operatorname{Pyr}(\mathcal{Q})}(z) &= 1 + \sum_{t \geq 1} L_{\operatorname{Pyr}(\mathcal{Q})}(t)\, z^t = 1 + \sum_{t \geq 1} \left(1 + \sum_{j=1}^{t} L_{\mathcal{Q}}(j) \right) z^t \\ &= \sum_{t \geq 0} z^t + \sum_{t \geq 1} \sum_{j=1}^{t} L_{\mathcal{Q}}(j)\, z^t = \frac{1}{1-z} + \sum_{j \geq 1} L_{\mathcal{Q}}(j) \sum_{t \geq j} z^t \\ &= \frac{1}{1-z} + \sum_{j \geq 1} L_{\mathcal{Q}}(j) \frac{z^j}{1-z} = \frac{1 + \sum_{j \geq 1} L_{\mathcal{Q}}(j)\, z^j}{1-z}. \end{aligned}$$

□

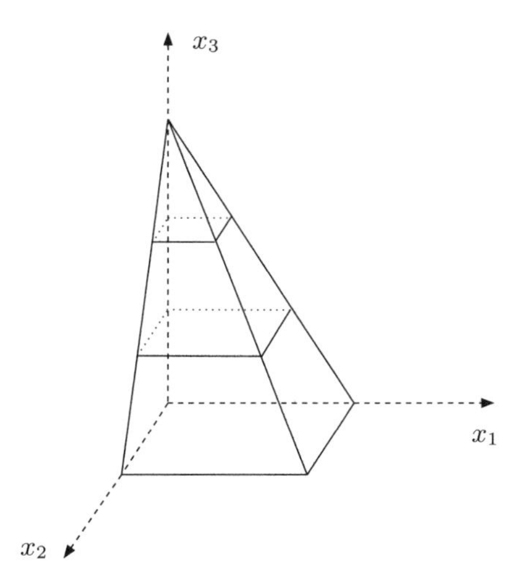

Fig. 2.4. Counting the lattice points in $t\,\mathrm{Pyr}(\mathcal{Q})$.

Our pyramid $\mathcal{P}$ that started this section is a pyramid over the unit $(d-1)$-cube, and so

$$\mathrm{Ehr}_{\mathcal{P}}(z) = \frac{1}{1-z}\,\frac{\sum_{k=1}^{d-1} A\,(d-1,k)\,z^{k-1}}{(1-z)^d} = \frac{\sum_{k=1}^{d-1} A\,(d-1,k)\,z^{k-1}}{(1-z)^{d+1}}\,. \tag{2.13}$$

Incidentally, this Ehrhart series gives rise to a generating function (different from (2.8)) for the Bernoulli polynomial B_d (see Exercise 2.22).

Let's summarize what we have proved for the pyramid over the unit cube.

Theorem 2.5. *Let $\mathcal{P}$ be the d-pyramid*

$$\mathcal{P} = \left\{ (x_1, x_2, \dots, x_d) \in \mathbb{R}^d : 0 \le x_1, x_2, \dots, x_{d-1} \le 1 - x_d \le 1 \right\}.$$

(a) *The lattice-point enumerator of $\mathcal{P}$ is the polynomial*

$$L_{\mathcal{P}}(t) = \frac{1}{d}\left(B_d(t+2) - B_d\right).$$

(b) *Its evaluation at negative integers yields $(-1)^d L_{\mathcal{P}}(-t) = L_{\mathcal{P}^\circ}(t)$.*

(c) *The Ehrhart series of $\mathcal{P}$ is $\mathrm{Ehr}_{\mathcal{P}}(z) = \frac{\sum_{k=1}^{d-1} A(d-1,k)z^{k-1}}{(1-z)^{d+1}}$.* □

Patterns are emerging...

2.5 The Lattice-Point Enumerators of the Cross-Polytopes

Consider the **cross-polytope** $\diamond$ in $\mathbb{R}^d$ given by the hyperplane description

$$\diamond := \left\{ (x_1, x_2, \ldots, x_d) \in \mathbb{R}^d : |x_1| + |x_2| + \cdots + |x_d| \le 1 \right\}. \tag{2.14}$$

Figure 2.5 shows the 3-dimensional instance of $\diamond$, an octahedron. The vertices of $\diamond$ are $(\pm 1, 0, \ldots, 0), (0, \pm 1, 0, \ldots, 0), \ldots, (0, \ldots, 0, \pm 1)$.

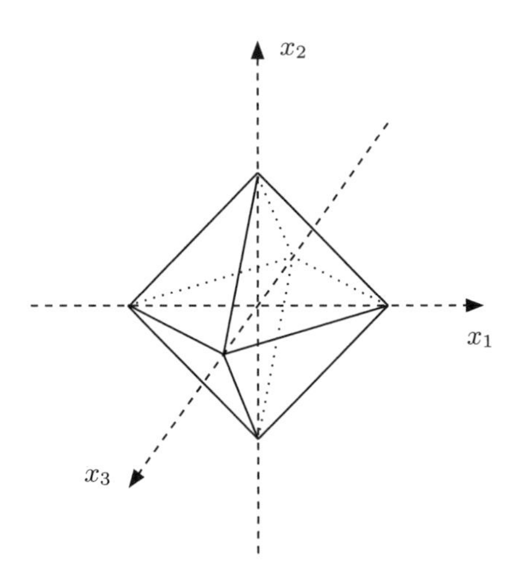

Fig. 2.5. The cross-polytope $\diamond$ in dimension 3.

To compute the discrete volume of $\diamond$, we use a process similar to that of Section 2.4. Namely, for a $(d-1)$-polytope $\mathcal{Q}$ with vertices $\mathbf{v}_1, \mathbf{v}_2, \ldots, \mathbf{v}_m$, define $\operatorname{BiPyr}(\mathcal{Q})$, the **bipyramid over** $\mathcal{Q}$, as the convex hull of

$$(\mathbf{v}_1, 0), (\mathbf{v}_2, 0), \ldots, (\mathbf{v}_m, 0), (0, \ldots, 0, 1), \quad \text{and} \quad (0, \ldots, 0, -1).$$

In our example above, the d-dimensional cross-polytope is the bipyramid over the $(d-1)$-dimensional cross-polytope. The number of integer points in $t \operatorname{BiPyr}(\mathcal{Q})$ is, by construction,

$$\begin{aligned} L_{\operatorname{BiPyr}(\mathcal{Q})}(t) &= 2 + 2L_{\mathcal{Q}}(1) + 2L_{\mathcal{Q}}(2) + \cdots + 2L_{\mathcal{Q}}(t-1) + L_{\mathcal{Q}}(t) \\ &= 2 + 2\sum_{j=1}^{t-1} L_{\mathcal{Q}}(j) + L_{\mathcal{Q}}(t). \end{aligned}$$

This identity allows us to compute the Ehrhart series of $\operatorname{BiPyr}(\mathcal{Q})$ from the Ehrhart series of $\mathcal{Q}$, in a manner similar to the proof of Theorem 2.4. We leave the proof of the following result as Exercise 2.23.

Theorem 2.6. *If $\mathcal{Q}$ contains the origin, then* $\mathrm{Ehr}_{\mathrm{BiPyr}(\mathcal{Q})}(z) = \frac{1+z}{1-z}\,\mathrm{Ehr}_{\mathcal{Q}}(z)$. □

This theorem allows us to compute the Ehrhart series of $\diamond$ effortlessly: The cross-polytope $\diamond$ in dimension 0 is the origin, with Ehrhart series $\frac{1}{1-z}$. The higher-dimensional cross-polytopes can be computed recursively through Theorem 2.6 as

$$\mathrm{Ehr}_\diamond(z) = \frac{(1+z)^d}{(1-z)^{d+1}}\,.$$

Since $\mathrm{Ehr}_\diamond(z) = 1 + \sum_{t \ge 1} L_\diamond(t)\, z^t$, we can retrieve $L_\diamond(t)$ by expanding $\mathrm{Ehr}_\diamond(z)$ into its power series at $z = 0$:

$$\begin{aligned}
\mathrm{Ehr}_\diamond(z) &= \frac{(1+z)^d}{(1-z)^{d+1}} = \frac{\sum_{k=0}^{d}\binom{d}{k} z^k}{(1-z)^{d+1}} \\
&= \sum_{k=0}^{d}\binom{d}{k} z^k \sum_{t\ge 0}\binom{t+d}{d} z^t = \sum_{k=0}^{d}\binom{d}{k}\sum_{t\ge k}\binom{t-k+d}{d} z^t \\
&= \sum_{k=0}^{d}\binom{d}{k}\sum_{t\ge 0}\binom{t-k+d}{d} z^t .
\end{aligned}$$

In the last step we used the fact that $\binom{t-k+d}{d} = 0$ for $0 \le t < k$. But then

$$1 + \sum_{t\ge 1} L_\diamond(t)\, z^t = \sum_{t\ge 0}\sum_{k=0}^{d}\binom{d}{k}\binom{t-k+d}{d} z^t ,$$

and hence $L_\diamond(t) = \sum_{k=0}^{d}\binom{d}{k}\binom{t-k+d}{d}$ for all $t \ge 1$.

We finish this section by counting the *interior* lattice points in $t\diamond$. We start by noticing, since t is an integer, that

$$\begin{aligned}
L_{\diamond^\circ}(t) &= \#\left\{(m_1, m_2, \dots, m_d) \in \mathbb{Z}^d : |m_1| + |m_2| + \dots + |m_d| < t\right\} \\
&= \#\left\{(m_1, m_2, \dots, m_d) \in \mathbb{Z}^d : |m_1| + |m_2| + \dots + |m_d| \le t-1\right\} \\
&= L_\diamond(t-1)\,.
\end{aligned}$$

On the other hand, we can use (2.7):

$$\begin{aligned}
L_\diamond(-t) &= \sum_{k=0}^{d}\binom{d}{k}\binom{-t-k+d}{d} \\
&= \sum_{k=0}^{d}\binom{d}{k}(-1)^d\binom{t-1+k}{d} \\
&= (-1)^d\sum_{k=0}^{d}\binom{d}{d-k}\binom{t-1+d-k}{d} \\
&= (-1)^d L_\diamond(t-1)\,.
\end{aligned}$$

Comparing the last two computations, we see that $(-1)^d L_\diamond(-t) = L_{\diamond^\circ}(t)$. Let us summarize:

Theorem 2.7. *Let $\diamond$ be the cross-polytope in $\mathbb{R}^d$.*

(a) *The lattice-point enumerator of $\diamond$ is the polynomial*

$$L_\diamond(t) = \sum_{k=0}^{d} \binom{d}{k} \binom{t-k+d}{d}.$$

(b) *Its evaluation at negative integers yields $(-1)^d L_\diamond(-t) = L_{\diamond^\circ}(t)$.*

(c) *The Ehrhart series of $\mathcal{P}$ is* $\operatorname{Ehr}_\diamond(z) = \frac{(1+z)^d}{(1-z)^{d+1}}$. □

2.6 Pick's Theorem

Going back to basic concepts, we now give a complete account of $L_\mathcal{P}$ for all integral convex polygons $\mathcal{P}$ in $\mathbb{R}^2$. Denote the number of integer points inside the polygon $\mathcal{P}$ by I, and the number of integer points on the boundary of $\mathcal{P}$ by B. The following result, called *Pick's theorem* in honor of its discoverer Georg Alexander Pick (1859–1942), presents the astonishing fact that the area A of $\mathcal{P}$ can be computed simply by counting lattice points:

Theorem 2.8 (Pick's theorem). *For an integral convex polygon,*

$$A = I + \frac{1}{2}B - 1.$$

Proof. We start by proving that Pick's identity has an additive character: we can decompose $\mathcal{P}$ into the union of two integral polygons $\mathcal{P}_1$ and $\mathcal{P}_2$ by joining two vertices of $\mathcal{P}$ with a line segment, as shown in Figure 2.6.

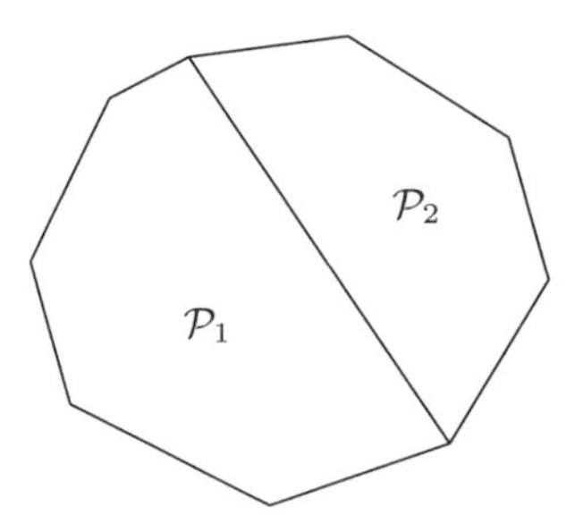

Fig. 2.6. Decomposition of a polygon into two.

We claim that the validity of Pick's identity for $\mathcal{P}$ follows from the validity of Pick's identity for $\mathcal{P}_1$ and $\mathcal{P}_2$. Denote the area, number of interior lattice points, and number of boundary lattice points of $\mathcal{P}_k$ by A_k, I_k, and B_k, respectively, for $k = 1, 2$. Clearly,

$$A = A_1 + A_2 \,.$$

Furthermore, if we denote the number of lattice points on the edge common to $\mathcal{P}_1$ and $\mathcal{P}_2$ by L, then

$$I = I_1 + I_2 + L - 2 \qquad \text{and} \qquad B = B_1 + B_2 - 2L + 2 \,.$$

Thus

$$\begin{aligned} I + \frac{1}{2}B - 1 &= I_1 + I_2 + L - 2 + \frac{1}{2}B_1 + \frac{1}{2}B_2 - L + 1 - 1 \\ &= I_1 + \frac{1}{2}B_1 - 1 + I_2 + \frac{1}{2}B_2 - 1 \,. \end{aligned}$$

This proves the claim. Note that our proof also shows that the validity of Pick's identity for $\mathcal{P}_1$ follows from the validity of Pick's identity for $\mathcal{P}$ and $\mathcal{P}_2$.

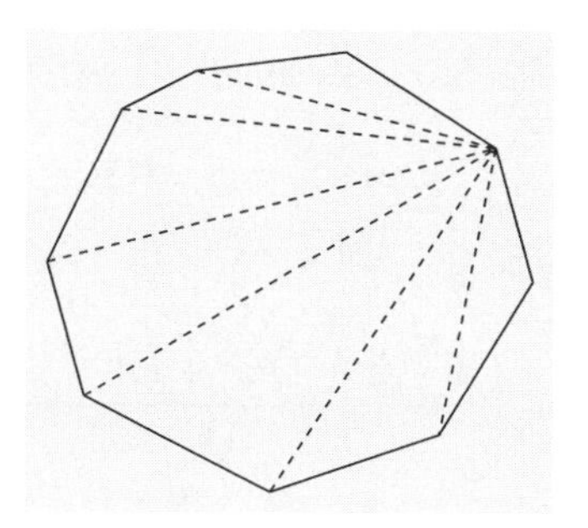

Fig. 2.7. Triangulation of a polygon.

Now, any convex polygon can be decomposed into triangles that share a common vertex, as illustrated in Figure 2.7. Hence it suffices to prove Pick's theorem for triangles. Further simplifying the picture, we can embed any integral triangle into an integral rectangle as suggested by Figure 2.8.

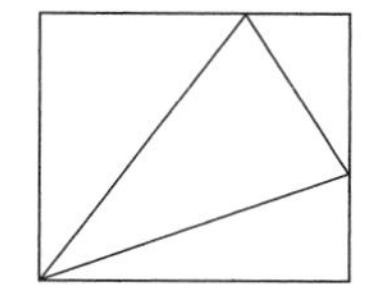

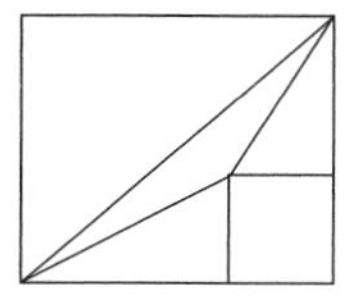

Fig. 2.8. Embedding a triangle in a rectangle.

This reduces the proof of Pick's theorem to proving the theorem for integral rectangles whose edges are parallel to the coordinate axes, and for rectangular triangles two of whose edges are parallel to the coordinate axes. These two cases are left to the reader as Exercise 2.24. □

Pick's theorem allows us not only to count the lattice points strictly inside the polygon $\mathcal{P}$ but also the total number of lattice points contained in $\mathcal{P}$, because this number is

$$I + B = A - \frac{1}{2}B + 1 + B = A + \frac{1}{2}B + 1 . \tag{2.15}$$

From this identity, it is now easy to describe the lattice-point enumerator $L_{\mathcal{P}}$:

Theorem 2.9. *Suppose $\mathcal{P}$ is an integral convex polygon with area A and B lattice points on its boundary.*

(a) *The lattice-point enumerator of $\mathcal{P}$ is the polynomial*

$$L_{\mathcal{P}}(t) = A\,t^2 + \frac{1}{2}B\,t + 1 .$$

(b) *Its evaluation at negative integers yields the relation*

$$L_{\mathcal{P}}(-t) = L_{\mathcal{P}^\circ}(t) .$$

(c) *The Ehrhart series of $\mathcal{P}$ is*

$$\mathrm{Ehr}_{\mathcal{P}}(z) = \frac{\left(A - \frac{B}{2} + 1\right) z^2 + \left(A + \frac{B}{2} - 2\right) z + 1}{(1-z)^3} .$$

Note that in the numerator of the Ehrhart series, the coefficient of z^2 is $L_{\mathcal{P}^\circ}(1)$, and the coefficient of z is $L_{\mathcal{P}}(1) - 3$.

Proof. Statement (a) follows from (2.15) if we can prove that the area of $t\mathcal{P}$ is At^2, and the number of boundary points on $t\mathcal{P}$ is Bt, which is the content of Exercise 2.25. Statement (b) follows with $L_{\mathcal{P}^\circ}(t) = L_{\mathcal{P}}(t) - Bt$. Finally, the Ehrhart series is

$$\begin{aligned}
\mathrm{Ehr}_{\mathcal{P}}(z) &= 1 + \sum_{t \ge 1} L_{\mathcal{P}}(t)\, z^t \\
&= \sum_{t \ge 0} \left(A\,t^2 + \frac{B}{2}\,t + 1 \right) z^t \\
&= A\,\frac{z^2 + z}{(1-z)^3} + \frac{B}{2}\,\frac{z}{(1-z)^2} + \frac{1}{1-z} \\
&= \frac{\left(A - \frac{B}{2} + 1\right) z^2 + \left(A + \frac{B}{2} - 2\right) z + 1}{(1-z)^3} .
\end{aligned}$$

□

2.7 Polygons with Rational Vertices

In this section we will establish formulas for the number of integer points in any *rational* convex polygon and its integral dilates.

A natural first step is to fix a triangulation of the polygon $\mathcal{P}$, which reduces our problem to that of counting integer points in rational *triangles*. However, this procedure merits some remarks. After counting lattice points in the triangles, we need to put those back together to form the polygon. But then we need to take care of the overcounting on line segments (where the triangles meet). Computing the number of lattice points on rational line segments is considerably easier than enumerating lattice points in 2-dimensional regions; however, it is still nontrivial (see Popoviciu's Theorem 1.5).

After triangulating $\mathcal{P}$, we can further simplify the picture by embedding an arbitrary rational triangle in a rational rectangle as in Figure 2.8. To compute lattice points in a triangle, we can first count the points in a rectangle with edges parallel to the coordinate axes, and then subtract the number of points in three right triangles, each with two edges are parallel to the axes, and possibly another rectangle, as shown in Figure 2.8. Since rectangles are easy to deal with (see Exercise 2.2), the problem reduces to finding a formula for a right triangle two of whose edges are parallel to the coordinate axes.

We now adjust and expand our generating-function machinery to these right triangles. Such a triangle $\mathcal{T}$ is a subset of $\mathbb{R}^2$ consisting of all points (x, y) satisfying

$$x \geq \frac{a}{d}, \quad y \geq \frac{b}{d}, \quad ex + fy \leq r$$

for some integers a, b, d, e, f, r (with $ea+fb \leq rd$; otherwise, the triangle would be empty). Because the lattice point count is invariant under horizontal and vertical integer translations and under flipping about the x- or y-axis, we may assume that $a, b, d, e, f, r \geq 0$ and $a, b < d$. (One should meditate about this fact for a minute.) Thus we arrive at the triangle $\mathcal{T}$ depicted in Figure 2.9.

To make our life a little easier, let's assume for the moment that e and f are relatively prime; we will deal with the general case in the exercises. So let

$$\mathcal{T} = \left\{ (x, y) \in \mathbb{R}^2 : \, x \geq \frac{a}{d}, \ y \geq \frac{b}{d}, \ ex + fy \leq r \right\}. \tag{2.16}$$

To derive a formula for

$$L_{\mathcal{T}}(t) = \# \left\{ (m, n) \in \mathbb{Z}^2 : \, m \geq \frac{ta}{d}, \ n \geq \frac{tb}{d}, \ em + fn \leq tr \right\}$$

we want to use methods similar to those in Chapter 1. As in Section 2.3, we introduce a slack variable s:

$$\begin{aligned} L_{\mathcal{T}}(t) &= \# \left\{ (m, n) \in \mathbb{Z}^2 : \, m \geq \frac{ta}{d}, \ n \geq \frac{tb}{d}, \ em + fn \leq tr \right\} \\ &= \# \left\{ (m, n, s) \in \mathbb{Z}^3 : \, m \geq \frac{ta}{d}, \ n \geq \frac{tb}{d}, \ s \geq 0, \ em + fn + s = tr \right\}. \end{aligned}$$

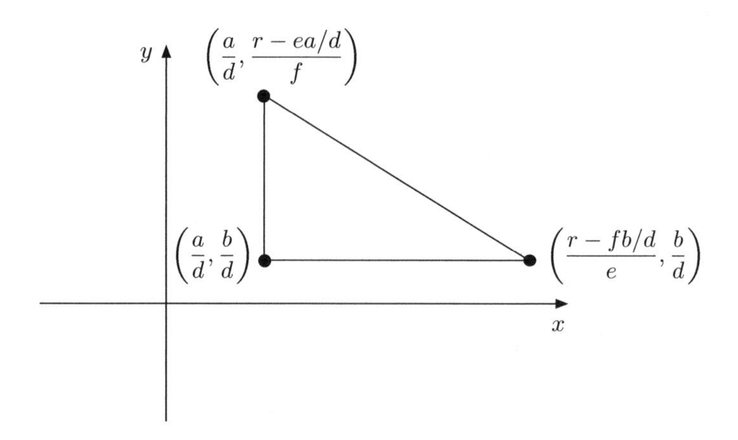

Fig. 2.9. A right rational triangle.

This counting function can now, as earlier, be interpreted as the coefficient of z^{tr} in the function

$$\left(\sum_{m\geq\frac{ta}{d}} z^{em}\right)\left(\sum_{n\geq\frac{tb}{d}} z^{fn}\right)\left(\sum_{s\geq 0} z^{s}\right).$$

Here the subscript (e.g., $m \geq \frac{ta}{d}$) under a summation sign means "sum over all integers satisfying this condition." For example, in the first sum we start with the least integer greater than or equal to $\frac{ta}{d}$, which is denoted by $\left\lceil\frac{ta}{d}\right\rceil$ (and is equal to $\left\lfloor\frac{ta-1}{d}\right\rfloor+1$ by Exercise 1.4 (j)). Hence the above generating function can be rewritten as

$$\begin{aligned}\left(\sum_{m\geq\left\lceil\frac{ta}{d}\right\rceil} z^{em}\right)\left(\sum_{n\geq\left\lceil\frac{tb}{d}\right\rceil} z^{fn}\right)\left(\sum_{s\geq 0} z^{s}\right) &= \frac{z^{\left\lceil\frac{ta}{d}\right\rceil e}}{1-z^e}\,\frac{z^{\left\lceil\frac{tb}{d}\right\rceil f}}{1-z^f}\,\frac{1}{1-z} \\ &= \frac{z^{u+v}}{(1-z^e)\left(1-z^f\right)(1-z)}, \qquad (2.17)\end{aligned}$$

where we have introduced, for ease of notation,

$$u := \left\lceil\frac{ta}{d}\right\rceil e \qquad \text{and} \qquad v := \left\lceil\frac{tb}{d}\right\rceil f. \qquad (2.18)$$

To extract the coefficient of z^{tr} of our generating function (2.17), we use familiar methods. As usual, we shift this coefficient to a constant term:

$$L_{\mathcal{T}}(t) = \operatorname{const}\left(\frac{z^{u+v-tr}}{(1-z^e)(1-z^f)(1-z)}\right)$$
$$= \operatorname{const}\left(\frac{1}{(1-z^e)(1-z^f)(1-z)z^{tr-u-v}}\right).$$

Before we apply the partial fraction machinery to this function, we should make sure that it is indeed a proper rational function, that is, that the total degree satisfies

$$u + v - tr - e - f - 1 < 0 \tag{2.19}$$

(see Exercise 2.31). Then we expand into partial fractions (here we're using our assumption that e and f do not have any common factors!):

$$\frac{1}{(1-z^e)(1-z^f)(1-z)z^{tr-u-v}}$$
$$= \sum_{j=1}^{e-1} \frac{A_j}{z-\xi_e^j} + \sum_{j=1}^{f-1} \frac{B_j}{z-\xi_f^j} + \sum_{k=1}^{3} \frac{C_k}{(z-1)^k} + \sum_{k=1}^{tr-u-v} \frac{D_k}{z^k}. \tag{2.20}$$

As numerous times before, the coefficients D_k do not contribute to the constant term, so that we obtain

$$L_{\mathcal{T}}(t) = -\sum_{j=1}^{e-1} \frac{A_j}{\xi_e^j} - \sum_{l=1}^{f-1} \frac{B_l}{\xi_f^l} - C_1 + C_2 - C_3. \tag{2.21}$$

We invite the reader to compute the coefficients appearing in this formula (Exercise 2.32):

$$A_j = -\frac{\xi_e^{j(v-tr+1)}}{e\left(1-\xi_e^{jf}\right)\left(1-\xi_e^j\right)},$$
$$B_l = -\frac{\xi_f^{l(u-tr+1)}}{f\left(1-\xi_f^{le}\right)\left(1-\xi_f^l\right)},$$
$$C_1 = -\frac{(u+v-tr)^2}{2ef} + \frac{u+v-tr}{2}\left(-\frac{1}{ef}+\frac{1}{e}+\frac{1}{f}\right) + \frac{1}{4}\left(\frac{1}{e}+\frac{1}{f}-1\right)$$
$$-\frac{1}{12}\left(\frac{e}{f}+\frac{1}{ef}+\frac{f}{e}\right), \tag{2.22}$$
$$C_2 = -\frac{u+v-tr+1}{ef} + \frac{1}{2e} + \frac{1}{2f},$$
$$C_3 = -\frac{1}{ef}.$$

Putting these ingredients into (2.21) yields the following formula for our lattice-point count.

Theorem 2.10. *For the rectangular rational triangle $\mathcal{T}$ given by (2.16), where e and f are relatively prime,*

$$\begin{aligned} L_{\mathcal{T}}(t) = &\frac{1}{2ef}(tr-u-v)^2 + \frac{1}{2}(tr-u-v)\left(\frac{1}{e}+\frac{1}{f}+\frac{1}{ef}\right) \\ &+\frac{1}{4}\left(1+\frac{1}{e}+\frac{1}{f}\right)+\frac{1}{12}\left(\frac{e}{f}+\frac{f}{e}+\frac{1}{ef}\right) \\ &+\frac{1}{e}\sum_{j=1}^{e-1}\frac{\xi_e^{j(v-tr)}}{\left(1-\xi_e^{jf}\right)\left(1-\xi_e^{j}\right)}+\frac{1}{f}\sum_{l=1}^{f-1}\frac{\xi_f^{l(u-tr)}}{\left(1-\xi_f^{le}\right)\left(1-\xi_f^{l}\right)}. \quad \square \end{aligned}$$

This identity can be rephrased in terms of the Fourier–Dedekind sum that we introduced in (1.13):

$$\begin{aligned} L_{\mathcal{T}}(t) = &\frac{1}{2ef}(tr-u-v)^2 + \frac{1}{2}(tr-u-v)\left(\frac{1}{e}+\frac{1}{f}+\frac{1}{ef}\right) \\ &+\frac{1}{4}\left(1+\frac{1}{e}+\frac{1}{f}\right)+\frac{1}{12}\left(\frac{e}{f}+\frac{f}{e}+\frac{1}{ef}\right) \\ &+s_{v-tr}(f,1;e)+s_{u-tr}(e,1;f). \end{aligned}$$

The general formula for $L_{\mathcal{T}}$—not assuming that e and f are relatively prime—is the content of Exercise 2.34.

Let us pause for a moment and study the nature of $L_{\mathcal{T}}$ as a function of t. Aside from the last two finite sums (which will be put in the spotlight in Chapter 8) and the appearance of u and v, $L_{\mathcal{T}}$ is a quadratic polynomial in t. And in those two sums, t appears only in the exponent of roots of unity, namely as the exponent of ξ_e and ξ_f. As a function of t, ξ_e^t is *periodic* with period e, and similarly ξ_f^t is periodic with period f. We should also remember that u and v are functions of t; but they can be easily written in terms of the fractional-part function, which again gives rise to periodic functions in t. So $L_{\mathcal{T}}(t)$ is a (quadratic) "polynomial" in t, whose coefficients are periodic functions in t. This is reminiscent of the counting functions of Chapter 1, which showed a similar periodic-polynomial behavior. Inspired by both examples, we define a **quasipolynomial** Q as an expression of the form $Q(t) = c_n(t)\,t^n + \cdots + c_1(t)\,t + c_0(t)$, where $c_0,\ldots,c_n$ are periodic functions in t. The **degree** of Q is n,[3] and the least common period of $c_0,\ldots,c_n$ is the **period** of Q. Alternatively, for a quasipolynomial Q, there exist a positive integer k and polynomials $p_0,p_1,\ldots,p_{k-1}$ such that

$$Q(t) = \begin{cases} Q(t) = p_0(t) & \text{if } t \equiv 0 \bmod k, \\ Q(t) = p_1(t) & \text{if } t \equiv 1 \bmod k, \\ \quad\vdots & \\ Q(t) = p_{k-1}(t) & \text{if } t \equiv k-1 \bmod k. \end{cases}$$

[3] Here we tacitly assume that c_n is not the zero function.

The minimal such k is the period of Q, and for this minimal k, the polynomials $p_0, p_1, \ldots, p_{k-1}$ are the **constituents** of Q.

By the triangulation and embedding-in-a-box arguments that started this section, we can now state a general structural result for rational polygons.

Theorem 2.11. *Let $\mathcal{P}$ be any rational polygon. Then $L_{\mathcal{P}}(t)$ is a quasipolynomial of degree 2. Its leading coefficient is the area of $\mathcal{P}$ (in particular, it is a constant).*

We have the technology at this point to also study the period of L_P; we let the reader enjoy the ensuing details (see Exercise 2.35).

Proof. By Exercises 2.2 and 2.34 (the general form of Theorem 2.10), the theorem holds for rational rectangles and right triangles whose edges are parallel to the axes. Now use the additivity of both degree-2 quasipolynomials and areas, and Popoviciu's theorem (Theorem 1.5). □

2.8 Euler's Generating Function for General Rational Polytopes

By now we have computed several instances of counting functions by setting up a generating function that fits the particular problem we're interested in. In this section, we set up such a generating function for the lattice-point enumerator of *any* rational polytope. Such a polytope is given by its hyperplane description as an intersection of half-spaces and hyperplanes. The half-spaces are algebraically given by linear inequalities, the hyperplanes by linear equations. If the polytope is rational, we can choose the coefficients of these inequalities and equations to be integers (Exercise 2.7). To unify both descriptions, we can introduce slack variables to turn the half-space inequalities into equalities. Furthermore, by translating our polytope into the nonnegative orthant (we can always shift a polytope by an integer vector without changing the lattice-point count), we may assume that all points in the polytope have nonnegative coordinates. In summary, after a harmless integer translation, we can describe any rational polytope $\mathcal{P}$ as

$$\mathcal{P} = \left\{ \mathbf{x} \in \mathbb{R}^d_{\geq 0} : \mathbf{A}\,\mathbf{x} = \mathbf{b} \right\} \tag{2.23}$$

for some integral matrix $\mathbf{A} \in \mathbb{Z}^{m \times d}$ and some integer vector $\mathbf{b} \in \mathbb{Z}^m$. (Note that d is not necessarily the dimension of $\mathcal{P}$.) To describe the tth dilate of $\mathcal{P}$, we simply scale a point $\mathbf{x} \in \mathcal{P}$ by $\frac{1}{t}$, or alternatively, multiply $\mathbf{b}$ by t:

$$t\mathcal{P} = \left\{ \mathbf{x} \in \mathbb{R}^d_{\geq 0} : \mathbf{A}\,\frac{\mathbf{x}}{t} = \mathbf{b} \right\} = \left\{ \mathbf{x} \in \mathbb{R}^d_{\geq 0} : \mathbf{A}\,\mathbf{x} = t\mathbf{b} \right\}.$$

Hence the lattice-point enumerator of $\mathcal{P}$ is the counting function

$$L_{\mathcal{P}}(t) = \# \left\{ \mathbf{x} \in \mathbb{Z}^d_{\geq 0} : \mathbf{A}\,\mathbf{x} = t\mathbf{b} \right\}. \tag{2.24}$$

Example 2.12. Suppose $\mathcal{P}$ is the quadrilateral with vertices $(0,0)$, $(2,0)$, $(1,1)$, and $\left(0,\frac{3}{2}\right)$:

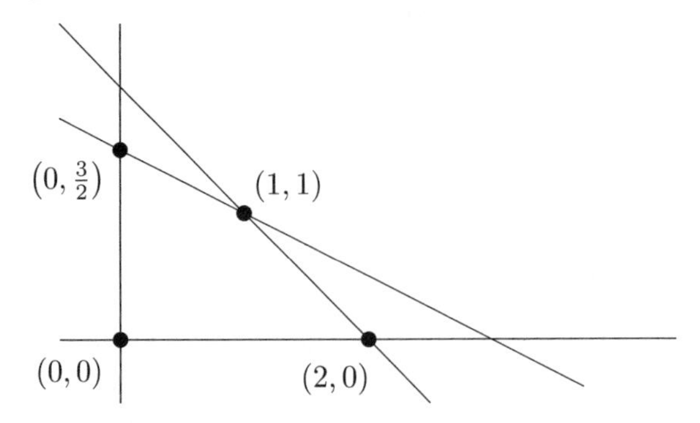

The half-space-inequality description of $\mathcal{P}$ is

$$\mathcal{P} = \left\{ (x_1, x_2) \in \mathbb{R}^2 : \, x_1, x_2 \geq 0, \, \begin{array}{c} x_1 + 2x_2 \leq 3, \\ x_1 + x_2 \leq 2 \end{array} \right\}.$$

Thus,

$$\begin{aligned} L_{\mathcal{P}}(t) &= \# \left\{ (x_1, x_2) \in \mathbb{Z}^2 : \, x_1, x_2 \geq 0, \, \begin{array}{c} x_1 + 2x_2 \leq 3t, \\ x_1 + x_2 \leq 2t \end{array} \right\} \\ &= \# \left\{ (x_1, x_2, x_3, x_4) \in \mathbb{Z}^4 : \, x_1, x_2, x_3, x_4 \geq 0, \, \begin{array}{c} x_1 + 2x_2 + x_3 = 3t, \\ x_1 + x_2 + x_4 = 2t \end{array} \right\} \\ &= \# \left\{ \mathbf{x} \in \mathbb{Z}^4_{\geq 0} : \begin{pmatrix} 1 & 2 & 1 & 0 \\ 1 & 1 & 0 & 1 \end{pmatrix} \mathbf{x} = \begin{pmatrix} 3t \\ 2t \end{pmatrix} \right\}. \end{aligned}$$

Using the ideas from Sections 1.3, 1.5, 2.3, and 2.7, we now construct a generating function for this counting function. In these previous sections, the lattice-point enumerator could be described with only *one* nontrivial linear equation, whereas now we have a system of such linear constraints. However, we can use the same approach of encoding the linear equation into geometric series; we just need more than one variable. When we expand the function

$$f(z_1, z_2) := \frac{1}{(1 - z_1 z_2)\left(1 - z_1^2 z_2\right)(1 - z_1)(1 - z_2) \, z_1^{3t} z_2^{2t}}$$

into geometric series, we have

$$\begin{aligned} f(z_1, z_2) &= \left(\sum_{n_1 \geq 0} (z_1 z_2)^{n_1} \right) \left(\sum_{n_2 \geq 0} \left(z_1^2 z_2\right)^{n_2} \right) \left(\sum_{n_3 \geq 0} z_1^{n_3} \right) \left(\sum_{n_4 \geq 0} z_2^{n_4} \right) \frac{1}{z_1^{3t} z_2^{2t}} \\ &= \sum_{n_1, \dots, n_4 \geq 0} z_1^{n_1 + 2n_2 + n_3 - 3t} z_2^{n_1 + n_2 + n_4 - 2t}. \end{aligned}$$

When we compute the constant term (in both z_1 and z_2), we are counting solutions $(n_1, n_2, n_3, n_4) \in \mathbb{Z}^4_{\geq 0}$ to

$$\begin{pmatrix} 1 & 2 & 1 & 0 \\ 1 & 1 & 0 & 1 \end{pmatrix} \begin{pmatrix} n_1 \\ n_2 \\ n_3 \\ n_4 \end{pmatrix} = \begin{pmatrix} 3t \\ 2t \end{pmatrix},$$

that is, the constant term of $f(z_1, z_2)$ counts the integer points in $\mathcal{P}$:

$$L_{\mathcal{P}}(t) = \text{const} \frac{1}{(1 - z_1 z_2)(1 - z_1^2 z_2)(1 - z_1)(1 - z_2) z_1^{3t} z_2^{2t}}.$$

We invite the reader to actually compute this constant term (Exercise 2.36). It turns out to be

$$\frac{7}{4}t^2 + \frac{5}{2}t + \frac{7 + (-1)^t}{8}. \qquad \square$$

Returning to the general case of a polytope $\mathcal{P}$ given by (2.23), we denote the columns of $\mathbf{A}$ by $\mathbf{c}_1, \mathbf{c}_2, \ldots, \mathbf{c}_d$. Let $\mathbf{z} = (z_1, z_2, \ldots, z_m)$ and expand the function

$$\frac{1}{(1 - \mathbf{z}^{\mathbf{c}_1})(1 - \mathbf{z}^{\mathbf{c}_2}) \cdots (1 - \mathbf{z}^{\mathbf{c}_d}) \mathbf{z}^{t\mathbf{b}}} \tag{2.25}$$

in terms of geometric series:

$$\left(\sum_{n_1 \geq 0} \mathbf{z}^{n_1 \mathbf{c}_1} \right) \left(\sum_{n_2 \geq 0} \mathbf{z}^{n_2 \mathbf{c}_2} \right) \cdots \left(\sum_{n_d \geq 0} \mathbf{z}^{n_d \mathbf{c}_d} \right) \frac{1}{\mathbf{z}^{t\mathbf{b}}}.$$

Here we use the abbreviating notation $\mathbf{z}^{\mathbf{c}} := z_1^{c_1} z_2^{c_2} \cdots z_m^{c_m}$ for the vectors $\mathbf{z} = (z_1, z_2, \ldots, z_m) \in \mathbb{C}^m$ and $\mathbf{c} = (c_1, c_2, \ldots, c_m) \in \mathbb{Z}^m$. When multiplying out everything, a typical term will look like

$$n_1 \mathbf{c}_1 + n_2 \mathbf{c}_2 + \cdots + n_d \mathbf{c}_d - t\mathbf{b} = \mathbf{A}\mathbf{n} - t\mathbf{b},$$

where $\mathbf{n} = (n_1, n_2, \ldots, n_d) \in \mathbb{Z}_{\geq 0}^d$. That is, if we take the constant term of our generating function (2.25), we're counting integer vectors $\mathbf{n} \in \mathbb{Z}_{\geq 0}^d$ satisfying

$$\mathbf{A}\mathbf{n} - t\mathbf{b} = 0, \qquad \text{that is,} \qquad \mathbf{A}\mathbf{n} = t\mathbf{b}.$$

So this constant term will pick up exactly the number of lattice points $\mathbf{n} \in \mathbb{Z}_{\geq 0}^d$ in $t\mathcal{P}$:

Theorem 2.13 (Euler's generating function). *Suppose the rational polytope $\mathcal{P}$ is given by (2.23). Then the Ehrhart quasipolynomial of $\mathcal{P}$ can be computed as follows:*

$$L_{\mathcal{P}}(t) = \text{const} \left(\frac{1}{(1 - \mathbf{z}^{\mathbf{c}_1})(1 - \mathbf{z}^{\mathbf{c}_2}) \cdots (1 - \mathbf{z}^{\mathbf{c}_d}) \mathbf{z}^{t\mathbf{b}}} \right). \qquad \square$$

We finish this section with rephrasing this constant-term identity in terms of Ehrhart series.

Corollary 2.14. *Suppose the rational polytope $\mathcal{P}$ is given by* (2.23). *Then the Ehrhart series of $\mathcal{P}$ can be computed as*

$$\mathrm{Ehr}_{\mathcal{P}}(x) = \mathrm{const}\left(\frac{1}{(1-\mathbf{z}^{\mathbf{c}_1})(1-\mathbf{z}^{\mathbf{c}_2})\cdots(1-\mathbf{z}^{\mathbf{c}_d})\left(1-\frac{x}{\mathbf{z}^{\mathbf{b}}}\right)} \right).$$

Proof. By Theorem 2.13,

$$\begin{aligned}\mathrm{Ehr}_{\mathcal{P}}(x) &= \sum_{t\geq 0} \mathrm{const}\left(\frac{1}{(1-\mathbf{z}^{\mathbf{c}_1})(1-\mathbf{z}^{\mathbf{c}_2})\cdots(1-\mathbf{z}^{\mathbf{c}_d})\,\mathbf{z}^{t\mathbf{b}}} \right) x^t \\ &= \mathrm{const}\left(\frac{1}{(1-\mathbf{z}^{\mathbf{c}_1})(1-\mathbf{z}^{\mathbf{c}_2})\cdots(1-\mathbf{z}^{\mathbf{c}_d})} \sum_{t\geq 0} \frac{x^t}{\mathbf{z}^{t\mathbf{b}}} \right) \\ &= \mathrm{const}\left(\frac{1}{(1-\mathbf{z}^{\mathbf{c}_1})(1-\mathbf{z}^{\mathbf{c}_2})\cdots(1-\mathbf{z}^{\mathbf{c}_d})} \frac{1}{1-\frac{x}{\mathbf{z}^{\mathbf{b}}}} \right). \end{aligned}$$

□

Notes

1. Convex polytopes are beautiful objects with a rich history and interesting theory, which we have only glimpsed here. For a good introduction to polytopes, we recommend [47, 90, 193]. Polytopes appear in a vast range of current research areas, including Gröbner bases and commutative algebra [174], combinatorial optimization [159], integral geometry [110], and geometry of numbers [163].

2. The distinction between the vertex and hyperplane description of a convex polytope leads to an interesting algorithmic question; namely, how quickly can we retrieve the first piece of data from the second and vice versa [159, 193]?

3. Ehrhart series are named after Eugène Ehrhart (1906–2000),[4] in anticipation of the theorems we will prove in Chapter 3. The Ehrhart series of a polytope belonging to the special class of *normal* polytopes equals another rational generating function, the *Hilbert–Poincaré series*. These series appear in the study of *graded algebras* (see, for example, [96, 171]). Ehrhart series also appear in the context of *toric varieties*, a vast and fruitful subject [64, 84].

4. The Eulerian numbers $A(d,k)$ are named after Leonhard Euler (1707–1783)[5] and arise naturally in the statistics of permutations: $A(d,k)$ counts permutations of $\{1,2,\ldots,d\}$ with $k-1$ ascents. For more on $A(d,k)$, see [62, Section 6.5].

[4] For more information about Ehrhart, see `http://icps.u-strasbg.fr/~clauss/Ehrhart.html`.

[5] For more information about Euler, see `http://www-groups.dcs.st-and.ac.uk/~history/Biographies/Euler.html`.

5. The pyramids of Section 2.4 have an interpretation as *order polytopes* [172]. A curious fact about the lattice-point enumerators of these pyramids is that they have arbitrarily large real roots as the dimension grows [24].

6. The counting function $L_\diamond$ for the cross-polytope can, incidentally, also be written as

$$\sum_{k=0}^{\min(d,t)} 2^k \binom{d}{k}\binom{t}{k}.$$

In particular, $L_\diamond$ is symmetric in d and t. The cross-polytope counting functions bear a connection to Laguerre polynomials, the d-dimensional harmonic oscillator, and the Riemann hypothesis. This connection appeared in [51], where Daniel Bump, Kwok-Kwong Choi, Pär Kurlberg, and Jeffrey Vaaler also found a curious fact about the roots of the polynomials $L_\diamond$: they all have real part $-\frac{1}{2}$ (an instance of a *local Riemann hypothesis*). This fact was proved independently by Peter Kirschenhofer, Attila Pethő, and Robert Tichy [109]; see also the Notes in Chapter 4.

7. Theorem 2.8 marks the beginning of the general study of lattice-point enumeration in polytopes. Its amazingly simple statement was discovered by Georg Alexander Pick (1859–1942)[6] in 1899 [143]. Pick's theorem holds also for a nonconvex polygon, provided its boundary forms a simple curve. In Chapter 12, we prove a generalization of Pick's theorem that includes nonconvex curves.

8. The results of Section 2.7 appeared in [29]. We will see in Chapter 8 that the finite sums over roots of unity can be rephrased in terms of *Dedekind–Rademacher sums*, which—as we will see in Chapter 8—can be computed very quickly. The theorems of Section 2.7 will then imply that the discrete volume of any rational polygon can be computed efficiently.

9. If we replace $t\mathbf{b}$ in (2.24) by a variable integer vector $\mathbf{v}$, the counting function

$$f(\mathbf{v}) = \#\left\{\mathbf{x} \in \mathbb{Z}^d_{\geq 0} : \mathbf{A}\,\mathbf{x} = \mathbf{v}\right\}$$

is called a *vector partition function*: it counts partitions of the vector $\mathbf{v}$ in terms of the columns of $\mathbf{A}$. Vector partition functions are the multivariate analogues of our lattice-point enumerators $L_\mathcal{P}(t)$, have many interesting properties, and give rise to intriguing open questions [20, 40, 63, 173, 178].

10. While Leonhard Euler most likely did not think of lattice-point enumeration in the sense of Ehrhart, we attribute Theorem 2.13 to him, since he certainly worked with generating functions of this type, probably thinking of

---

[6] For more information about Pick, see
`http://www-groups.dcs.st-and.ac.uk/~history/Biographies/Pick.html`.

them as vector partition functions. The potential of Euler's generating function for Ehrhart polynomials was already realized by Ehrhart [79, 81]. Several modern approaches to computing Ehrhart polynomials are based on Theorem 2.13 (see, for example, [19, 46, 119]).

Exercises

2.1. ♣ Fix positive integers a, b, c, d such that $a/b < c/d$, and let $\mathcal{P}$ be the interval $\left[\frac{a}{b}, \frac{c}{d}\right]$ (so $\mathcal{P}$ is a 1-dimensional rational convex polytope). Compute $L_{\mathcal{P}}(t) = \#\left(t\mathcal{P} \cap \mathbb{Z}\right)$ and $L_{\mathcal{P}^\circ}(t)$ and show directly that $L_{\mathcal{P}}(t)$ and $L_{\mathcal{P}^\circ}(t)$ are quasipolynomials with period $\operatorname{lcm}(b, d)$ that satisfy

$$L_{\mathcal{P}^\circ}(-t) = -L_{\mathcal{P}}(t)\,.$$

(*Hint:* Exercise 1.4 (i).)

2.2. ♣ Fix positive rational numbers a_1, b_1, a_2, b_2 and let $\mathcal{R}$ be the rectangle with vertices (a_1, b_1), (a_2, b_1), (a_2, b_2), and (a_1, b_2). Compute $L_{\mathcal{R}}(t)$ and $\operatorname{Ehr}_{\mathcal{R}}(z)$.

2.3. Fix positive integers a and b, and let $\mathcal{T}$ be a triangle with vertices $(0,0)$, $(a,0)$, and $(0,b)$.

(a) Compute $L_{\mathcal{T}}(t)$ and $\operatorname{Ehr}_{\mathcal{T}}(z)$.
(b) Use (a) to derive the following formula for the greatest common divisor of a and b:

$$\gcd(a,b) = 2\sum_{k=1}^{b-1}\left\lfloor \frac{ka}{b} \right\rfloor + a + b - ab\,.$$

(*Hint:* Exercise 1.12.)

2.4. Prove that for two polytopes $\mathcal{P} \subset \mathbb{R}^m$ and $\mathcal{Q} \subset \mathbb{R}^n$,

$$\#\left((\mathcal{P} \times \mathcal{Q}) \cap \mathbb{Z}^{m+n}\right) = \#\left(\mathcal{P} \cap \mathbb{Z}^m\right) \cdot \#\left(\mathcal{Q} \cap \mathbb{Z}^n\right).$$

Hence, $L_{\mathcal{P}\times\mathcal{Q}}(t) = L_{\mathcal{P}}(t)\, L_{\mathcal{Q}}(t)$.

2.5. Prove that if $\mathcal{F}$ is a face of $\mathcal{P}$ and $\mathcal{G}$ is a face of $\mathcal{F}$, then $\mathcal{G}$ is also a face of $\mathcal{P}$. (That is, the face relation is transitive.)

2.6. ♣ Suppose Δ is a d-simplex with vertices $V = \{\mathbf{v}_1, \mathbf{v}_2, \dots, \mathbf{v}_{d+1}\}$. Prove that for any nonempty subset $W \subseteq V$, $\operatorname{conv} W$ is a face of Δ, and conversely, that any face of Δ is of the form $\operatorname{conv} W$ for some $W \subseteq V$. Conclude the following corollaries from this characterization of the faces of a simplex:

(a) A face of any simplex is again a simplex.
(b) The intersection of two faces of a simplex Δ is again a face of Δ.

2.7. ♣ Prove that a rational convex polytope can be described by a system of linear inequalities and equations with *integral* coefficients.

2.8. ♣ Prove the properties (2.3) of the Eulerian numbers for all integers $1 \leq k \leq d$, namely:

(a) $A(d,k) = A(d, d+1-k)$;
(b) $A(d,k) = (d-k+1)A(d-1,k-1) + kA(d-1,k)$;
(c) $\sum_{k=0}^{d} A(d,k) = d!$;
(d) $A(d,k) = \sum_{j=0}^{k} (-1)^j \binom{d+1}{j} (k-j)^d$.

2.9. ♣ Prove (2.6); namely, for $d \geq 0$, $\frac{1}{(1-z)^{d+1}} = \sum_{k\geq 0} \binom{d+k}{d} z^k$.

2.10. ♣ Prove (2.7): For $t, k \in \mathbb{Z}$ and $d \in \mathbb{Z}_{>0}$,

$$(-1)^d \binom{-t+k}{d} = \binom{t+d-1-k}{d}.$$

2.11. The *Stirling numbers of the first kind*, $\operatorname{stirl}(n,m)$, are defined through the finite generating function

$$x(x-1)\cdots(x-n+1) = \sum_{m=0}^{n} \operatorname{stirl}(n,m)\, x^m .$$

(See also [165, Sequence A008275].) Prove that

$$\frac{1}{d!} \sum_{k=0}^{d} (-1)^{d-k} \operatorname{stirl}(d+1, k+1)\, t^k = \binom{d+t}{d},$$

the lattice-point enumerator for the standard d-simplex.

2.12. Give a direct proof that the number of solutions $(m_1, m_2, \dots, m_{d+1}) \in \mathbb{Z}_{\geq 0}^{d+1}$ to $m_1 + m_2 + \cdots + m_{d+1} = t$ equals $\binom{d+t}{d}$. (*Hint:* think of t objects lined up and separated by d walls.)

2.13. Compute $L_{\mathcal{P}}(t)$, where $\mathcal{P}$ is the regular tetrahedron with vertices $(0,0,0), (1,1,0), (1,0,1), (0,1,1)$.

2.14. ♣ Prove that the power series

$$\sum_{k\geq 0} \frac{B_k}{k!} z^k$$

that defines the Bernoulli numbers has radius of convergence 2π.

2.15. ♣ Prove (2.11); namely, $B_d(1-x) = (-1)^d B_d(x)$.

2.16. ♣ Prove (2.12); namely, $B_d = 0$ for all odd $d \geq 3$.

2.17. Show that for each positive integer n,

$$n\, x^{n-1} = \sum_{k=1}^{n} \binom{n}{k} B_{n-k}(x) \,.$$

This gives us a change of basis for the polynomials of degree $\leq n$, allowing us to represent any polynomial as a sum of Bernoulli polynomials.

2.18. As a complement to the previous exercise, show that we also have a change of basis in the other direction. Namely, we can represent a single Bernoulli polynomial in terms of the monomials as follows:

$$B_n(x) = \sum_{k=0}^{n} \binom{n}{k} B_k\, x^{n-k} .$$

2.19. Show that for all positive integers m, n, and any $x \in \mathbb{R}$,

$$\frac{1}{m} \sum_{k=0}^{m-1} B_n\left(x + \frac{k}{m}\right) = m^{-n} B_n(mx) \,.$$

(This is a *Hecke-operator*-type identity, originally found by Joseph Ludwig Raabe in 1851.)

2.20. Show that $B_n(x+1) - B_n(x) = n\, x^{n-1}$.

2.21. An alternative way to define the Bernoulli polynomials is to give elementary properties that uniquely characterize them. Show that the following three properties uniquely determine the Bernoulli polynomials, as defined in the text by (2.8):

(a) $B_0(x) = 1$.
(b) $\frac{dB_n(x)}{dx} = n\, B_{n-1}(x)$, for all $n \geq 1$.
(c) $\int_0^1 B_n(x)\, dx = 0$, for all $n \geq 1$.

2.22. Prove that for $d = 1$,

$$\sum_{t \geq 0} B_1(t)\, z^t = \frac{3z - 1}{2(1-z)^2} \,,$$

and for $d \geq 2$,

$$\sum_{t \geq 0} B_d(t)\, z^t = \frac{d \sum_{k=1}^{d-1} A\,(d-1, k)\, z^{k+1}}{(1-z)^{d+1}} + \frac{B_d}{1-z} \,.$$

2.23. ♣ Prove Theorem 2.6: $\operatorname{Ehr}_{\operatorname{BiPyr}(\mathcal{Q})}(z) = \frac{1+z}{1-z} \operatorname{Ehr}_{\mathcal{Q}}(z)$.

2.24. ♣ Let $\mathcal{R}$ be an integral rectangle whose edges are parallel to the coordinate axes, and let $\mathcal{T}$ be a rectangular triangle two of whose edges are parallel to the coordinate axes. Show that Pick's theorem holds for $\mathcal{R}$ and $\mathcal{T}$.

2.25. ♣ Suppose $\mathcal{P}$ is an integral polygon with area A and B lattice points on its boundary. Show that the area of $t\mathcal{P}$ is At^2, and the number of boundary points on $t\mathcal{P}$ is Bt. (*Hint:* Exercise 1.12.)

2.26. Let $\mathcal{P}$ be the self-intersecting polygon defined by the line segments $[(0,0),(4,2)]$, $[(4,2),(4,0)]$, $[(4,0),(0,2)]$, and $[(0,2),(0,0)]$. Show that Pick's theorem does not hold for $\mathcal{P}$.

2.27. Suppose that $\mathcal{P}$ and $\mathcal{Q}$ are integral polygons, and that $\mathcal{Q}$ lies entirely inside $\mathcal{P}$. Then the area bounded by the boundaries of $\mathcal{P}$ and $\mathcal{Q}$, denoted by $\mathcal{P} - \mathcal{Q}$, is a "doubly connected polygon." Find and prove the analogue of Pick's theorem for $\mathcal{P} - \mathcal{Q}$. Generalize your formula to a polygon with n "holes" (instead of one).

2.28. Consider the rhombus

$$\mathcal{R} = \{(x,y) : a|x| + b|y| \le ab\},$$

where a and b are fixed positive integers. Find a formula for $L_{\mathcal{R}}(t)$.

2.29. We define the nth *Farey sequence* as all the rational numbers $\frac{a}{b}$ in the interval $[0,1]$ when a and b are coprime and $b \le n$. For instance, the sixth Farey sequence is $\frac{0}{1}, \frac{1}{6}, \frac{1}{5}, \frac{1}{4}, \frac{1}{3}, \frac{2}{5}, \frac{1}{2}, \frac{3}{5}, \frac{2}{3}, \frac{3}{4}, \frac{4}{5}, \frac{5}{6}, \frac{1}{1}$.

(a) For two consecutive fractions $\frac{a}{b}$ and $\frac{c}{d}$ in a Farey sequence, prove that $bc - ad = 1$.

(b) For three consecutive fractions $\frac{a}{b}$, $\frac{c}{d}$, and $\frac{e}{f}$ in a Farey sequence, show that $\frac{c}{d} = \frac{a+e}{b+f}$.

2.30. Let $\lceil x \rceil$ denote the smallest integer larger than or equal to x. Prove that for all positive integers a and b,

$$a + (-1)^b \sum_{m=0}^{a} (-1)^{\left\lceil \frac{bm}{a} \right\rceil} \equiv b + (-1)^a \sum_{n=0}^{b} (-1)^{\left\lceil \frac{an}{b} \right\rceil} \mod 4 .$$

(*Hint:* This is a variation of Exercise 1.6. One way to obtain this identity is by counting lattice points in a certain triangle, keeping track only of the parity.)

2.31. ♣ Verify (2.19).

2.32. ♣ Compute the partial fraction coefficients (2.22).

2.33. Let a, b be positive integers. Show that

$$\frac{1}{1-z^{ab}} = -\frac{\xi_a^k}{ab}\left(z-\xi_a^k\right)^{-1} + \frac{ab-1}{2ab} + \text{terms with positive powers of } \left(z-\xi_a^k\right).$$

2.34. ♣ Let $\mathcal{T}$ be given by (2.16), and let $c = \gcd(e, f)$. Prove that

$$\begin{aligned} L_{\mathcal{T}}(t) &= \frac{1}{2ef}(tr-u-v)^2 + \frac{1}{2}(tr-u-v)\left(\frac{1}{e}+\frac{1}{f}+\frac{1}{ef}\right) \\ &+ \frac{1}{4}\left(1+\frac{1}{e}+\frac{1}{f}\right) + \frac{1}{12}\left(\frac{e}{f}+\frac{f}{e}+\frac{1}{ef}\right) \\ &+ \left(\frac{1}{2e}+\frac{1}{2f}-\frac{u+v-tr}{ef}\right)\sum_{k=1}^{c-1}\frac{\xi_c^{-ktr}}{1-\xi_c^k} - \frac{1}{ef}\sum_{k=1}^{c-1}\frac{\xi_c^{k(-tr+1)}}{(1-\xi_c)^2} \\ &+ \frac{1}{e}\sum_{\substack{j=1 \\ \frac{e}{c} \nmid j}}^{e-1}\frac{\xi_e^{j(v-tr)}}{\left(1-\xi_e^{jf}\right)\left(1-\xi_e^j\right)} + \frac{1}{f}\sum_{\substack{l=1 \\ \frac{f}{c} \nmid l}}^{f-1}\frac{\xi_f^{l(u-tr)}}{\left(1-\xi_f^{le}\right)\left(1-\xi_f^l\right)}. \end{aligned}$$

2.35. Let $\mathcal{P}$ be a rational polygon, and let d be the least common multiple of the denominators of the vertices of $\mathcal{P}$. Prove directly (using Exercise 2.34) that the period of $L_{\mathcal{P}}$ divides d.

2.36. ♣ Finish the calculation in Example 2.12, that is, compute

$$\operatorname{const} \frac{1}{\left(1-z_1z_2\right)\left(1-z_1^2z_2\right)\left(1-z_1\right)\left(1-z_2\right)z_1^{3t}z_2^{2t}}.$$

2.37. Compute the vector partition function of the quadrilateral given in Example 2.12, that is, compute the counting function

$$f\left(v_1, v_2\right) := \#\left\{\mathbf{x} \in \mathbb{Z}_{\geq 0}^4 : \begin{pmatrix} 1 & 2 & 1 & 0 \\ 1 & 1 & 0 & 1 \end{pmatrix}\mathbf{x} = \begin{pmatrix} v_1 \\ v_2 \end{pmatrix}\right\}$$

for $v_1, v_2 \in \mathbb{Z}$. (This function depends on the relationship between v_1 and v_2.)

2.38. Search on the Internet for the program `polymake` [86]. You can download it for free. Experiment.

Open Problems

2.39. Pick $d+1$ of the 2^d vertices of the unit d-cube $\square$, and let Δ be the simplex defined by their convex hull.

(a) Which choice of vertices maximizes $\operatorname{vol}\Delta$?
(b) What is the maximum volume of such a Δ?

2.40. Find classes of integer d-polytopes $(\mathcal{P}_d)_{d\geq 1}$ for which $L_{\mathcal{P}_d}(t)$ is symmetric in d and t. (The standard simplices Δ and the cross-polytopes $\diamond$ form two such classes.)

2.41. We mentioned already in the Notes that all the roots of the polynomials $L_\diamond$ have real part $-\frac{1}{2}$ [51, 109]. Find other classes of polytopes whose lattice-point enumerator exhibits such a special behavior.

3

Counting Lattice Points in Polytopes: The Ehrhart Theory

Ubi materia, ibi geometria.

Johannes Kepler (1571–1630)

Given the profusion of examples that gave rise to the polynomial behavior of the integer-point counting function $L_{\mathcal{P}}(t)$ for special polytopes $\mathcal{P}$, we now ask whether there is a general structure theorem. As the ideas unfold, the reader is invited to look back at Chapters 1 and 2 as appetizers and indeed as special cases of the theorems developed below.

3.1 Triangulations and Pointed Cones

Because most of the proofs that follow work like a charm for a simplex, we first dissect a polytope into simplices. This dissection is captured by the following definition.

A **triangulation** of a convex d-polytope $\mathcal{P}$ is a finite collection T of d-simplices with the following properties:

- $\mathcal{P} = \bigcup_{\Delta \in T} \Delta$.
- For any $\Delta_1, \Delta_2 \in T$, $\Delta_1 \cap \Delta_2$ is a face common to both Δ_1 and Δ_2.

We say that $\mathcal{P}$ can be **triangulated using no new vertices** if there exists a triangulation T such that the vertices of any $\Delta \in T$ are vertices of $\mathcal{P}$.

Theorem 3.1 (Existence of triangulations). *Every convex polytope can be triangulated using no new vertices.*

This theorem seems intuitively obvious but is not entirely trivial to prove. We carefully work out a proof in Appendix B.

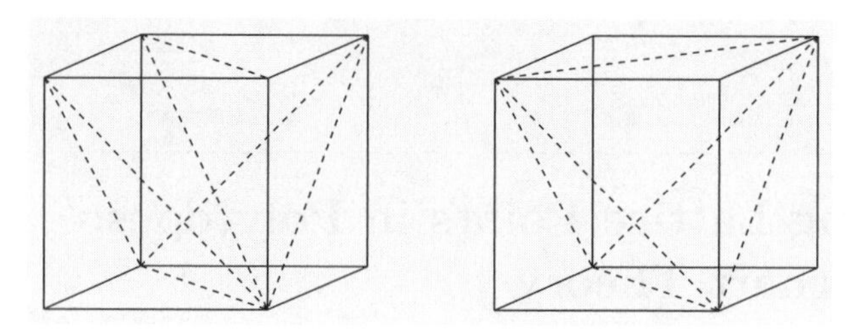

Fig. 3.1. Two (very different) triangulations of the 3-cube.

A **pointed cone** $\mathcal{K} \subseteq \mathbb{R}^d$ is a set of the form

$$\mathcal{K} = \{\mathbf{v} + \lambda_1 \mathbf{w}_1 + \lambda_2 \mathbf{w}_2 + \cdots + \lambda_m \mathbf{w}_m : \lambda_1, \lambda_2, \ldots, \lambda_m \geq 0\},$$

where $\mathbf{v}, \mathbf{w}_1, \mathbf{w}_2, \ldots, \mathbf{w}_m \in \mathbb{R}^d$ are such that there exists a hyperplane H for which $H \cap \mathcal{K} = \{\mathbf{v}\}$; that is, $\mathcal{K} \setminus \{\mathbf{v}\}$ lies strictly on one side of H. The vector $\mathbf{v}$ is called the **apex** of $\mathcal{K}$, and the $\mathbf{w}_k$'s are the **generators** of $\mathcal{K}$. The cone is **rational** if $\mathbf{v}, \mathbf{w}_1, \mathbf{w}_2, \ldots, \mathbf{w}_m \in \mathbb{Q}^d$, in which case we may choose $\mathbf{w}_1, \mathbf{w}_2, \ldots, \mathbf{w}_m \in \mathbb{Z}^d$ by clearing denominators. The **dimension** of $\mathcal{K}$ is the dimension of the affine space spanned by $\mathcal{K}$; if $\mathcal{K}$ is of dimension d, we call it a d-cone. The d-cone $\mathcal{K}$ is **simplicial** if $\mathcal{K}$ has precisely d linearly independent generators.

Just as polytopes have a description as an intersection of half-spaces, so do pointed cones: A rational pointed d-cone is the d-dimensional intersection of finitely many half-spaces of the form

$$\left\{\mathbf{x} \in \mathbb{R}^d : a_1 x_1 + a_2 x_2 + \cdots + a_d x_d \leq b\right\}$$

with integral parameters $a_1, a_2, \ldots, a_d, b \in \mathbb{Z}$ such that the corresponding hyperplanes of the form

$$\left\{\mathbf{x} \in \mathbb{R}^d : a_1 x_1 + a_2 x_2 + \cdots + a_d x_d = b\right\}$$

meet in exactly one point.

Cones are important for many reasons. The most practical for us is a process called *coning over a polytope.* Given a convex polytope $\mathcal{P} \subset \mathbb{R}^d$ with vertices $\mathbf{v}_1, \mathbf{v}_2, \ldots, \mathbf{v}_n$, we lift these vertices into $\mathbb{R}^{d+1}$ by adding a 1 as their last coordinate. So, let

$$\mathbf{w}_1 = (\mathbf{v}_1, 1), \ \mathbf{w}_2 = (\mathbf{v}_2, 1), \ \ldots, \ \mathbf{w}_n = (\mathbf{v}_n, 1).$$

Now we define the **cone over** $\mathcal{P}$ as

$$\operatorname{cone}(\mathcal{P}) = \{\lambda_1 \mathbf{w}_1 + \lambda_2 \mathbf{w}_2 + \cdots + \lambda_n \mathbf{w}_n : \lambda_1, \lambda_2, \ldots, \lambda_n \geq 0\} \subset \mathbb{R}^{d+1}.$$

This pointed cone has the origin as apex, and we can recover our original polytope $\mathcal{P}$ (strictly speaking, the translated set $\{(\mathbf{x}, 1) : \mathbf{x} \in \mathcal{P}\}$) by cutting $\operatorname{cone}(\mathcal{P})$ with the hyperplane $x_{d+1} = 1$, as shown in Figure 3.2.

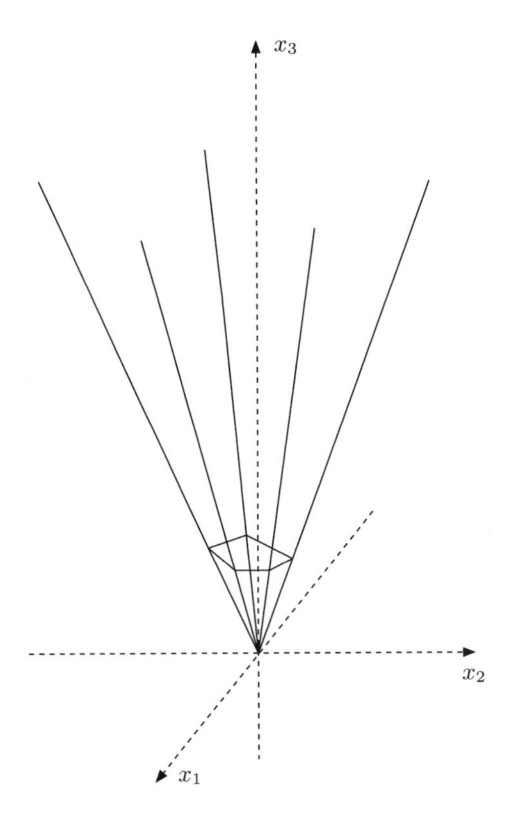

Fig. 3.2. Coning over a polytope.

By analogy with the language of polytopes, we say that the hyperplane $H = \{\mathbf{x} \in \mathbb{R}^d : \mathbf{a} \cdot \mathbf{x} = b\}$ is a **supporting hyperplane** of the pointed d-cone $\mathcal{K}$ if $\mathcal{K}$ lies entirely on one side of H, that is,

$$\mathcal{K} \subset \left\{ \mathbf{x} \in \mathbb{R}^d : \mathbf{a} \cdot \mathbf{x} \leq b \right\} \qquad \text{or} \qquad \mathcal{K} \subset \left\{ \mathbf{x} \in \mathbb{R}^d : \mathbf{a} \cdot \mathbf{x} \geq b \right\}.$$

A **face** of $\mathcal{K}$ is a set of the form $\mathcal{K} \cap H$, where H is a supporting hyperplane of $\mathcal{K}$. The $(d-1)$-dimensional faces are called **facets** and the 1-dimensional faces **edges** of $\mathcal{K}$. The apex of $\mathcal{K}$ is its unique 0-dimensional face.

Just as polytopes can be triangulated into simplices, pointed cones can be triangulated into simplicial cones. So, a collection T of simplicial d-cones is a **triangulation** of the d-cone $\mathcal{K}$ if it satisfies:

- $\mathcal{K} = \bigcup_{\mathcal{S} \in T} \mathcal{S}$.
- For any $\mathcal{S}_1, \mathcal{S}_2 \in T$, $\mathcal{S}_1 \cap \mathcal{S}_2$ is a face common to both $\mathcal{S}_1$ and $\mathcal{S}_2$.

We say that $\mathcal{K}$ can be **triangulated using no new generators** if there exists a triangulation T such that the generators of any $\mathcal{S} \in T$ are generators of $\mathcal{P}$.

Theorem 3.2. *Any pointed cone can be triangulated into simplicial cones using no new generators.*

Proof. This theorem is really a corollary to Theorem 3.1. Given a pointed d-cone $\mathcal{K}$, there exists a hyperplane H that intersects $\mathcal{K}$ only at the apex. Now translate H "into" the cone, so that $H \cap \mathcal{K}$ consists of more than just one point. This intersection is a $(d-1)$-polytope $\mathcal{P}$, whose vertices are determined by the generators of $\mathcal{K}$. Now triangulate $\mathcal{P}$ using no new vertices. The cone over each simplex of the triangulation is a simplicial cone. These simplicial cones, by construction, triangulate $\mathcal{K}$. □

3.2 Integer-Point Transforms for Rational Cones

We want to encode the information contained by the lattice points in a set $S \subset \mathbb{R}^d$. It turns out that the following multivariate generating function allows us to do this in an efficient way if S is a rational cone or polytope:

$$\sigma_S(\mathbf{z}) = \sigma_S(z_1, z_2, \ldots, z_d) := \sum_{\mathbf{m} \in S \cap \mathbb{Z}^d} \mathbf{z}^{\mathbf{m}}.$$

The generating function σ_S simply lists all integer points in S in a special form: not as a list of vectors, but as a formal sum of monomials. We call σ_S the **integer-point transform** of S; the function σ_S also goes by the name *moment generating function* or simply *generating function* of S. The integer-point transform σ_S opens the door to both algebraic and analytic techniques.

Example 3.3. As a warm-up example, consider the 1-dimensional cone $\mathcal{K} = [0, \infty)$. Its integer-point transform is our old friend

$$\sigma_{\mathcal{K}}(z) = \sum_{m \in [0,\infty) \cap \mathbb{Z}} z^m = \sum_{m \geq 0} z^m = \frac{1}{1-z}.$$ □

Example 3.4. Now we consider the two-dimensional cone

$$\mathcal{K} := \{\lambda_1(1,1) + \lambda_2(-2,3) : \lambda_1, \lambda_2 \geq 0\} \subset \mathbb{R}^2$$

depicted in Figure 3.3. To obtain the integer-point transform $\sigma_{\mathcal{K}}$, we tile $\mathcal{K}$ by copies of the *fundamental parallelogram*

$$\Pi := \{\lambda_1(1,1) + \lambda_2(-2,3) : 0 \leq \lambda_1, \lambda_2 < 1\} \subset \mathbb{R}^2.$$

More precisely, we translate Π by nonnegative integer linear combinations of the generators $(1,1)$ and $(-2,3)$, and these translates will exactly cover

$\mathcal{K}$. How can we list the integer points in $\mathcal{K}$ as monomials? Let's first list all vertices of the translates of Π. These are nonnegative integer combinations of the generators $(1,1)$ and $(-2,3)$, so we can list them using geometric series:

$$\sum_{\substack{\mathbf{m}=j(1,1)+k(-2,3)\\ j,k\geq 0}} \mathbf{z}^{m} = \sum_{j\geq 0}\sum_{k\geq 0} \mathbf{z}^{j(1,1)+k(-2,3)} = \frac{1}{\left(1-z_1z_2\right)\left(1-z_1^{-2}z_2^{3}\right)} \, .$$

We now use the integer points $(m,n) \in \Pi$ to generate a subset of $\mathbb{Z}^2$ by adding to (m,n) nonnegative linear integer combinations of the generators $(1,1)$ and $(-2,3)$. Namely, we let

$$\mathcal{L}_{(m,n)} := \left\{(m,n) + j(1,1) + k(-2,3) : \, j,k \in \mathbb{Z}_{\geq 0}\right\}.$$

It is immediate that $\mathcal{K}\cap\mathbb{Z}^2$ is the disjoint union of the subsets $\mathcal{L}_{(m,n)}$ as (m,n) ranges over $\Pi \cap \mathbb{Z}^2 = \{(0,0),(0,1),(0,2),(-1,2),(-1,3)\}$. Hence

$$\begin{aligned}\sigma_{\mathcal{K}}(\mathbf{z}) &= \left(1+z_2+z_2^2+z_1^{-1}z_2^2+z_1^{-1}z_2^3\right) \sum_{\substack{\mathbf{m}=j(1,1)+k(-2,3)\\ j,k\geq 0}} \mathbf{z}^{m} \\ &= \frac{1+z_2+z_2^2+z_1^{-1}z_2^2+z_1^{-1}z_2^3}{\left(1-z_1z_2\right)\left(1-z_1^{-2}z_2^{3}\right)} \, . \end{aligned}$$

□

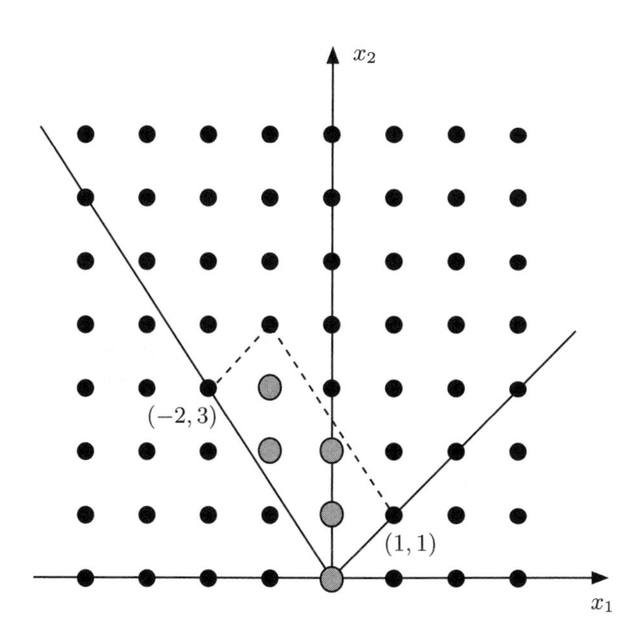

Fig. 3.3. The cone $\mathcal{K}$ and its fundamental parallelogram.

Similar geometric series suffice to describe integer-point transforms for simplicial d-cones. The following result utilizes the geometric series in several directions simultaneously.

Theorem 3.5. *Suppose*

$$\mathcal{K} := \{\lambda_1 \mathbf{w}_1 + \lambda_2 \mathbf{w}_2 + \cdots + \lambda_d \mathbf{w}_d : \lambda_1, \lambda_2, \ldots, \lambda_d \geq 0\}$$

is a simplicial d-cone, where $\mathbf{w}_1, \mathbf{w}_2, \ldots, \mathbf{w}_d \in \mathbb{Z}^d$. *Then for* $\mathbf{v} \in \mathbb{R}^d$, *the integer-point transform* $\sigma_{\mathbf{v}+\mathcal{K}}$ *of the shifted cone* $\mathbf{v}+\mathcal{K}$ *is the rational function*

$$\sigma_{\mathbf{v}+\mathcal{K}}(\mathbf{z}) = \frac{\sigma_{\mathbf{v}+\Pi}(\mathbf{z})}{(1 - \mathbf{z}^{\mathbf{w}_1})(1 - \mathbf{z}^{\mathbf{w}_2}) \cdots (1 - \mathbf{z}^{\mathbf{w}_d})},$$

where Π *is the half-open parallelepiped*

$$\Pi := \{\lambda_1 \mathbf{w}_1 + \lambda_2 \mathbf{w}_2 + \cdots + \lambda_d \mathbf{w}_d : 0 \leq \lambda_1, \lambda_2, \ldots, \lambda_d < 1\}.$$

The half-open parallelepiped Π is called the **fundamental parallelepiped** of $\mathcal{K}$.

Proof. In $\sigma_{\mathbf{v}+\mathcal{K}}(\mathbf{z}) = \sum_{\mathbf{m} \in (\mathbf{v}+\mathcal{K}) \cap \mathbb{Z}^d} \mathbf{z}^{\mathbf{m}}$, we list each integer point $\mathbf{m}$ in $\mathbf{v}+\mathcal{K}$ as the monomial $\mathbf{z}^{\mathbf{m}}$. Such a lattice point can, by definition, be written as

$$\mathbf{m} = \mathbf{v} + \lambda_1 \mathbf{w}_1 + \lambda_2 \mathbf{w}_2 + \cdots + \lambda_d \mathbf{w}_d$$

for some numbers $\lambda_1, \lambda_2, \ldots, \lambda_d \geq 0$. Because the $\mathbf{w}_k$'s form a basis of $\mathbb{R}^d$, this representation is unique. Let's write each of the λ_k's in terms of their integer and fractional part: $\lambda_k = \lfloor \lambda_k \rfloor + \{\lambda_k\}$. So

$$\mathbf{m} = \mathbf{v} + (\{\lambda_1\} \mathbf{w}_1 + \{\lambda_2\} \mathbf{w}_2 + \cdots + \{\lambda_d\} \mathbf{w}_d) + \lfloor \lambda_1 \rfloor \mathbf{w}_1 + \lfloor \lambda_2 \rfloor \mathbf{w}_2 + \cdots + \lfloor \lambda_d \rfloor \mathbf{w}_d,$$

and we should note that, since $0 \leq \{\lambda_k\} < 1$, the vector

$$\mathbf{p} := \mathbf{v} + \{\lambda_1\} \mathbf{w}_1 + \{\lambda_2\} \mathbf{w}_2 + \cdots + \{\lambda_d\} \mathbf{w}_d$$

is in $\mathbf{v}+\Pi$. In fact, $\mathbf{p} \in \mathbb{Z}^d$, since $\mathbf{m}$ and $\lfloor \lambda_k \rfloor \mathbf{w}_k$ are all integer vectors. Again the representation of $\mathbf{p}$ in terms of the $\mathbf{w}_k$'s is unique. In summary, we have proved that any $\mathbf{m} \in \mathbf{v} + \mathcal{K} \cap \mathbb{Z}^d$ can be uniquely written as

$$\mathbf{m} = \mathbf{p} + k_1 \mathbf{w}_1 + k_2 \mathbf{w}_2 + \cdots + k_d \mathbf{w}_d \tag{3.1}$$

for some $\mathbf{p} \in (\mathbf{v}+\Pi) \cap \mathbb{Z}^d$ and some integers $k_1, k_2, \ldots, k_d \geq 0$. On the other hand, let us write the rational function on the right-hand side of the theorem as a product of series:

$$\frac{\sigma_{\mathbf{v}+\Pi}(\mathbf{z})}{(1 - \mathbf{z}^{\mathbf{w}_1}) \cdots (1 - \mathbf{z}^{\mathbf{w}_d})} = \left(\sum_{\mathbf{p} \in (\mathbf{v}+\Pi) \cap \mathbb{Z}^d} \mathbf{z}^{\mathbf{p}} \right) \left(\sum_{k_1 \geq 0} \mathbf{z}^{k_1 \mathbf{w}_1} \right) \cdots \left(\sum_{k_d \geq 0} \mathbf{z}^{k_d \mathbf{w}_d} \right).$$

If we multiply everything out, a typical exponent will look exactly like (3.1). □

Our proof contains a crucial geometric idea. Namely, we *tile* the cone $\mathbf{v}+\mathcal{K}$ with translates of $\mathbf{v}+\Pi$ by nonnegative integral combinations of the $\mathbf{w}_k$'s. It is this tiling that gives rise to the nice integer-point transform in Theorem 3.5. Computationally, we therefore favor cones over polytopes due to our ability to tile a simplicial cone with copies of the fundamental domain, as above. Another reason for favoring cones over polytopes appears in Brion's theorem in Chapter 9.

Theorem 3.5 shows that the real complexity of computing the integer-point transform $\sigma_{\mathbf{v}+\mathcal{K}}$ is embedded in the location of the lattice points in the parallelepiped $\mathbf{v}+\Pi$.

By mildly strengthening the hypothesis of Theorem 3.5, we obtain a slightly easier generating function, a result we shall need in Section 3.4 and Chapter 4.

Corollary 3.6. *Suppose*

$$\mathcal{K} := \{\lambda_1 \mathbf{w}_1 + \lambda_2 \mathbf{w}_2 + \cdots + \lambda_d \mathbf{w}_d : \lambda_1, \lambda_2, \ldots, \lambda_d \geq 0\}$$

is a simplicial d-cone, where $\mathbf{w}_1, \mathbf{w}_2, \ldots, \mathbf{w}_d \in \mathbb{Z}^d$, *and* $\mathbf{v} \in \mathbb{R}^d$, *such that the boundary of* $\mathbf{v}+\mathcal{K}$ *contains no integer point. Then*

$$\sigma_{\mathbf{v}+\mathcal{K}}(\mathbf{z}) = \frac{\sigma_{\mathbf{v}+\Pi}(\mathbf{z})}{(1-\mathbf{z}^{\mathbf{w}_1})(1-\mathbf{z}^{\mathbf{w}_2})\cdots(1-\mathbf{z}^{\mathbf{w}_d})},$$

where Π *is the open parallelepiped*

$$\Pi := \{\lambda_1 \mathbf{w}_1 + \lambda_2 \mathbf{w}_2 + \cdots + \lambda_d \mathbf{w}_d : 0 < \lambda_1, \lambda_2, \ldots, \lambda_d < 1\}.$$

Proof. The proof of Theorem 3.5 goes through almost verbatim, except that $\mathbf{v}+\Pi$ now has no boundary lattice points, so that there is no harm in choosing Π to be open. □

Since a general pointed cone can always be triangulated into simplicial cones, the integer-point transforms add up in an inclusion–exclusion manner (note that the intersection of simplicial cones in a triangulation is again a simplicial cone, by Exercise 3.2). Hence we have the following corollary.

Corollary 3.7. *Given any pointed cone*

$$\mathcal{K} = \{\mathbf{v} + \lambda_1 \mathbf{w}_1 + \lambda_2 \mathbf{w}_2 + \cdots + \lambda_m \mathbf{w}_m : \lambda_1, \lambda_2, \ldots, \lambda_m \geq 0\}$$

with $\mathbf{v} \in \mathbb{R}^d$, $\mathbf{w}_1, \mathbf{w}_2, \ldots, \mathbf{w}_m \in \mathbb{Z}^d$, *the integer-point transform* $\sigma_{\mathcal{K}}(\mathbf{z})$ *evaluates to a rational function in the coordinates of* $\mathbf{z}$. □

Philosophizing some more, one can show that the original infinite sum $\sigma_{\mathcal{K}}(\mathbf{z})$ converges only for $\mathbf{z}$ in a subset of $\mathbb{C}^d$, whereas the rational function that represents $\sigma_{\mathcal{K}}$ gives us its meromorphic continuation. Later, in Chapters 4 and 9, we make use of this continuation.

3.3 Expanding and Counting Using Ehrhart's Original Approach

Here is *the* fundamental theorem concerning the lattice-point count in an integral convex polytope.

Theorem 3.8 (Ehrhart's theorem). *If $\mathcal{P}$ is an integral convex d-polytope, then $L_{\mathcal{P}}(t)$ is a polynomial in t of degree d.*

This result is due to Eugène Ehrhart, in whose honor we call $L_{\mathcal{P}}$ the **Ehrhart polynomial** of $\mathcal{P}$. Naturally, there is an extension of Ehrhart's theorem to rational polytopes, which we will discuss in Section 3.7.

Our proof of Ehrhart's theorem uses generating functions of the form $\sum_{t\geq 0} f(t)\, z^t$, similar in spirit to the ones discussed in the beginning of Chapter 1. If f is a polynomial, this power series takes on a special form, which we invite the reader to prove (Exercise 3.8):

Lemma 3.9. *If*

$$\sum_{t\geq 0} f(t)\, z^t = \frac{g(z)}{(1-z)^{d+1}},$$

then f is a polynomial of degree d if and only if g is a polynomial of degree at most d and $g(1) \neq 0$. □

The reason we introduced generating functions of the form $\sigma_S(\mathbf{z}) = \sum_{\mathbf{m}\in S\cap\mathbb{Z}^d} \mathbf{z}^{\mathbf{m}}$ in Section 3.2 is that they are extremely handy for lattice-point problems. The connection to lattice points is evident, since we are summing over them. If we're interested in the lattice-point *count*, we simply evaluate σ_S at $\mathbf{z} = (1, 1, \dots, 1)$:

$$\sigma_S(1, 1, \dots, 1) = \sum_{\mathbf{m}\in S\cap\mathbb{Z}^d} \mathbf{1}^{\mathbf{m}} = \sum_{\mathbf{m}\in S\cap\mathbb{Z}^d} 1 = \#\left(S\cap\mathbb{Z}^d\right).$$

(Here we denote by $\mathbf{1}$ a vector all of whose components are 1.) Naturally, we should make this evaluation only if S is bounded; Theorem 3.5 already tells us that it's no fun evaluating $\sigma_{\mathcal{K}}(\mathbf{1})$ if $\mathcal{K}$ is a cone.

But the magic of the generating function σ_S does not stop there. To literally take it to the next level, we cone over a convex polytope $\mathcal{P}$. If $\mathcal{P} \subset \mathbb{R}^d$ has the vertices $\mathbf{v}_1, \mathbf{v}_2, \dots, \mathbf{v}_n \in \mathbb{Z}^d$, recall that we lift these vertices into $\mathbb{R}^{d+1}$, by adding a 1 as their last coordinate. So let

$$\mathbf{w}_1 = (\mathbf{v}_1, 1),\ \mathbf{w}_2 = (\mathbf{v}_2, 1),\ \dots,\ \mathbf{w}_n = (\mathbf{v}_n, 1).$$

Then

$$\operatorname{cone}(\mathcal{P}) = \{\lambda_1\mathbf{w}_1 + \lambda_2\mathbf{w}_2 + \cdots + \lambda_n\mathbf{w}_n : \lambda_1, \lambda_2, \dots, \lambda_n \geq 0\} \subset \mathbb{R}^{d+1}.$$

Recall that we can recover our original polytope $\mathcal{P}$ by cutting cone($\mathcal{P}$) with the hyperplane $x_{d+1} = 1$. We can recover more than just the original polytope in cone($\mathcal{P}$): By cutting the cone with the hyperplane $x_{d+1} = 2$, we obtain a copy of $\mathcal{P}$ dilated by a factor of 2. (The reader should meditate on why this cut is a 2-dilate of $\mathcal{P}$.) More generally, we can cut the cone with the hyperplane $x_{d+1} = t$ and obtain $t\mathcal{P}$, as suggested by Figure 3.4.

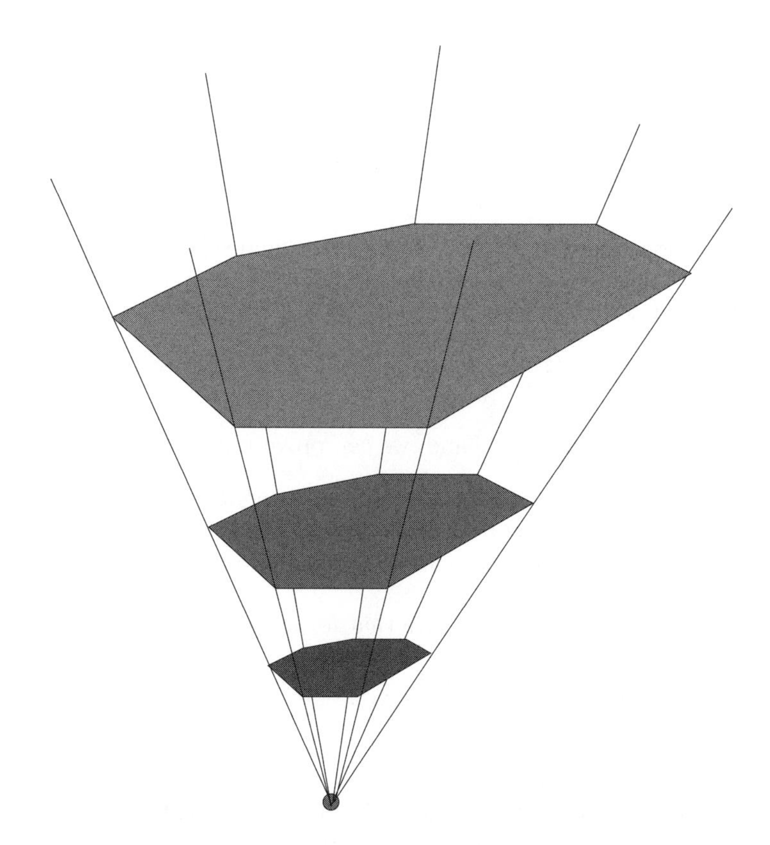

Fig. 3.4. Recovering dilates of $\mathcal{P}$ in cone($\mathcal{P}$).

Now let's form the integer-point transform $\sigma_{\text{cone}(\mathcal{P})}$ of cone($\mathcal{P}$). By what we just said, we should look at different powers of z_{d+1}: there is one term (namely, 1), with z_{d+1}^0, corresponding to the origin; the terms with z_{d+1}^1 correspond to lattice points in $\mathcal{P}$ (listed as monomials in $z_1, z_2, \ldots, z_d$), the terms with z_{d+1}^2 correspond to points in $2\mathcal{P}$, etc. In short,

$$\begin{aligned}
&\sigma_{\operatorname{cone}(\mathcal{P})}\left(z_1, z_2, \ldots, z_{d+1}\right) \\
&\quad = 1 + \sigma_{\mathcal{P}}\left(z_1, \ldots, z_d\right) z_{d+1} + \sigma_{2\mathcal{P}}\left(z_1, \ldots, z_d\right) z_{d+1}^2 + \cdots \\
&\quad = 1 + \sum_{t \geq 1} \sigma_{t\mathcal{P}}\left(z_1, \ldots, z_d\right) z_{d+1}^t .
\end{aligned}$$

Specializing further for enumeration purposes, we recall that $\sigma_{\mathcal{P}}\left(1, 1, \ldots, 1\right) = \#\left(\mathcal{P} \cap \mathbb{Z}^d\right)$, and so

$$\begin{aligned}
\sigma_{\operatorname{cone}(\mathcal{P})}\left(1, 1, \ldots, 1, z_{d+1}\right) &= 1 + \sum_{t \geq 1} \sigma_{t\mathcal{P}}\left(1, 1, \ldots, 1\right) z_{d+1}^t \\
&= 1 + \sum_{t \geq 1} \#\left(t\mathcal{P} \cap \mathbb{Z}^d\right) z_{d+1}^t .
\end{aligned}$$

But by definition, the enumerators on the right-hand side are just evaluations of Ehrhart's counting function, that is, $\sigma_{\operatorname{cone}(\mathcal{P})}\left(1, 1, \ldots, 1, z_{d+1}\right)$ is nothing but the Ehrhart series of $\mathcal{P}$:

Lemma 3.10. $\sigma_{\operatorname{cone}(\mathcal{P})}\left(1, 1, \ldots, 1, z\right) = 1 + \sum_{t \geq 1} L_{\mathcal{P}}(t)\, z^t = \operatorname{Ehr}_{\mathcal{P}}(z)$. □

With all this machinery at hand, we can prove Ehrhart's theorem.

Proof of Theorem 3.8. It suffices to prove the theorem for *simplices*, because we can triangulate any integral polytope into integral simplices, using no new vertices. Note that these simplices will intersect in lower-dimensional integral simplices.

By Lemma 3.9, it suffices to prove that for an integral d-simplex Δ,

$$\operatorname{Ehr}_{\Delta}(z) = 1 + \sum_{t \geq 1} L_{\Delta}(t)\, z^t = \frac{g(z)}{(1 - z)^{d+1}}$$

for some polynomial g of degree at most d with $g(1) \neq 0$. In Lemma 3.10 we showed that the Ehrhart series of Δ equals $\sigma_{\operatorname{cone}(\Delta)}\left(1, 1, \ldots, 1, z\right)$, so let's study the integer-point transform attached to $\operatorname{cone}(\Delta)$.

The simplex Δ has $d+1$ vertices $\mathbf{v}_1, \mathbf{v}_2, \ldots, \mathbf{v}_{d+1}$, and so $\operatorname{cone}(\Delta) \subset \mathbb{R}^{d+1}$ is simplicial, with apex the origin and generators

$$\mathbf{w}_1 = \left(\mathbf{v}_1, 1\right), \ \mathbf{w}_2 = \left(\mathbf{v}_2, 1\right), \ \ldots, \ \mathbf{w}_{d+1} = \left(\mathbf{v}_{d+1}, 1\right) \in \mathbb{Z}^{d+1}.$$

Now we use Theorem 3.5:

$$\sigma_{\operatorname{cone}(\Delta)}\left(z_1, z_2, \ldots, z_{d+1}\right) = \frac{\sigma_{\Pi}\left(z_1, z_2, \ldots, z_{d+1}\right)}{\left(1 - \mathbf{z}^{\mathbf{w}_1}\right)\left(1 - \mathbf{z}^{\mathbf{w}_2}\right) \cdots \left(1 - \mathbf{z}^{\mathbf{w}_{d+1}}\right)},$$

where $\Pi = \left\{\lambda_1 \mathbf{w}_1 + \lambda_2 \mathbf{w}_2 + \cdots + \lambda_{d+1} \mathbf{w}_{d+1} : 0 \leq \lambda_1, \lambda_2, \ldots, \lambda_{d+1} < 1\right\}$. This parallelepiped is bounded, so the attached generating function σ_{Π} is a Laurent polynomial in $z_1, z_2, \ldots, z_{d+1}$.

We claim that the z_{d+1}-degree of σ_Π is at most d. In fact, since the x_{d+1}-coordinate of each $\mathbf{w}_k$ is 1, the x_{d+1}-coordinate of a point in Π is $\lambda_1+\lambda_2+\cdots+\lambda_{d+1}$ for some $0 \le \lambda_1, \lambda_2, \ldots, \lambda_{d+1} < 1$. But then $\lambda_1+\lambda_2+\cdots+\lambda_{d+1} < d+1$, so if this sum is an integer it is at most d, which implies that the z_{d+1}-degree of $\sigma_\Pi\left(z_1, z_2, \ldots, z_{d+1}\right)$ is at most d. Consequently, $\sigma_\Pi\left(1,1,\ldots,1,z_{d+1}\right)$ is a polynomial in z_{d+1} of degree at most d. The evaluation $\sigma_\Pi\left(1,1,1,\ldots,1\right)$ of this polynomial at $z_{d+1}=1$ is not zero, because $\sigma_\Pi\left(1,1,1,\ldots,1\right) = \#\left(\Pi \cap \mathbb{Z}^{d+1}\right)$ and the origin is a lattice point in Π.

Finally, if we specialize $\mathbf{z}^{\mathbf{w}_k}$ to $z_1 = z_2 = \cdots = z_d = 1$, we obtain z_{d+1}^1, so that

$$\sigma_{\operatorname{cone}(\Delta)}\left(1,1,\ldots,1,z_{d+1}\right) = \frac{\sigma_\Pi\left(1,1,\ldots,1,z_{d+1}\right)}{\left(1-z_{d+1}\right)^{d+1}}.$$

The left–hand side is $\operatorname{Ehr}_\Delta\left(z_{d+1}\right) = 1+\sum_{t\ge 1} L_\Delta(t)\, z_{d+1}^t$ by Lemma 3.10. □

3.4 The Ehrhart Series of an Integral Polytope

We can actually take our proof of Ehrhart's theorem one step further by studying the polynomial $\sigma_\Pi\left(1,1,\ldots,1,z_{d+1}\right)$. As mentioned above, the coefficient of z_{d+1}^k simply counts the integer points in the parallelepiped Π cut with the hyperplane $x_{d+1} = k$. Let us record this.

Corollary 3.11. *Suppose Δ is an integral d-simplex with vertices $\mathbf{v}_1, \mathbf{v}_2, \ldots,$ $\mathbf{v}_{d+1}$, and let $\mathbf{w}_j = (\mathbf{v}_j, 1)$. Then*

$$\operatorname{Ehr}_\Delta(z) = 1 + \sum_{t\ge 1} L_\Delta(t)\, z^t = \frac{h_d\, z^d + h_{d-1}\, z^{d-1} + \cdots + h_1\, z + h_0}{(1-z)^{d+1}},$$

where h_k equals the number of integer points in

$$\left\{\lambda_1\mathbf{w}_1 + \lambda_2\mathbf{w}_2 + \cdots + \lambda_{d+1}\mathbf{w}_{d+1} : 0 \le \lambda_1, \lambda_2, \ldots, \lambda_{d+1} < 1\right\}$$

with last variable equal to k. □

This result can actually be used to compute $\operatorname{Ehr}_\Delta$, and therefore the Ehrhart polynomial, of an integral simplex Δ in low dimensions very quickly (a fact that the reader may discover in some of the exercises). We remark, however, that things are not as simple for arbitrary integral polytopes. Not only is triangulation a nontrivial task in general, but one would also have to deal with overcounting where simplices of a triangulation meet.

Corollary 3.11 implies that the numerator of the Ehrhart series of an integral simplex has nonnegative coefficients, since those coefficients count something. Although the latter cannot be said of the coefficients of the Ehrhart series of a general integral polytope, the nonnegativity property magically survives.

Theorem 3.12 (Stanley's nonnegativity theorem). *Suppose $\mathcal{P}$ is an integral convex d-polytope with Ehrhart series*

$$\mathrm{Ehr}_{\mathcal{P}}(z) = \frac{h_d\, z^d + h_{d-1}\, z^{d-1} + \cdots + h_0}{(1-z)^{d+1}} .$$

Then $h_0, h_1, \dots, h_d \geq 0$.

Proof. Triangulate $\mathrm{cone}(\mathcal{P}) \subset \mathbb{R}^{d+1}$ into the simplicial cones $\mathcal{K}_1, \mathcal{K}_2, \dots, \mathcal{K}_m$. Now Exercise 3.14 ensures that there exists a vector $\mathbf{v} \in \mathbb{R}^{d+1}$ such that

$$\mathrm{cone}(\mathcal{P}) \cap \mathbb{Z}^d = (\mathbf{v} + \mathrm{cone}(\mathcal{P})) \cap \mathbb{Z}^d$$

(that is, we neither lose nor gain any lattice points when shifting $\mathrm{cone}(\mathcal{P})$ by $\mathbf{v}$) and neither the facets of $\mathbf{v} + \mathrm{cone}(\mathcal{P})$ nor the triangulation hyperplanes contain any lattice points. This implies that every lattice point in $\mathbf{v} + \mathrm{cone}(\mathcal{P})$ belongs to exactly one simplicial cone $\mathbf{v} + \mathcal{K}_j$:

$$\mathrm{cone}(\mathcal{P}) \cap \mathbb{Z}^d = (\mathbf{v} + \mathrm{cone}(\mathcal{P})) \cap \mathbb{Z}^d = \bigcup_{j=1}^{m} \left((\mathbf{v} + \mathcal{K}_j) \cap \mathbb{Z}^d \right) , \qquad (3.2)$$

and this union is a *disjoint* union. If we translate the last identity into generating-function language, it becomes

$$\sigma_{\mathrm{cone}(\mathcal{P})} \left(z_1, z_2, \dots, z_{d+1} \right) = \sum_{j=1}^{m} \sigma_{\mathbf{v}+\mathcal{K}_j} \left(z_1, z_2, \dots, z_{d+1} \right) .$$

But now we recall that the Ehrhart series of $\mathcal{P}$ is just a special evaluation of $\sigma_{\mathrm{cone}(\mathcal{P})}$ (Lemma 3.10):

$$\mathrm{Ehr}_{\mathcal{P}}(z) = \sigma_{\mathrm{cone}(\mathcal{P})} \left(1, 1, \dots, 1, z \right) = \sum_{j=1}^{m} \sigma_{\mathbf{v}+\mathcal{K}_j} \left(1, 1, \dots, 1, z \right) . \qquad (3.3)$$

It suffices to show that the rational generating functions $\sigma_{\mathbf{v}+\mathcal{K}_j} \left(1, 1, \dots, 1, z \right)$ for the simplicial cones $\mathbf{v} + \mathcal{K}_j$ have nonnegative numerator. But this fact follows from evaluating the rational function in Corollary 3.6 at $(1, 1, \dots, 1, z)$. □

This proof shows a little more: Since the origin is in precisely *one* simplicial cone on the right-hand side of (3.2), we get on the right-hand side of (3.3) precisely *one* term that contributes $1/(1-z)^{d+1}$ to $\mathrm{Ehr}_{\mathcal{P}}$; all other terms contribute to higher powers of the numerator polynomial of $\mathrm{Ehr}_{\mathcal{P}}$. That is, the constant term h_0 equals 1. The reader might feel that we are chasing our tail at this point, since we assumed from the very beginning that the constant term of the infinite series $\mathrm{Ehr}_{\mathcal{P}}$ is 1, and hence h_0 has to be 1, as a quick look at the expansion of the rational function representing $\mathrm{Ehr}_{\mathcal{P}}$ shows. The above argument shows merely that this convention is geometrically sound. Let us record this:

Lemma 3.13. *Suppose $\mathcal{P}$ is an integral convex d-polytope with Ehrhart series*

$$\operatorname{Ehr}_{\mathcal{P}}(z) = \frac{h_d\, z^d + h_{d-1}\, z^{d-1} + \cdots + h_0}{(1-z)^{d+1}} .$$

Then $h_0 = 1$. □

For a general integral polytope $\mathcal{P}$, the reader has probably already discovered how to extract the Ehrhart polynomial of $\mathcal{P}$ from its Ehrhart series:

Lemma 3.14. *Suppose $\mathcal{P}$ is an integral convex d-polytope with Ehrhart series*

$$\operatorname{Ehr}_{\mathcal{P}}(z) = 1 + \sum_{t \ge 1} L_{\mathcal{P}}(t)\, z^t = \frac{h_d\, z^d + h_{d-1}\, z^{d-1} + \cdots + h_1\, z + 1}{(1-z)^{d+1}} .$$

Then

$$L_{\mathcal{P}}(t) = \binom{t+d}{d} + h_1 \binom{t+d-1}{d} + \cdots + h_{d-1} \binom{t+1}{d} + h_d \binom{t}{d} .$$

Proof. Expand into a binomial series:

$$\begin{aligned}
\operatorname{Ehr}_{\mathcal{P}}(z) &= \frac{h_d\, z^d + h_{d-1}\, z^{d-1} + \cdots + h_1\, z + 1}{(1-z)^{d+1}} \\
&= \left(h_d\, z^d + h_{d-1}\, z^{d-1} + \cdots + h_1\, z + 1\right) \sum_{t \ge 0} \binom{t+d}{d} z^t \\
&= h_d \sum_{t \ge 0} \binom{t+d}{d} z^{t+d} + h_{d-1} \sum_{t \ge 0} \binom{t+d}{d} z^{t+d-1} + \cdots \\
&\quad + h_1 \sum_{t \ge 0} \binom{t+d}{d} z^{t+1} + \sum_{t \ge 0} \binom{t+d}{d} z^t \\
&= h_d \sum_{t \ge d} \binom{t}{d} z^t + h_{d-1} \sum_{t \ge d-1} \binom{t+1}{d} z^t + \cdots \\
&\quad + h_1 \sum_{t \ge 1} \binom{t+d-1}{d} z^t + \sum_{t \ge 0} \binom{t+d}{d} z^t .
\end{aligned}$$

In all infinite sums on the right-hand side, we can start the index t with 0 without changing the sums, by the definition (2.1) of the binomial coefficient. Hence

$$\begin{aligned}
&\operatorname{Ehr}_{\mathcal{P}}(z) \\
&= \sum_{t \ge 0} \left(h_d \binom{t}{d} + h_{d-1} \binom{t+1}{d} + \cdots + h_1 \binom{t+d-1}{d} + \binom{t+d}{d} \right) z^t .
\end{aligned}$$

□

The representation of the polynomial $L_{\mathcal{P}}(t)$ in terms of the coefficients of $\mathrm{Ehr}_{\mathcal{P}}$ can be interpreted as the Ehrhart polynomial expressed in the basis $\binom{t}{d}, \binom{t+1}{d}, \dots, \binom{t+d}{d}$ (see Exercise 3.9). This representation is very useful, as the following results show.

Corollary 3.15. *If $\mathcal{P}$ is an integral convex d-polytope, then the constant term of the Ehrhart polynomial $L_{\mathcal{P}}$ is* 1.

Proof. Use the expansion of Lemma 3.14. The constant term is

$$L_{\mathcal{P}}(0) = \binom{d}{d} + h_1 \binom{d-1}{d} + \cdots + h_{d-1} \binom{1}{d} + h_d \binom{0}{d} = \binom{d}{d} = 1 . \qquad \square$$

This proof is exciting, because it marks the first instance where we extend the domain of an Ehrhart polynomial beyond the positive integers, for which the lattice-point enumerator was initially defined. More precisely, Ehrhart's theorem (Theorem 3.8) implies that the counting function

$$L_{\mathcal{P}}(t) = \# \left(t\mathcal{P} \cap \mathbb{Z}^d \right),$$

originally defined for positive integers t, can be extended to all real or even complex arguments t (as a polynomial). A natural question arises: are there nice *interpretations* of $L_{\mathcal{P}}(t)$ for arguments t that are not positive integers? Corollary 3.15 gives such an interpretation for $t = 0$. In Chapter 4, we will give interpretations of $L_{\mathcal{P}}(t)$ for *negative* integers t.

Corollary 3.16. *Suppose $\mathcal{P}$ is an integral convex d-polytope with Ehrhart series*

$$\mathrm{Ehr}_{\mathcal{P}}(z) = \frac{h_d\, z^d + h_{d-1}\, z^{d-1} + \cdots + h_1\, z + 1}{(1-z)^{d+1}} .$$

Then $h_1 = L_{\mathcal{P}}(1) - d - 1 = \# \left(\mathcal{P} \cap \mathbb{Z}^d \right) - d - 1$.

Proof. Use the expansion of Lemma 3.14 with $t = 1$:

$$L_{\mathcal{P}}(1) = \binom{d+1}{d} + h_1 \binom{d}{d} + \cdots + h_{d-1} \binom{2}{d} + h_d \binom{1}{d} = d + 1 + h_1 . \qquad \square$$

The proof of Corollary 3.16 suggests that there are also formulas for $h_2, h_3, \dots$ in terms of the evaluations $L_{\mathcal{P}}(1), L_{\mathcal{P}}(2), \dots$, and we invite the reader to find them (Exercise 3.10).

A final corollary to Theorem 3.12 and Lemma 3.14 states how large the denominators of the Ehrhart coefficients can be:

Corollary 3.17. *Suppose $\mathcal{P}$ is an integral polytope with Ehrhart polynomial $L_{\mathcal{P}}(t) = c_d\, t^d + c_{d-1}\, t^{d-1} + \cdots + c_1\, t + 1$. Then all coefficients satisfy $d!\, c_k \in \mathbb{Z}$.*

Proof. By Theorem 3.12 and Lemma 3.14,

$$L_{\mathcal{P}}(t) = \binom{t+d}{d} + h_1 \binom{t+d-1}{d} + \cdots + h_{d-1} \binom{t+1}{d} + h_d \binom{t}{d},$$

where the h_k's are integers. Hence multiplying out this expression yields a polynomial in t whose coefficients can be written as rational numbers with denominator $d!$. □

We finish this section with a general result that gives relations between negative integer roots of a polynomial and its generating function. This theorem will become handy in Chapter 4, in which we find an interpretation for the evaluation of an Ehrhart polynomial at negative integers.

Theorem 3.18. *Suppose p is a degree-d polynomial with the rational generating function*

$$\sum_{t \geq 0} p(t)\, z^t = \frac{h_d\, z^d + h_{d-1}\, z^{d-1} + \cdots + h_1\, z + h_0}{(1-z)^{d+1}}.$$

Then $h_d = h_{d-1} = \cdots = h_{k+1} = 0$ and $h_k \neq 0$ if and only if $p(-1) = p(-2) = \cdots = p\,(-(d-k)) = 0$ and $p\,(-(d-k+1)) \neq 0$.

Proof. Suppose $h_d = h_{d-1} = \cdots = h_{k+1} = 0$ and $h_k \neq 0$. Then the proof of Lemma 3.14 gives

$$p(t) = h_0 \binom{t+d}{d} + \cdots + h_{k-1} \binom{t+d-k+1}{d} + h_k \binom{t+d-k}{d}.$$

All the binomial coefficients are zero for $t = -1, -2, \ldots, -d+k$, so those are roots of p. On the other hand, all binomial coefficients but the last one are zero for $t = -d+k-1$, and since $h_k \neq 0$, $-d+k-1$ is not a root of p.

Conversely, suppose $p(-1) = p(-2) = \cdots = p\,(-(d-k)) = 0$ and $p\,(-(d-k+1)) \neq 0$. The first root -1 of p gives

$$0 = p(-1) = h_0 \binom{d-1}{d} + h_1 \binom{d-2}{d} + \cdots + h_{d-1} \binom{0}{d} + h_d \binom{-1}{d} = h_d \binom{-1}{d},$$

so we must have $h_d = 0$. The next root -2 forces $h_{d-1} = 0$, and so on, up to the root $-d+k$, which forces $h_{k+1} = 0$. It remains to show that $h_k \neq 0$. But if h_k were zero then, by a similar line of reasoning as in the first part of the proof, $p(-d+k-1) = 0$, a contradiction. □

3.5 From the Discrete to the Continuous Volume of a Polytope

Given a geometric object $S \subset \mathbb{R}^d$, its **volume**, defined by the integral $\operatorname{vol} S := \int_S dx$, is one of the fundamental data of S. By the definition of the integral, say in the Riemannian sense, we can think of computing $\operatorname{vol} S$ by approximating S with d-dimensional boxes that get smaller and smaller. To be precise, if we take the boxes with side length $1/t$ then they each have volume $1/t^d$. We might further think of the boxes as filling out the space between grid points in the lattice $\left(\frac{1}{t}\mathbb{Z}\right)^d$. This means that volume computation can be approximated by counting boxes, or equivalently, lattice points in $\left(\frac{1}{t}\mathbb{Z}\right)^d$:

$$\operatorname{vol} S = \lim_{t \to \infty} \frac{1}{t^d} \cdot \# \left(S \cap \left(\frac{1}{t}\mathbb{Z} \right)^d \right).$$

It is a short step to counting integer points in dilates of S, because

$$\# \left(S \cap \left(\frac{1}{t}\mathbb{Z} \right)^d \right) = \# \left(tS \cap \mathbb{Z}^d \right).$$

Let us summarize:

Lemma 3.19. *Suppose $S \subset \mathbb{R}^d$ is d-dimensional. Then*

$$\operatorname{vol} S = \lim_{t \to \infty} \frac{1}{t^d} \cdot \# \left(tS \cap \mathbb{Z}^d \right). \qquad \square$$

We emphasize here that S is d-dimensional, because otherwise (since S could be lower-dimensional although living in d-space), by our current definition $\operatorname{vol} S = 0$. (We will extend our volume definition in Chapter 5 to give nonzero *relative* volume to objects that are not full-dimensional.)

Part of the magic of Ehrhart's theorem lies in the fact that for an integral d-polytope $\mathcal{P}$, we do not have to take a limit to compute $\operatorname{vol} \mathcal{P}$; we need to compute "only" the $d+1$ coefficients of a polynomial.

Corollary 3.20. *Suppose $\mathcal{P} \subset \mathbb{R}^d$ is a integral convex d-polytope with Ehrhart polynomial $c_d\, t^d + c_{d-1}\, t^{d-1} + \cdots + c_1\, t + 1$. Then $c_d = \operatorname{vol} \mathcal{P}$.*

Proof. By Lemma 3.19,

$$\operatorname{vol} \mathcal{P} = \lim_{t \to \infty} \frac{c_d\, t^d + c_{d-1}\, t^{d-1} + \cdots + c_1\, t + 1}{t^d} = c_d\,. \qquad \square$$

On the one hand, this should not come as a surprise, because counting integer points in some object should grow roughly like the volume of the object as we make it bigger and bigger. On the other hand, the fact that we can compute the volume as one term of a polynomial should be very surprising: the polynomial is a counting function and as such is something *discrete*, yet by computing it (and its leading term), we derive some *continuous* data. Even more, we can—at least theoretically—compute this continuous datum (the volume) of the object by calculating a few values of the polynomial and then interpolating; this can be described as a completely discrete operation!

We finish this section by showing how to retrieve the continuous volume of an integer polytope from its Ehrhart series.

Corollary 3.21. *Suppose $\mathcal{P} \subset \mathbb{R}^d$ is an integral convex d-polytope, and*

$$\operatorname{Ehr}_{\mathcal{P}}(z) = \frac{h_d\, z^d + h_{d-1}\, z^{d-1} + \cdots + h_1\, z + 1}{(1-z)^{d+1}} .$$

Then $\operatorname{vol} \mathcal{P} = \frac{1}{d!} \left(h_d + h_{d-1} + \cdots + h_1 + 1 \right).$

Proof. Use the expansion of Lemma 3.14. The leading coefficient is

$$\frac{1}{d!} \left(h_d + h_{d-1} + \cdots + h_1 + 1 \right). \qquad \square$$

3.6 Interpolation

We now use the polynomial behavior of the discrete volume $L_{\mathcal{P}}$ of an integral polytope $\mathcal{P}$ to compute the continuous volume $\operatorname{vol} \mathcal{P}$ and the discrete volume $L_{\mathcal{P}}$ from finite data.

Two points uniquely determine a line. There exists a unique quadratic passing through any three given points. In general, a degree-d polynomial p is determined by $d+1$ points $(x, p(x)) \in \mathbb{R}^2$. Namely, evaluating $p(x) = c_d x^d + c_{d-1} x^{d-1} + \cdots + c_0$ at distinct inputs $x_1, x_2, \dots, x_{d+1}$ gives

$$\begin{pmatrix} p(x_1) \\ p(x_2) \\ \vdots \\ p(x_{d+1}) \end{pmatrix} = \mathbf{V} \begin{pmatrix} c_d \\ c_{d-1} \\ \vdots \\ c_0 \end{pmatrix}, \tag{3.4}$$

where

$$\mathbf{V} = \begin{pmatrix} x_1^d & x_1^{d-1} & \cdots & x_1 & 1 \\ x_2^d & x_2^{d-1} & \cdots & x_2 & 1 \\ \vdots & \vdots & & \vdots & \vdots \\ x_{d+1}^d & x_{d+1}^{d-1} & \cdots & x_{d+1} & 1 \end{pmatrix},$$

so that

$$\begin{pmatrix} c_d \\ c_{d-1} \\ \vdots \\ c_0 \end{pmatrix} = \mathbf{V}^{-1} \begin{pmatrix} p(x_1) \\ p(x_2) \\ \vdots \\ p(x_{d+1}) \end{pmatrix}. \tag{3.5}$$

(Exercise 3.16 makes sure that $\mathbf{V}$ is invertible.) The identity (3.5) gives the famous *Lagrange interpolation formula.*

This gives us an efficient way to compute $L_{\mathcal{P}}$, at least when $\dim \mathcal{P}$ is not too large. The continuous volume of $\mathcal{P}$ will follow instantly, since it is the leading coefficient c_d of $L_{\mathcal{P}}$. In the case of an Ehrhart polynomial $L_{\mathcal{P}}$, we know that $L_{\mathcal{P}}(0) = 1$, so that (3.4) simplifies to

$$\begin{pmatrix} L_{\mathcal{P}}(x_1) - 1 \\ L_{\mathcal{P}}(x_2) - 1 \\ \vdots \\ L_{\mathcal{P}}(x_d) - 1 \end{pmatrix} = \begin{pmatrix} x_1^d & x_1^{d-1} & \cdots & x_1 \\ x_2^d & x_2^{d-1} & \cdots & x_2 \\ \vdots & \vdots & & \vdots \\ x_d^d & x_d^{d-1} & \cdots & x_d \end{pmatrix} \begin{pmatrix} c_d \\ c_{d-1} \\ \vdots \\ c_1 \end{pmatrix}.$$

Example 3.22 (Reeve's tetrahedron). Let $\mathcal{T}_h$ be the tetrahedron with vertices $(0,0,0)$, $(1,0,0)$, $(0,1,0)$, and $(1,1,h)$, where h is a positive integer (see Figure 3.5).

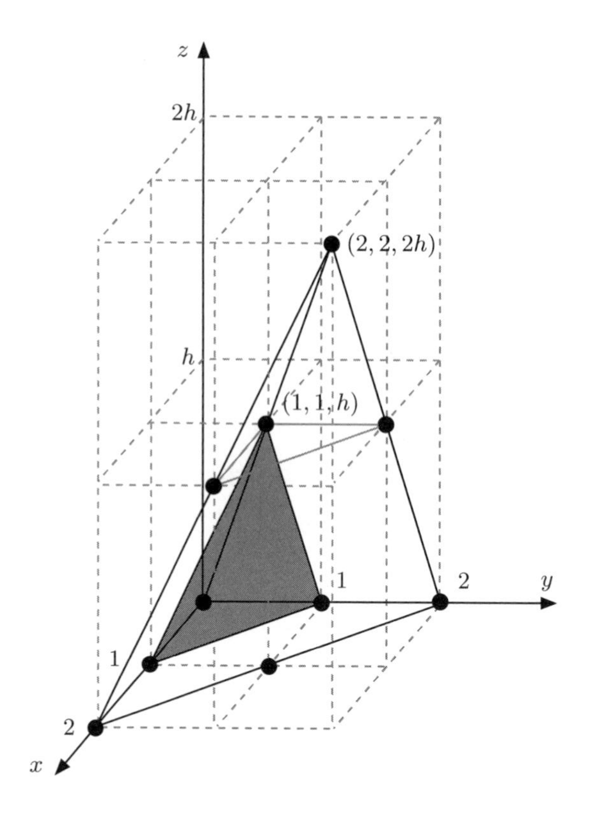

Fig. 3.5. Reeve's tetrahedron $\mathcal{T}_h$ (and $2\mathcal{T}_h$).

To interpolate the Ehrhart polynomial $L_{\mathcal{T}_h}(t)$ from its values at various points, we use Figure 3.5 to deduce the following:

$$4 = L_{\mathcal{T}_h}(1) = \operatorname{vol}(\mathcal{T}_h) + c_2 + c_1 + 1,$$
$$h + 9 = L_{\mathcal{T}_h}(2) = \operatorname{vol}(\mathcal{T}_h) \cdot 2^3 + c_2 \cdot 2^2 + c_1 \cdot 2 + 1.$$

Using the volume formula for a pyramid, we know that

$$\mathrm{vol}\,(\mathcal{T}_h) = \frac{1}{3}(\text{base area})(\text{height}) = \frac{h}{6}\,.$$

Thus $h + 1 = h + 2c_2 - 1$, which gives us $c_2 = 1$ and $c_1 = 2 - \frac{h}{6}$. Therefore

$$L_{\mathcal{T}_h}(t) = \frac{h}{6}\,t^3 + t^2 + \left(2 - \frac{h}{6}\right) t + 1\,. \qquad \square$$

3.7 Rational Polytopes and Ehrhart Quasipolynomials

We do not have to change much to study lattice-point enumeration for *rational* polytopes, and most of this section will consist of exercises for the reader. The structural result paralleling Theorem 3.8 is as follows.

Theorem 3.23 (Ehrhart's theorem for rational polytopes). *If $\mathcal{P}$ is a rational convex d-polytope, then $L_{\mathcal{P}}(t)$ is a quasipolynomial in t of degree d. Its period divides the least common multiple of the denominators of the coordinates of the vertices of $\mathcal{P}$.*

We will call the least common multiple of the denominators of the coordinates of the vertices of $\mathcal{P}$ the **denominator** of $\mathcal{P}$. Theorem 3.23, also due to Ehrhart, extends Theorem 3.8, because the denominator of an integral polytope $\mathcal{P}$ is one. Exercises 3.21 and 3.22 show that the word "divides" in Theorem 3.23 is far from being replaceable by "equals."

We start the path toward a proof of Theorem 3.23 by stating the analogue of Lemma 3.9 for quasipolynomials (see Exercise 3.19):

Lemma 3.24. *If*

$$\sum_{t \ge 0} f(t)\, z^t = \frac{g(z)}{h(z)}\,,$$

then f is a quasipolynomial of degree d with period dividing p if and only if g and h are polynomials such that $\deg(g) < \deg(h)$, all roots of h are p^{th} roots of unity of multiplicity at most $d+1$, and there is a root of multiplicity equal to $d+1$ (all of this assuming that g/h has been reduced to lowest terms). $\square$

Our goal is now evident: we will prove that if $\mathcal{P}$ is a rational convex d-polytope with denominator p, then

$$\mathrm{Ehr}_{\mathcal{P}}(z) = 1 + \sum_{t \ge 1} L_{\mathcal{P}}(t)\, z^t = \frac{g(z)}{\left(1 - z^p\right)^{d+1}}\,,$$

for some polynomial g of degree less than $p(d+1)$. As in Section 3.3, we will have to prove this only for the case of a rational *simplex*. So suppose the d-simplex Δ has vertices $\mathbf{v}_1, \mathbf{v}_2, \dots, \mathbf{v}_{d+1} \in \mathbb{Q}^d$, and the denominator of Δ is p. Again we will cone over Δ: let

$$\mathbf{w}_1 = (\mathbf{v}_1, 1), \mathbf{w}_2 = (\mathbf{v}_2, 1), \dots, \mathbf{w}_{d+1} = (\mathbf{v}_{d+1}, 1);$$

then

$$\operatorname{cone}(\Delta) = \{\lambda_1 \mathbf{w}_1 + \lambda_2 \mathbf{w}_2 + \cdots + \lambda_{d+1} \mathbf{w}_{d+1} : \lambda_1, \lambda_2, \dots, \lambda_{d+1} \geq 0\} \subset \mathbb{R}^{d+1}.$$

To be able to use Theorem 3.5, we first have to ensure that we have a description of cone(Δ) with integral generators. But since the denominator of Δ is p, we can replace each generator $\mathbf{w}_k$ by $p\mathbf{w}_k \in \mathbb{Z}^{d+1}$, and we're ready to apply Theorem 3.5. From this point, the proof of Theorem 3.23 proceeds exactly like that of Theorem 3.8, and we invite the reader to finish it up (Exercise 3.20).

Although the proofs of Theorem 3.23 and Theorem 3.8 are almost identical, the arithmetic structure of Ehrhart quasipolynomials is much more subtle and less well known than that of Ehrhart polynomials.

3.8 Reflections on the Coin-Exchange Problem and the Gallery of Chapter 2

At this point, we encourage the reader to look back at the first two chapters in light of the basic Ehrhart-theory results. Popoviciu's theorem (Theorem 1.5) and its higher-dimensional analogue give a special set of Ehrhart quasipolynomials. On the other hand, in Chapter 2 we encountered many integral polytopes. Ehrhart's theorem (Theorem 3.8) explains why their lattice-point enumeration functions were all polynomials.

Notes

1. Triangulations of polytopes and manifolds are an active source of research with many interesting open problems; see, e.g., [69].

2. Eugène Ehrhart laid the foundation for the central theme of this book in the 1960s, starting with the proof of Theorem 3.8 in 1962 [79]. The proof we give here follows Ehrhart's original lines of thought. An interesting fact is that he did his most beautiful work as a teacher at a *lycée* in Strasbourg (France), receiving his doctorate at age 60 on the urging of some colleagues.

3. Given any d linearly independent vectors $\mathbf{w}_1, \mathbf{w}_2, \dots, \mathbf{w}_d \in \mathbb{R}^d$, the **lattice** generated by them is the set of all integer linear combinations of $\mathbf{w}_1, \mathbf{w}_2, \dots, \mathbf{w}_d$. Alternatively, one can define a lattice as a discrete subgroup of $\mathbb{R}^d$, and these two notions can be shown to be equivalent. One might wonder whether replacing the lattice $\mathbb{Z}^d$ by an arbitrary lattice $\mathcal{L}$ throughout the statements of the theorems—requiring now that the vertices of a polytope be in $\mathcal{L}$—gives us any different results. The fact that the theorems of this chapter remain the same follows from the observation that any lattice can be mapped to $\mathbb{Z}^d$ by an invertible linear transformation.

4. Richard Stanley developed much of the theory of Ehrhart (quasi-)polynomials, initially from a commutative-algebra point of view. Theorem 3.12 is due to him [170]. The proof we give here appeared in [30]. An extension of Theorem 3.12 was found by Takayuki Hibi; he proved that if $h_d > 0$ then $h_k \geq h_1$ for all $1 \leq k \leq d-1$ (using the notation of Theorem 3.12) [98].

5. The tetrahedron $\mathcal{T}_h$ of Example 3.22 was used by John Reeve to show that Pick's theorem does not hold in $\mathbb{R}^3$ (see Exercise 3.18) [154]. Incidentally, the formula for $L_{\mathcal{T}_h}$ also proves that the coefficients of an Ehrhart polynomial (of a closed polytope) are not always positive.

6. There are several interesting questions (some of which are still open) regarding the periods of Ehrhart quasipolynomials. Some particularly nice examples about what can happen with periods were given by Tyrrell McAllister and Kevin Woods [126].

7. Most of the results remain true if we replace "convex polytope" by "polytopal complex," that is, a finite union of polytopes. One important exception is Corollary 3.15: the constant term of an "Ehrhart polynomial" of an integral polytopal complex C is the *Euler characteristric* of C.

8. The reader might wonder why we do not discuss polytopes with *irrational* vertices. The answer is simple: nobody has yet found a theory that would parallel the results in this chapter, even in dimension two. One notable exception is [11], in which irrational extensions of Brion's theorem are given; we will study the rational case of Brion's theorem in Chapter 9. On the other hand, Ehrhart theory has been extended to functions other than strict lattice-point counting; one instance is described in Chapter 11.

Exercises

3.1. To any permutation $\pi \in S_d$ on d elements, we associate the simplex

$$\Delta_\pi := \operatorname{conv}\left\{\mathbf{0}, \mathbf{e}_{\pi(1)}, \mathbf{e}_{\pi(1)} + \mathbf{e}_{\pi(2)}, \dots, \mathbf{e}_{\pi(1)} + \mathbf{e}_{\pi(2)} + \cdots + \mathbf{e}_{\pi(d)}\right\},$$

where $\mathbf{e}_1, \mathbf{e}_2, \dots, \mathbf{e}_d$ denote the unit vectors in $\mathbb{R}^d$.

(a) Prove that $\{\Delta_\pi : \pi \in S_d\}$ is a triangulation of the unit d-cube $[0,1]^d$.
(b) Prove that all Δ_π are congruent to each other, that is, each one can be obtained from any other by reflections, translations, and rotations.
(c) Show that for all $\pi \in S_d$, $L_{\Delta_\pi}(t) = \binom{d+t}{d}$.

3.2. ♣ Suppose T is a triangulation of a pointed cone. Prove that the intersection of two simplicial cones in T is again a simplicial cone.

3.3. Find the generating function $\sigma_{\mathcal{K}}(\mathbf{z})$ for the following cones:

(a) $\mathcal{K} = \{\lambda_1(0,1) + \lambda_2(1,0) : \lambda_1, \lambda_2 \geq 0\}$;
(b) $\mathcal{K} = \{\lambda_1(0,1) + \lambda_2(1,1) : \lambda_1, \lambda_2 \geq 0\}$;
(c) $\mathcal{K} = \{(3,4) + \lambda_1(0,1) + \lambda_2(2,1) : \lambda_1, \lambda_2 \geq 0\}$.

3.4. ♣ Let $S \subseteq \mathbb{R}^m$ and $T \subseteq \mathbb{R}^n$. Show that $\sigma_{S\times T}(z_1, z_2, \ldots, z_{m+n}) = \sigma_S(z_1, z_2, \ldots, z_m)\, \sigma_T(z_{m+1}, z_{m+2}, \ldots, z_{m+n})$.

3.5. ♣ Let $\mathcal{K}$ be a rational d-cone, and let $\mathbf{m} \in \mathbb{Z}^d$. Show that $\sigma_{\mathbf{m}+\mathcal{K}}(\mathbf{z}) = \mathbf{z}^{\mathbf{m}} \sigma_{\mathcal{K}}(\mathbf{z})$.

3.6. ♣ For a set $S \subset \mathbb{R}^d$, let $-S := \{-x : x \in S\}$. Prove that

$$\sigma_{-S}(z_1, z_2, \ldots, z_d) = \sigma_S\left(\frac{1}{z_1}, \frac{1}{z_2}, \ldots, \frac{1}{z_d}\right).$$

3.7. Given a pointed cone $\mathcal{K} \subset \mathbb{R}^d$ with apex at the origin, let $S := \mathcal{K} \cap \mathbb{Z}^d$. Show that if $\mathbf{x}, \mathbf{y} \in S$ then $\mathbf{x} + \mathbf{y} \in S$. (In algebraic terms, S is a *semigroup*, since $\mathbf{0} \in S$ and associativity of the addition in S follows trivially from associativity in $\mathbb{R}^d$.)

3.8. ♣ Prove Lemma 3.9: If

$$\sum_{t\geq 0} f(t)\, z^t = \frac{g(z)}{(1-z)^{d+1}},$$

then f is a polynomial of degree d if and only if g is a polynomial of degree at most d and $g(1) \neq 0$.

3.9. Prove that $\binom{x+n}{n}, \binom{x+n-1}{n}, \ldots, \binom{x}{n}$ is a basis for the vector space Pol_n of polynomials (in the variable x) of degree less than or equal to n.

3.10. For a polynomial $p(t) = c_d t^d + c_{d-1} t^{d-1} + \cdots + c_0$, let $H_p(z)$ be defined by

$$\sum_{t\geq 0} p(t)\, z^t = \frac{H_p(z)}{(1-z)^{d+1}}.$$

Consider the map $\phi_d \colon \mathrm{Pol}_d \to \mathrm{Pol}_d,\ p \mapsto H_p$.

(a) Show that ϕ_d is a linear transformation.
(b) Compute the matrix describing ϕ_d for $d = 0, 1, 2, \ldots$.
(c) Deduce formulas for $h_2, h_3, \ldots,$ similar to the one in Corollary 3.16.

3.11. Compute the Ehrhart polynomials and the Ehrhart series of the simplices with the following vertices:

(a) $(0,0,0)$, $(1,0,0)$, $(0,2,0)$, and $(0,0,3)$;
(b) $(0,0,0,0)$, $(1,0,0,0)$, $(0,2,0,0)$, $(0,0,3,0)$, and $(0,0,0,4)$.

3.12. Define the **hypersimplex** $\Delta(d,k)$ as the convex hull of

$$\{\mathbf{e}_{j_1} + \mathbf{e}_{j_2} + \cdots + \mathbf{e}_{j_k} : 1 \leq j_1 < j_2 < \cdots < j_k \leq d\},$$

where $\mathbf{e}_1, \mathbf{e}_2, \ldots, \mathbf{e}_d$ are the standard basis vectors in $\mathbb{R}^d$. For example, $\Delta(d,1)$ and $\Delta(d,d-1)$ are regular $(d-1)$-simplices. Compute the Ehrhart polynomial and the Ehrhart series of $\Delta(d,k)$.

3.13. ♣ Suppose H is the hyperplane given by

$$H = \left\{\mathbf{x} \in \mathbb{R}^d : a_1x_1 + a_2x_2 + \cdots + a_dx_d = 0\right\}$$

for some $a_1, a_2, \ldots, a_d \in \mathbb{Z}$, which we may assume to have no common factor. Prove that there exists $\mathbf{v} \in \mathbb{Z}^d$ such that $\bigcup_{n\in\mathbb{Z}} \left((n\mathbf{v} + H) \cap \mathbb{Z}^d\right) = \mathbb{Z}^d$. (This implies, in particular, that the points in $\mathbb{Z}^d \setminus H$ are all at least some minimal distance away from H; this minimal distance is essentially given by the dot product of $\mathbf{v}$ with $(a_1, a_2, \ldots, a_d)$.)

3.14. ♣ A hyperplane H is **rational** if it can be written in the form

$$H = \left\{\mathbf{x} \in \mathbb{R}^d : a_1x_1 + a_2x_2 + \cdots + a_dx_d = b\right\}$$

for some $a_1, a_2, \ldots, a_d, b \in \mathbb{Z}$. A **hyperplane arrangement** in $\mathbb{R}^d$ is a finite set of hyperplanes in $\mathbb{R}^d$. Prove that a rational hyperplane arrangement $\mathcal{H}$ can be translated so that no hyperplane in $\mathcal{H}$ contains any integer points.

3.15. The conclusion of the previous exercise can be strengthened: Prove that a rational hyperplane arrangement $\mathcal{H}$ can be translated such that no hyperplane in $\mathcal{H}$ contains any *rational* points.

3.16. Show that, given distinct numbers $x_1, x_2, \ldots, x_{d+1}$, the matrix

$$\mathbf{V} = \begin{pmatrix} x_1^d & x_1^{d-1} & \cdots & x_1 & 1 \\ x_2^d & x_2^{d-1} & \cdots & x_2 & 1 \\ \vdots & \vdots & & \vdots & \vdots \\ x_{d+1}^d & x_{d+1}^{d-1} & \cdots & x_{d+1} & 1 \end{pmatrix}$$

is not singular. (V is known as the *Vandermonde matrix*.)

3.17. Let $\mathcal{P}$ be an integral d-polytope. Show that

$$\operatorname{vol} \mathcal{P} = \frac{1}{d!}\left((-1)^d + \sum_{k=1}^{d} \binom{d}{k} (-1)^{d-k} L_{\mathcal{P}}(k)\right).$$

3.18. As in Example 3.22, let $\mathcal{T}_n$ be the tetrahedron with vertices $(0,0,0)$, $(1,0,0)$, $(0,1,0)$, and $(1,1,n)$, where n is a positive integer. Show that the volume of $\mathcal{T}_n$ is unbounded as $n \to \infty$, yet for all n, $\mathcal{T}_n$ has no interior and four boundary lattice points. This example proves that Pick's theorem does not hold for a three-dimensional integral polytope $\mathcal{P}$, in the sense that there is no linear relationship among $\operatorname{vol}\mathcal{P}$, $L_{\mathcal{P}}(1)$, and $L_{\mathcal{P}^\circ}(1)$.

3.19. ♣ Prove Lemma 3.24: If

$$\sum_{t\geq 0} f(t)\, z^t = \frac{g(z)}{h(z)} \,,$$

then f is a quasipolynomial of degree d with period dividing p if and only if g and h are polynomials such that $\deg(g) < \deg(h)$, all roots of h are p^{th} roots of unity of multiplicity at most $d+1$, and there is a root of multiplicity equal to $d+1$ (all of this assuming that g/h has been reduced to lowest terms).

3.20. ♣ Provide the details for the proof of Theorem 3.23: If $\mathcal{P}$ is a rational convex d-polytope, then $L_{\mathcal{P}}(t)$ is a quasipolynomial in t of degree d. Its period divides the least common multiple of the coordinates of the vertices of $\mathcal{P}$.

3.21. Let $\mathcal{T}$ be the rational triangle with vertices $(0,0)$, $\left(1, \frac{p-1}{p}\right)$, and $(p,0)$, where p is a fixed integer ≥ 2. Show that $L_{\mathcal{T}}(t) = \frac{p-1}{2}\, t^2 + \frac{p+1}{2}\, t + 1$; in particular, $L_{\mathcal{T}}$ is a *polynomial.*

3.22. Prove that for any $d \geq 2$ and any $p \geq 1$, there exists a d-polytope $\mathcal{P}$ whose Ehrhart quasipolynomial is a *polynomial* (i.e., it has period 1), yet $\mathcal{P}$ has a vertex with denominator p.

3.23. Prove that the period of the Ehrhart quasipolynomial of a 1-dimensional polytope is *always* equal to the lcm of the denominators of its vertices.

3.24. Let $\mathcal{T}$ be the triangle with vertices $\left(-\frac{1}{2}, -\frac{1}{2}\right)$, $\left(\frac{1}{2}, -\frac{1}{2}\right)$, and $\left(0, \frac{3}{2}\right)$. Show that $L_{\mathcal{T}}(t) = t^2 + c(t)\, t + 1$, where

$$c(t) = \begin{cases} 1 & \text{if } t \text{ is even,} \\ 0 & \text{if } t \text{ is odd.} \end{cases}$$

(This shows that the periods of the "coefficients" of an Ehrhart quasipolynomial do not necessarily increase with decreasing power.) Find the Ehrhart series of $\mathcal{T}$.

3.25. Prove the following extension of Theorem 3.12: Suppose $\mathcal{P}$ is a rational d-polytope with denominator p. Then

$$\operatorname{Ehr}_{\mathcal{P}}(z) = \frac{f(z)}{\left(1 - z^p\right)^{d+1}} \,,$$

where f is a polynomial with nonnegative integral coefficients.

3.26. Find and prove a statement that extends Lemma 3.14 to Ehrhart quasipolynomials.

3.27. Prove the following extension of Corollary 3.15 to rational polytopes. Namely, the Ehrhart quasipolynomial $L_{\mathcal{P}}$ of the rational convex polytope $\mathcal{P} \subset \mathbb{R}^d$ satisfies $L_{\mathcal{P}}(0) = 1$.

3.28. Prove the following analogue of Corollary 3.17 for rational polytopes: Suppose $\mathcal{P}$ is a rational polytope with Ehrhart quasipolynomial $L_{\mathcal{P}}(t) = c_d(t)\, t^d + c_{d-1}(t)\, t^{d-1} + \cdots + c_1(t)\, t + c_0(t)$. Then for all $t \in \mathbb{Z}$ and $0 \le k \le d$, we have $d!\, c_k(t) \in \mathbb{Z}$.

3.29. ♣ Prove that Corollary 3.20 also holds for rational polytopes: Suppose $\mathcal{P} \subset \mathbb{R}^d$ is a rational convex d-polytope with Ehrhart quasipolynomial $c_d(t)\, t^d + c_{d-1}(t)\, t^{d-1} + \cdots + c_0(t)$. Then $c_d(t)$ equals the volume of $\mathcal{P}$; in particular, $c_d(t)$ is constant.

3.30. Suppose $\mathcal{P}$ is a rational convex polytope. Show that as rational functions,

$$\mathrm{Ehr}_{2\mathcal{P}}(z) = \frac{1}{2}\left(\mathrm{Ehr}_{\mathcal{P}}\left(\sqrt{z}\right) - \mathrm{Ehr}_{\mathcal{P}}\left(-\sqrt{z}\right)\right).$$

3.31. Suppose f and g are quasipolynomials. Prove that the **convolution**

$$F(t) := \sum_{s=0}^{t} f(s)\, g(t-s)$$

is also a quasipolynomial. What can you say about the degree and the period of F, given the degrees and periods of f and g?

3.32. Given two positive, relatively prime integers a and b, let

$$f(t) := \begin{cases} 1 & \text{if } a|t, \\ 0 & \text{otherwise,} \end{cases} \qquad \text{and} \qquad g(t) := \begin{cases} 1 & \text{if } b|t, \\ 0 & \text{otherwise.} \end{cases}$$

Form the convolution of f and g. What function is it?

3.33. Suppose $\mathcal{P} \subset \mathbb{R}^m$ and $\mathcal{Q} \subset \mathbb{R}^n$ are rational polytopes. Prove that the convolution of $L_{\mathcal{P}}$ and $L_{\mathcal{Q}}$ equals the Ehrhart quasipolynomial of the polytope given by the convex hull of $\mathcal{P} \times \{\mathbf{0}_n\} \times \{0\}$ and $\{\mathbf{0}_m\} \times \mathcal{Q} \times \{1\}$. Here $\mathbf{0}_d$ denotes the origin in $\mathbb{R}^d$.

3.34. We define the **unimodular group** $\mathrm{SL}_d(\mathbb{Z})$ as the set of all $d \times d$ matrices with integer entries and determinant ± 1.

(a) Show that $\mathrm{SL}_d(\mathbb{Z})$ acts on the integer lattice $\mathbb{Z}^d$ as a one-to-one, onto map. That is, fix any $\mathbf{A} \in \mathrm{SL}_d(\mathbb{Z})$. Then $\mathbf{A}$ maps $\mathbb{Z}^d$ to itself in a bijective fashion.

(b) For any open simplex $\Delta^\circ \subset \mathbb{R}^d$ and any $\mathbf{A} \in \mathrm{SL}_d(\mathbb{Z})$, consider the image of Δ° under $\mathbf{A}$, defined by $\mathbf{A}(\Delta^\circ) := \{\mathbf{A}\,\mathbf{x} : \mathbf{x} \in \Delta^\circ\}$. Show that

$$\#\left\{\Delta^\circ \cap \mathbb{Z}^d\right\} = \#\left\{\mathbf{A}(\Delta^\circ) \cap \mathbb{Z}^d\right\}.$$

(c) Let $\mathcal{P}$ be an integral polytope, and let $\mathcal{Q} := \mathbf{A}(\mathcal{P})$, where $\mathbf{A} \in \mathrm{SL}_d(\mathbb{Z})$, so that $\mathcal{P}$ and $\mathcal{Q}$ are unimodular images of each other. Show that $L_{\mathcal{P}}(t) = L_{\mathcal{Q}}(t)$. (*Hint:* Write $\mathcal{P}$ as the disjoint union of open simplices.)

3.35. Search on the Internet for the program `LattE`: Lattice-Point Enumeration [66, 115]. You can download it for free. Experiment.

Open Problems

3.36. How many triangulations are there for a given polytope?

3.37. What is the minimal number of simplices needed to triangulate the unit d-cube? (These numbers are known for $d \le 7$.)

3.38. Classify the polynomials of a fixed degree d that are Ehrhart polynomials. This is completely done for $d = 2$ [160] and partially known for $d = 3$ and 4 [24, Section 3].

3.39. Study the roots of Ehrhart polynomials of integral polytopes in a fixed dimension [24, 37, 42, 94]. Study the roots of the numerators of Ehrhart series.

3.40. Come up with an efficient algorithm that computes the period of an Ehrhart quasipolynomial. (See [188], in which Woods describes an efficient algorithm that checks whether a given integer is a period of an Ehrhart quasipolynomial.)

3.41. Suppose $\mathcal{P}$ and $\mathcal{Q}$ are integer polytopes with the same Ehrhart polynomial, that is, $L_{\mathcal{P}}(t) = L_{\mathcal{Q}}(t)$. What additional conditions on $\mathcal{P}$ and $\mathcal{Q}$ do we need to ensure that integer translates of $\mathcal{P}$ and $\mathcal{Q}$ are unimodular images of each other? That is, when is $\mathcal{Q} = \mathbf{A}(\mathcal{P}) + \mathbf{m}$ for some $\mathbf{A} \in \mathrm{SL}_d(\mathbb{Z})$ and $\mathbf{m} \in \mathbb{Z}^d$?

3.42. Find an "Ehrhart theory" for irrational polytopes.

4

Reciprocity

In mathematics you don't understand things. You just get used to them.

John von Neumann (1903–1957)

While Exercise 1.4 (i) gave us the elementary identity

$$\left\lfloor \frac{t-1}{a} \right\rfloor = -\left\lfloor \frac{-t}{a} \right\rfloor - 1 \tag{4.1}$$

for $t \in \mathbb{Z}$ and $a \in \mathbb{Z}_{>0}$, this fact is a special instance of a more general theme. Namely, (4.1) marks the simplest (one-dimensional) case of a *reciprocity theorem* that is central to Ehrhart theory. Let $\mathcal{I} := [0, 1/a] \subset \mathbb{R}$, a rational

Fig. 4.1. Lattice points in $t\mathcal{I}$.

1-polytope (see Figure 4.1). Its discrete volume is (recalling Exercise 1.3)

$$L_{\mathcal{I}}(t) = \left\lfloor \frac{t}{a} \right\rfloor + 1 .$$

The lattice-point enumerator for the *interior* $\mathcal{I}^\circ = (0, 1/a)$, on the other hand, is

$$L_{\mathcal{I}^\circ}(t) = \left\lfloor \frac{t-1}{a} \right\rfloor \tag{4.2}$$

(see Exercise 4.1). The identity (4.1) says that algebraically,

$$L_{\mathcal{I}^\circ}(t) = -L_{\mathcal{I}}(-t) .$$

This chapter is devoted to proving that a similar identity holds for rational polytopes in any dimension:

Theorem 4.1 (Ehrhart–Macdonald reciprocity). *Suppose $\mathcal{P}$ is a convex rational polytope. Then the evaluation of the quasipolynomial $L_{\mathcal{P}}$ at negative integers yields*

$$L_{\mathcal{P}}(-t) = (-1)^{\dim \mathcal{P}} L_{\mathcal{P}^\circ}(t) \, .$$

This theorem belongs to a class of famous *reciprocity theorems.* A common theme in combinatorics is to begin with an interesting object P, and

1. define a counting function $f(t)$ attached to P that makes physical sense for positive integer values of t;
2. recognize the function f as a polynomial in t;
3. substitute negative integral values of t into the counting function f, and recognize $f(-t)$ as a counting function of a new mathematical object Q.

For us, P is the closure of a polytope and Q is its interior.

4.1 Generating Functions for Somewhat Irrational Cones

Our approach to proving Theorem 4.1 parallels the steps of Chapter 3: we deduce Theorem 4.1 from an identity for rational cones. We start with a reciprocity theorem for simplicial cones.

Theorem 4.2. *Fix linearly independent vectors $\mathbf{w}_1, \mathbf{w}_2, \dots, \mathbf{w}_d \in \mathbb{Z}^d$, and let $\mathcal{K} = \{\lambda_1 \mathbf{w}_1 + \lambda_2 \mathbf{w}_2 + \cdots + \lambda_d \mathbf{w}_d : \lambda_1, \dots, \lambda_d \geq 0\}$, the simplicial cone generated by the $\mathbf{w}_j$'s. Then for those $\mathbf{v} \in \mathbb{R}^d$ for which the boundary of the shifted simplicial cone $\mathbf{v} + \mathcal{K}$ contains no integer point,*

$$\sigma_{\mathbf{v}+\mathcal{K}} \left(\frac{1}{z_1}, \frac{1}{z_2}, \dots, \frac{1}{z_d} \right) = (-1)^d \, \sigma_{-\mathbf{v}+\mathcal{K}} \left(z_1, z_2, \dots, z_d \right) .$$

Remark. This theorem is meaningless on the level of formal power series; however, the identity holds at the level of rational functions. We will establish that $\sigma_{\mathbf{v}+\mathcal{K}}$ is a rational function in the process of proving the theorem.

Proof. As in the proofs of Theorem 3.5 and Corollary 3.6, we have the formula

$$\sigma_{\mathbf{v}+\mathcal{K}}(\mathbf{z}) = \frac{\sigma_{\mathbf{v}+\Pi}(\mathbf{z})}{\left(1 - \mathbf{z}^{\mathbf{w}_1}\right)\left(1 - \mathbf{z}^{\mathbf{w}_2}\right) \cdots \left(1 - \mathbf{z}^{\mathbf{w}_d}\right)} ,$$

where Π is the open parallelepiped

$$\Pi = \{\lambda_1 \mathbf{w}_1 + \lambda_2 \mathbf{w}_2 + \cdots + \lambda_d \mathbf{w}_d : 0 < \lambda_1, \lambda_2, \dots, \lambda_d < 1\} . \tag{4.3}$$

This also proves that $\sigma_{\mathbf{v}+\mathcal{K}}$ is a rational function. Note that by assumption, $\mathbf{v} + \Pi$ contains no integer points on its boundary. Naturally,

$$\sigma_{-\mathbf{v}+\mathcal{K}}(\mathbf{z}) = \frac{\sigma_{-\mathbf{v}+\Pi}(\mathbf{z})}{(1-\mathbf{z}^{\mathbf{w}_1})(1-\mathbf{z}^{\mathbf{w}_2})\cdots(1-\mathbf{z}^{\mathbf{w}_d})},$$

so we need to relate the parallelepipeds $\mathbf{v}+\Pi$ and $-\mathbf{v}+\Pi$. This relation is illustrated in Figure 4.2 for the case $d=2$; the identity for general d is (see Exercise 4.2)

$$\mathbf{v}+\Pi = -(-\mathbf{v}+\Pi) + \mathbf{w}_1 + \mathbf{w}_2 + \cdots + \mathbf{w}_d\,. \tag{4.4}$$

Now we translate the geometry of (4.4) into generating functions:

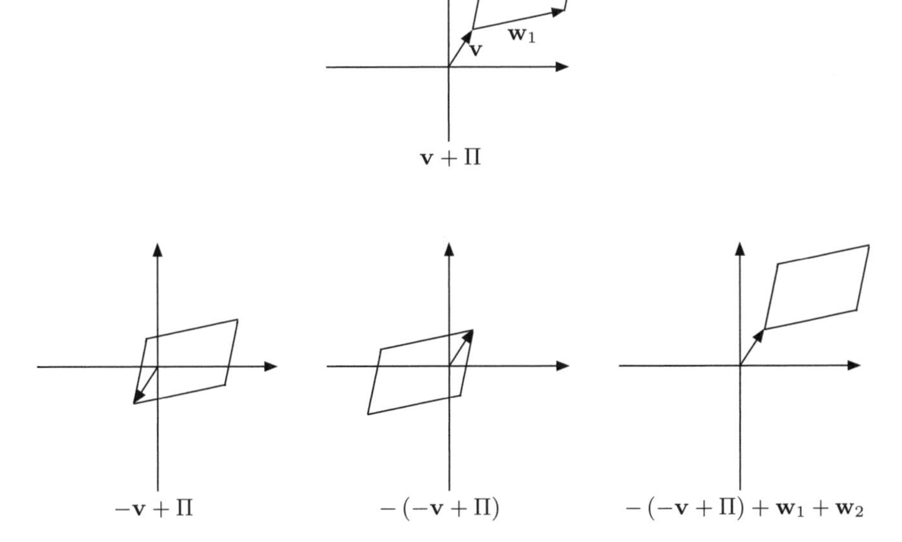

Fig. 4.2. From $-\mathbf{v}+\Pi$ to $\mathbf{v}+\Pi$.

$$\begin{aligned}\sigma_{\mathbf{v}+\Pi}(\mathbf{z}) &= \sigma_{-(-\mathbf{v}+\Pi)}(\mathbf{z})\,\mathbf{z}^{\mathbf{w}_1}\mathbf{z}^{\mathbf{w}_2}\cdots\mathbf{z}^{\mathbf{w}_d}\\ &= \sigma_{-\mathbf{v}+\Pi}\left(\tfrac{1}{z_1},\tfrac{1}{z_2},\dots,\tfrac{1}{z_d}\right)\mathbf{z}^{\mathbf{w}_1}\mathbf{z}^{\mathbf{w}_2}\cdots\mathbf{z}^{\mathbf{w}_d}\end{aligned}$$

(the last equation follows from Exercise 3.6). Let's abbreviate the vector $\left(\frac{1}{z_1},\frac{1}{z_2},\dots,\frac{1}{z_d}\right)$ by $\frac{1}{\mathbf{z}}$. Then the last identity is equivalent to

$$\sigma_{\mathbf{v}+\Pi}\left(\frac{1}{\mathbf{z}}\right) = \sigma_{-\mathbf{v}+\Pi}(\mathbf{z})\,\mathbf{z}^{-\mathbf{w}_1}\mathbf{z}^{-\mathbf{w}_2}\cdots\mathbf{z}^{-\mathbf{w}_d},$$

whence

$$\begin{aligned}
\sigma_{\mathbf{v}+\mathcal{K}}\left(\frac{1}{\mathbf{z}}\right) &= \frac{\sigma_{\mathbf{v}+\Pi}\left(\frac{1}{\mathbf{z}}\right)}{\left(1-\mathbf{z}^{-\mathbf{w}_1}\right)\left(1-\mathbf{z}^{-\mathbf{w}_2}\right)\cdots\left(1-\mathbf{z}^{-\mathbf{w}_d}\right)} \\
&= \frac{\sigma_{-\mathbf{v}+\Pi}(\mathbf{z})\,\mathbf{z}^{-\mathbf{w}_1}\mathbf{z}^{-\mathbf{w}_2}\cdots\mathbf{z}^{-\mathbf{w}_d}}{\left(1-\mathbf{z}^{-\mathbf{w}_1}\right)\left(1-\mathbf{z}^{-\mathbf{w}_2}\right)\cdots\left(1-\mathbf{z}^{-\mathbf{w}_d}\right)} \\
&= \frac{\sigma_{-\mathbf{v}+\Pi}(\mathbf{z})}{\left(\mathbf{z}^{\mathbf{w}_1}-1\right)\left(\mathbf{z}^{\mathbf{w}_2}-1\right)\cdots\left(\mathbf{z}^{\mathbf{w}_d}-1\right)} \\
&= (-1)^d \frac{\sigma_{-\mathbf{v}+\Pi}(\mathbf{z})}{\left(1-\mathbf{z}^{\mathbf{w}_1}\right)\left(1-\mathbf{z}^{\mathbf{w}_2}\right)\cdots\left(1-\mathbf{z}^{\mathbf{w}_d}\right)} \\
&= (-1)^d \sigma_{-\mathbf{v}+\mathcal{K}}(\mathbf{z})\,.
\end{aligned}$$

□

4.2 Stanley's Reciprocity Theorem for Rational Cones

For the general reciprocity theorem for cones, we patch the simplicial cones of a triangulation together, in a manner very similar to our proof of Theorem 3.12.

Theorem 4.3 (Stanley reciprocity). *Suppose $\mathcal{K}$ is a rational d-cone with the origin as apex. Then*

$$\sigma_{\mathcal{K}}\left(\frac{1}{z_1}, \frac{1}{z_2}, \ldots, \frac{1}{z_d}\right) = (-1)^d \sigma_{\mathcal{K}^\circ}\left(z_1, z_2, \ldots, z_d\right).$$

Proof. Triangulate $\mathcal{K}$ into the simplicial cones $\mathcal{K}_1, \mathcal{K}_2, \ldots, \mathcal{K}_m$. Now Exercise 3.14 ensures that there exists a vector $\mathbf{v} \in \mathbb{R}^d$ such that the shifted cone $\mathbf{v} + \mathcal{K}$ contains exactly the *interior* lattice points of $\mathcal{K}$,

$$\mathcal{K}^\circ \cap \mathbb{Z}^d = (\mathbf{v} + \mathcal{K}) \cap \mathbb{Z}^d, \tag{4.5}$$

and there are no boundary lattice points on any of the triangulation cones:

$$\partial\left(\mathbf{v} + \mathcal{K}_j\right) \cap \mathbb{Z}^d = \varnothing \qquad \text{for all } j = 1, \ldots, m, \tag{4.6}$$

as well as

$$\partial\left(-\mathbf{v} + \mathcal{K}_j\right) \cap \mathbb{Z}^d = \varnothing \qquad \text{for all } j = 1, \ldots, m. \tag{4.7}$$

We invite the reader (Exercise 4.3) to realize that (4.5)–(4.7) imply

$$\mathcal{K} \cap \mathbb{Z}^d = (-\mathbf{v} + \mathcal{K}) \cap \mathbb{Z}^d. \tag{4.8}$$

Now by Theorem 4.2,

$$\begin{aligned}
\sigma_{\mathcal{K}}\left(\tfrac{1}{\mathbf{z}}\right) &= \sigma_{-\mathbf{v}+\mathcal{K}}\left(\tfrac{1}{\mathbf{z}}\right) = \sum_{j=1}^{m} \sigma_{-\mathbf{v}+\mathcal{K}_j}\left(\tfrac{1}{\mathbf{z}}\right) = \sum_{j=1}^{m} (-1)^d \sigma_{\mathbf{v}+\mathcal{K}_j}(\mathbf{z}) \\
&= (-1)^d \sigma_{\mathbf{v}+\mathcal{K}}(\mathbf{z}) = (-1)^d \sigma_{\mathcal{K}^\circ}(\mathbf{z})\,.
\end{aligned}$$

Note that the second and fourth equalities are true because of the validity of (4.7) and (4.6), respectively. □

4.3 Ehrhart–Macdonald Reciprocity for Rational Polytopes

In preparation for the proof of Theorem 4.1, we define the **Ehrhart series** for the *interior* of the rational polytope $\mathcal{P}$ as

$$\mathrm{Ehr}_{\mathcal{P}^\circ}(z) := \sum_{t \ge 1} L_{\mathcal{P}^\circ}(t)\, z^t .$$

Our convention of beginning the series with $t = 1$ stems from the fact that this generating function is a special evaluation of the integer-point transform of the open cone $(\mathrm{cone}(\mathcal{P}))^\circ$: Much in sync with Lemma 3.10, we have

$$\mathrm{Ehr}_{\mathcal{P}^\circ}(z) = \sigma_{(\mathrm{cone}(\mathcal{P}))^\circ}\,(1, 1, \dots, 1, z) . \tag{4.9}$$

We are now ready to prove the Ehrhart-series analogue of Theorem 4.1.

Theorem 4.4. *Suppose $\mathcal{P}$ is a convex rational polytope. Then the evaluation of the rational function* $\mathrm{Ehr}_{\mathcal{P}}$ *at* $1/z$ *yields*

$$\mathrm{Ehr}_{\mathcal{P}}\left(\frac{1}{z}\right) = (-1)^{\dim \mathcal{P}+1}\, \mathrm{Ehr}_{\mathcal{P}^\circ}(z) .$$

Proof. Suppose $\mathcal{P}$ is a d-polytope. We recall Lemma 3.10, which states that the generating function of the Ehrhart polynomial of $\mathcal{P}$ is an evaluation of the generating function of $\mathrm{cone}(\mathcal{P})$:

$$\mathrm{Ehr}_{\mathcal{P}}(z) = \sum_{t \ge 0} L_{\mathcal{P}}(t)\, z^t = \sigma_{\mathrm{cone}(\mathcal{P})}\,(1, 1, \dots, 1, z) .$$

Equation (4.9) above gives the analogous evaluation of $\sigma_{(\mathrm{cone}(\mathcal{P}))^\circ}$ that yields $\mathrm{Ehr}_{\mathcal{P}^\circ}$. Now we apply Theorem 4.3 to the $(d+1)$-cone $\mathcal{K} = \mathrm{cone}(\mathcal{P})$:

$$\sigma_{(\mathrm{cone}(\mathcal{P}))^\circ}\,(1, 1, \dots, 1, z) \;=\; (-1)^{d+1}\, \sigma_{\mathrm{cone}(\mathcal{P})}\left(1, 1, \dots, 1, \frac{1}{z}\right) . \qquad \square$$

Theorem 4.1 now follows like a breeze.

Proof of Ehrhart–Macdonald reciprocity (Theorem 4.1). We first apply Exercise 4.5 to the Ehrhart series of $\mathcal{P}$: namely, as rational functions,

$$\mathrm{Ehr}_{\mathcal{P}}\left(\frac{1}{z}\right) = \sum_{t \le 0} L_{\mathcal{P}}(-t)\, z^t = -\sum_{t \ge 1} L_{\mathcal{P}}(-t)\, z^t .$$

Now we combine this identity with Theorem 4.4 to obtain

$$\sum_{t \ge 1} L_{\mathcal{P}^\circ}(t)\, z^t = (-1)^{d+1}\, \mathrm{Ehr}_{\mathcal{P}}\left(\frac{1}{z}\right) = (-1)^d \sum_{t \ge 1} L_{\mathcal{P}}(-t)\, z^t .$$

Comparing the coefficients of the two power series yields the reciprocity theorem. $\square$

With Ehrhart–Macdonald reciprocity, we can now restate Theorem 3.18 in terms of Ehrhart polynomials:

Theorem 4.5. *Suppose $\mathcal{P}$ is an integral d-polytope with Ehrhart series*

$$\operatorname{Ehr}_{\mathcal{P}}(z) = \frac{h_d\, z^d + h_{d-1}\, z^{d-1} + \cdots + h_1\, z + 1}{(1-z)^{d+1}} .$$

Then $h_d = h_{d-1} = \cdots = h_{k+1} = 0$ and $h_k \neq 0$ if and only if $(d-k+1)\mathcal{P}$ is the smallest integer dilate of $\mathcal{P}$ that contains an interior lattice point.

Proof. Theorem 3.18 says that h_k is the highest nonzero coefficient if and only if $L_{\mathcal{P}}(-1) = L_{\mathcal{P}}(-2) = \cdots = L_{\mathcal{P}}(-(d-k)) = 0$ and $L_{\mathcal{P}}(-(d-k+1)) \neq 0$. Now use Ehrhart–Macdonald reciprocity (Theorem 4.1). □

The largest k for which $h_k \neq 0$ is called the **degree** of $\mathcal{P}$. The above theorem says that the degree of $\mathcal{P}$ is k precisely if $(d-k+1)\mathcal{P}$ is the smallest integer dilate of $\mathcal{P}$ that contains an interior lattice point.

4.4 The Ehrhart Series of Reflexive Polytopes

As an application of Theorem 4.4, we now study a special class of integral polytopes whose Ehrhart series have an additional symmetry structure. We call a polytope $\mathcal{P}$ that contains the origin in its interior **reflexive** if it is integral and has the half-space description

$$\mathcal{P} = \left\{ \mathbf{x} \in \mathbb{R}^d : \mathbf{A}\,\mathbf{x} \leq \mathbf{1} \right\},$$

where $\mathbf{A}$ is an integral matrix. (Here $\mathbf{1}$ denotes a vector all of whose coordinates are 1.) The following theorem gives a characterization of reflexive polytopes through their Ehrhart series.

Theorem 4.6 (Hibi's palindromic theorem). *Suppose $\mathcal{P}$ is an integral d-polytope that contains the origin in its interior and that has the Ehrhart series*

$$\operatorname{Ehr}_{\mathcal{P}}(z) = \frac{h_d\, z^d + h_{d-1}\, z^{d-1} + \cdots + h_1\, z + h_0}{(1-z)^{d+1}} .$$

Then $\mathcal{P}$ is reflexive if and only if $h_k = h_{d-k}$ for all $0 \leq k \leq \frac{d}{2}$.

The two main ingredients for the proof of this result are Theorem 4.4 and the following:

Lemma 4.7. *Suppose $a_1, a_2, \dots, a_d, b \in \mathbb{Z}$ satisfy $\gcd(a_1, a_2, \dots, a_d, b) = 1$ and $b > 1$. Then there exist positive integers c and t such that $tb < c < (t+1)b$ and $\left\{ (m_1, m_2, \dots, m_d) \in \mathbb{Z}^d : a_1 m_1 + a_2 m_2 + \cdots + a_d m_d = c \right\} \neq \varnothing$.*

Proof. Let $g = \gcd(a_1, a_2, \dots, a_d)$; by our assumption, $\gcd(g, b) = 1$, so one can find integers k and t such that

$$kg - tb = 1\,. \tag{4.10}$$

Furthermore, we can choose k and t in such a way that $t > 0$. Let $c = kg$; equation (4.10) and the condition $b > 1$ imply that $tb < c < (t+1)b$. Finally, since $g = \gcd(a_1, a_2, \dots, a_d)$ there exist $m_1, m_2, \dots, m_d \in \mathbb{Z}$ such that

$$a_1 m_1 + a_2 m_2 + \cdots + a_d m_d = kg = c\,. \qquad \square$$

Proof of Theorem 4.6. We recall that $\mathcal{P}$ is reflexive if and only if

$$\mathcal{P} = \left\{ \mathbf{x} \in \mathbb{R}^d : \mathbf{A}\,\mathbf{x} \le \mathbf{1} \right\} \quad \text{for some integral matrix } \mathbf{A}. \tag{4.11}$$

We claim that $\mathcal{P}$ has such a half-space description if and only if

$$\mathcal{P}^\circ \cap \mathbb{Z}^d = \{\mathbf{0}\} \quad \text{and for all } t \in \mathbb{Z}_{>0},\ \ (t+1)\mathcal{P}^\circ \cap \mathbb{Z}^d = t\mathcal{P} \cap \mathbb{Z}^d. \tag{4.12}$$

This condition means that the only lattice points that we gain when passing from $t\mathcal{P}$ to $(t+1)\mathcal{P}$ are those on the boundary of $(t+1)\mathcal{P}$. The fact that (4.11) implies (4.12) is the content of Exercise 4.11. Conversely, if $\mathcal{P}$ satisfies (4.12) then there are no lattice points between tH and $(t+1)H$ for any facet hyperplane H of $\mathcal{P}$ (Exercise 4.12). That is, if a facet hyperplane is given by $H = \left\{ \mathbf{x} \in \mathbb{R}^d : a_1 x_1 + a_2 x_2 + \cdots + a_d x_d = b \right\}$, where we may assume $\gcd(a_1, a_2, \dots, a_d, b) = 1$, then

$$\left\{ \mathbf{x} \in \mathbb{Z}^d : tb < a_1 x_1 + a_2 x_2 + \cdots + a_d x_d < (t+1)b \right\} = \varnothing\,.$$

But then Lemma 4.7 implies that $b = 1$, and so $\mathcal{P}$ has a half-space description of the form (4.11).

Thus we have established that $\mathcal{P}$ is reflexive if and only if it satisfies (4.12). Now by Theorem 4.4,

$$\mathrm{Ehr}_{\mathcal{P}^\circ}(z) = (-1)^{d+1}\,\mathrm{Ehr}_{\mathcal{P}}\left(\frac{1}{z}\right) = \frac{h_0\, z^{d+1} + h_1\, z^d + \cdots + h_{d-1}\, z^2 + h_d\, z}{(1-z)^{d+1}}\,.$$

By condition (4.12), $\mathcal{P}$ is reflexive if and only if this rational function is equal to

$$\begin{aligned} \sum_{t \ge 1} L_{\mathcal{P}}(t-1)\, z^t &= z \sum_{t \ge 0} L_{\mathcal{P}}(t)\, z^t = z\,\mathrm{Ehr}_{\mathcal{P}}(z) \\ &= \frac{h_d\, z^{d+1} + h_{d-1}\, z^d + \cdots + h_1\, z^2 + h_0\, z}{(1-z)^{d+1}}\,, \end{aligned}$$

that is, if and only if $h_k = h_{d-k}$ for all $0 \le k \le \frac{d}{2}$. $\square$

4.5 More "Reflections" on the Coin-Exchange Problem and the Gallery of Chapter 2

We have already encountered special cases of Ehrhart–Macdonald reciprocity several times. Note that Theorem 4.1 allows us to conclude that counting the number of *interior* lattice points in a rational polytope is tantamount to counting lattice points in its *closure*. Exercises 1.31, 2.1, and 2.7, as well as part (b) of each theorem in the gallery of Chapter 2, confirm that

$$L_{\mathcal{P}}(-t) = (-1)^{\dim \mathcal{P}} L_{\mathcal{P}^\circ}(t) .$$

Notes

1. Ehrhart–Macdonald reciprocity (Theorem 4.1) had been conjectured (and proved in several special cases) by Eugène Ehrhart for about a decade before I. G. Macdonald found a general proof in 1971 [123]. One can actually relax the condition of Ehrhart–Macdonald reciprocity: it holds for polytopal complexes that are homeomorphic to a d-manifold. The proof we give here (including the proof of Theorem 4.3) appeared in [30].

2. Theorem 4.3 is due to Richard Stanley [169], who proved more general versions of this theorem. The reader might recall that the rational function representing the Ehrhart series of a rational cone can be thought of as its meromorphic continuation. Stanley reciprocity (Theorem 4.3) gives a functional identity for such meromorphic continuations.

3. The term *reflexive polytope* was coined by Victor Batyrev, who found exciting applications of these polytopes to mirror symmetry in physical string theory [16]. Batyrev proved that the toric variety $X_{\mathcal{P}}$ defined by a reflexive polytope $\mathcal{P}$ is *Fano*, and that every generic hypersurface of $X_{\mathcal{P}}$ is *Calabi–Yau*. That the Ehrhart series of a reflexive polytope exhibits an unexpected symmetry (Theorem 4.6) was discovered by Takayuki Hibi [97]. The number of reflexive polytopes in dimension d is known for $d \leq 4$ [117, 118]; for example, there are precisely 16 reflexive polytopes in dimension 2, up to symmetries (see also [165, Sequence A090045]). A striking result is that the sum of the numbers of lattice points on the boundaries of a reflexive polygon and its dual is always 12 [147]. A similar result holds in dimension 3 (with 12 replaced by 24) [18], but no elementary proof of the latter fact is known [22, Section 4].

4. There is an equivalent definition for reflexive polytopes: $\mathcal{P}$ is reflexive if and only if both $\mathcal{P}$ and its dual $\mathcal{P}^*$ are integral polytopes. The *dual polytope* of $\mathcal{P}$ (often also called the *polar polytope*) is defined as $\mathcal{P}^* := \left\{ \mathbf{x} \in \mathbb{R}^d : \mathbf{x} \cdot \mathbf{y} \leq 1 \text{ for all } \mathbf{y} \in \mathcal{P} \right\}$. The concept of (polar) duality is not confined to polytopes but can be defined for any nonempty subset of $\mathbb{R}^d$. Duality is a crucial chapter in the theory of polytopes, and one of its applications is the equivalence of the vertex and hyperplane description of a polytope. For more about (polar) duality, the reader might consult [12, Chapter IV].

5. The cross-polytopes $\diamond$ from Section 2.5 form a special class of reflexive polytopes. We mentioned in the Notes of Chapter 2 that the roots of the Ehrhart polynomials $L_\diamond$ all have real part $-\frac{1}{2}$ [51, 109]. Christian Bey, Martin Henk, and Jörg Wills proved in [37] that if all roots of $L_\mathcal{P}$, for some integral polytope $\mathcal{P}$, have real part $-\frac{1}{2}$, then $\mathcal{P}$ is the unimodular image of a reflexive polytope.

Exercises

4.1. ♣ Prove (4.2): For $a \in \mathbb{Z}_{>0}$, $L_{(0,1/a)}(t) = \left\lfloor \frac{t-1}{a} \right\rfloor$.

4.2. ♣ Explain (4.4): If $\mathbf{w}_1, \mathbf{w}_2, \dots, \mathbf{w}_d \in \mathbb{R}^d$ are linear independent and

$$\Pi = \{\lambda_1 \mathbf{w}_1 + \lambda_2 \mathbf{w}_2 + \cdots + \lambda_d \mathbf{w}_d : 0 < \lambda_1, \lambda_2, \dots, \lambda_d < 1\},$$

then $\mathbf{v} + \Pi = -(-\mathbf{v} + \Pi) + \mathbf{w}_1 + \mathbf{w}_2 + \cdots + \mathbf{w}_d$.

4.3. ♣ Prove that (4.5)–(4.7) imply (4.8); that is, if $\mathcal{K}$ is a rational pointed d-cone $\mathcal{K}$ with the origin as apex, and $\mathbf{v} \in \mathbb{R}^d$ is such that

$$\mathcal{K}^\circ \cap \mathbb{Z}^d = (\mathbf{v} + \mathcal{K}) \cap \mathbb{Z}^d,$$
$$\partial(\mathbf{v} + \mathcal{K}_j) \cap \mathbb{Z}^d = \varnothing \qquad \text{for all } j = 1, \dots, m,$$

and

$$\partial(-\mathbf{v} + \mathcal{K}_j) \cap \mathbb{Z}^d = \varnothing \qquad \text{for all } j = 1, \dots, m,$$

then

$$\mathcal{K} \cap \mathbb{Z}^d = (-\mathbf{v} + \mathcal{K}) \cap \mathbb{Z}^d.$$

4.4. Prove the following generalization of Theorem 4.3 to rational pointed cones with arbitrary apex: Suppose $\mathcal{K}$ is a rational pointed d-cone $\mathcal{K}$ with the origin as apex, and $\mathbf{v} \in \mathbb{R}^d$. Then the integer-point transform $\sigma_{\mathbf{v}+\mathcal{K}}(\mathbf{z})$ of the pointed d-cone $\mathbf{v} + \mathcal{K}$ is a rational function that satisfies

$$\sigma_{\mathbf{v}+\mathcal{K}}\left(\frac{1}{\mathbf{z}}\right) = (-1)^d \sigma_{(-\mathbf{v}+\mathcal{K})^\circ}(\mathbf{z}).$$

4.5. ♣ Suppose $Q\colon \mathbb{Z} \to \mathbb{C}$ is a quasipolynomial. We know that $R_Q^+(z) := \sum_{t \geq 0} Q(t)\, z^t$ evaluates to a rational function.

(a) Prove that $R_Q^-(z) := \sum_{t<0} Q(t)\, z^t$ also evaluates to a rational function.
(b) Let $Q(t) = 1$. Prove that as rational functions, $R_Q^+(z) + R_Q^-(z) = 0$.
(c) Suppose Q is a polynomial. Prove that as rational functions, $R_Q^+(z) + R_Q^-(z) = 0$.
(d) Suppose Q is a quasipolynomial. Prove that as rational functions, $R_Q^+(z) + R_Q^-(z) = 0$.

4.6. ♣ Suppose that $\mathcal{P}$ is a rational d-polytope for which

$$L_{\mathcal{P}^\circ}(t) = L_{\mathcal{P}}(t-k) \qquad \text{and} \qquad L_{\mathcal{P}^\circ}(1) = L_{\mathcal{P}^\circ}(2) = \cdots = L_{\mathcal{P}^\circ}(k-1) = 0$$

for some integer k. (This situation applies to some of the polytopes in the gallery of Chapter 2.) Prove that

$$\operatorname{Ehr}_{\mathcal{P}}\left(\frac{1}{z}\right) = (-1)^{d+1} z^k \operatorname{Ehr}_{\mathcal{P}}(z)\,.$$

4.7. Suppose $\mathcal{P}$ is an integral d-polytope with Ehrhart series

$$\operatorname{Ehr}_{\mathcal{P}}(z) = \frac{h_d\, z^d + h_{d-1}\, z^{d-1} + \cdots + h_1\, z + 1}{(1-z)^{d+1}}\,.$$

Prove that $h_d = L_{\mathcal{P}^\circ}(1)$.

4.8. Suppose $\mathcal{P}$ is a convex integral d-polytope. Show that the dilate $(d+1)\mathcal{P}$ contains an interior lattice point.

4.9. Suppose $\mathcal{P}$ is a convex integral polytope. Denote the boundary of $\mathcal{P}$ by $\partial\mathcal{P}$. Prove that $L_{\partial\mathcal{P}}(t)$ is a polynomial that is either even or odd. Determine its constant term.

4.10. Recall the restricted partition function

$$p_{\{a_1,a_2,\dots,a_n\}}(t) := \#\left\{(m_1,\dots,m_d) \in \mathbb{Z}_{\geq 0}^d : \, m_1 a_1 + \cdots + m_d a_d = t\right\}$$

from Chapter 1. Prove that as quasipolynomials,

$$p_{\{a_1,a_2,\dots,a_n\}}(-t - a_1 - a_2 - \cdots - a_n) = (-1)^{n-1}\, p_{\{a_1,a_2,\dots,a_n\}}(t)$$

and that

$$\begin{aligned} p_{\{a_1,a_2,\dots,a_n\}}(-1) &= p_{\{a_1,a_2,\dots,a_n\}}(-2) = \cdots \\ &= p_{\{a_1,a_2,\dots,a_n\}}(-a_1 - a_2 - \cdots - a_n + 1) = 0\,. \end{aligned}$$

4.11. ♣ Prove that (4.11) implies (4.12), that is, show that if the polytope $\mathcal{P}$ is given by $\mathcal{P} = \left\{\mathbf{x} \in \mathbb{R}^d : \, \mathbf{A}\,\mathbf{x} \leq \mathbf{1}\right\}$ for an integral matrix $\mathbf{A}$, then $\mathcal{P}^\circ \cap \mathbb{Z}^d = \{\mathbf{0}\}$ and for all $t \in \mathbb{Z}_{>0}$, $(t+1)\mathcal{P}^\circ \cap \mathbb{Z}^d = t\mathcal{P} \cap \mathbb{Z}^d$.

4.12. ♣ Suppose $\mathcal{P}$ is an integral polytope that satisfies (4.12): $\mathcal{P}^\circ \cap \mathbb{Z}^d = \{\mathbf{0}\}$ and for all $t \in \mathbb{Z}_{>0}$, $(t+1)\mathcal{P}^\circ \cap \mathbb{Z}^d = t\mathcal{P} \cap \mathbb{Z}^d$. Then for any $t \in \mathbb{Z}$, there are no lattice points between tH and $(t+1)H$ for any facet hyperplane H of $\mathcal{P}$.

Open Problems

4.13. Prove the following conjecture of Batyrev [17]: For an integer polytope $\mathcal{P}$ of degree j, the volume of $\mathcal{P}$ is bounded by a constant that depends only on h_j.

4.14. Suppose $\mathcal{P}$ is a 3-dimensional reflexive polytope. Denote by e^* the edge in the dual polytope $\mathcal{P}^*$ that corresponds to the edge e in $\mathcal{P}$. Give an elementary proof that

$$\sum_{e \text{ edge of } \mathcal{P}} \operatorname{length}(e) \cdot \operatorname{length}(e^*) = 24 \, .$$

4.15. Find the number of reflexive polytopes in dimension $d \geq 5$.

5

Face Numbers and the Dehn–Sommerville Relations in Ehrhartian Terms

"Data! Data! Data!" he cried, impatiently. "I can't make bricks without clay."

Sherlock Holmes ("The Adventure of the Copper Beeches," by Arthur Conan Doyle, 1859–1930)

Our goal in this chapter is twofold, or rather, there is one goal in two different guises. The first one is to prove a set of fascinating identities, which give linear relations among the face numbers f_k. They are called *Dehn–Sommerville relations*, in honor of their discoverers Max Wilhelm Dehn (1878–1952)[1] and Duncan MacLaren Young Sommerville (1879–1934).[2] Our second goal is to unify the Dehn–Sommerville relations (Theorem 5.1 below) with Ehrhart–Macdonald reciprocity (Theorem 4.1).

5.1 Face It!

We denote the number of k-dimensional faces of $\mathcal{P}$ by the symbol f_k. As k varies from 0 to d, the **face numbers** f_k encode intrinsic information about the polytope $\mathcal{P}$. The d-polytope $\mathcal{P}$ is **simple** if each vertex of $\mathcal{P}$ lies on precisely d edges of $\mathcal{P}$.

Theorem 5.1 (Dehn–Sommerville relations). *If $\mathcal{P}$ is a simple d-polytope and $0 \leq k \leq d$, then*

$$f_k = \sum_{j=0}^{k} (-1)^j \binom{d-j}{d-k} f_j \, .$$

[1] For more information about Dehn, see http://www-groups.dcs.st-and.ac.uk/~history/Biographies/Dehn.html.

[2] For more information about Sommerville, see http://www-groups.dcs.st-and.ac.uk/~history/Biographies/Sommerville.html.

This theorem takes on a particularly nice form for $k = d$, namely the famous *Euler relation*, which holds for any polytope (not just simple ones).

Theorem 5.2 (Euler relation). *If $\mathcal{P}$ is a convex d-polytope, then*

$$\sum_{j=0}^{d} (-1)^j f_j = 1 .$$

This identity is less trivial than it might look. We give a quick proof for *rational* polytopes, for which we can use Ehrhart–Macdonald reciprocity (Theorem 4.1).

Proof of Theorem 5.2, assuming $\mathcal{P}$ is rational. Let us count the integer points in $t\mathcal{P}$ according to the (relatively) open faces that contain them:[3]

$$L_{\mathcal{P}}(t) = \sum_{\mathcal{F} \subseteq \mathcal{P}} L_{\mathcal{F}^\circ}(t) = \sum_{\mathcal{F} \subseteq \mathcal{P}} (-1)^{\dim \mathcal{F}} L_{\mathcal{F}}(-t) .$$

Here, and in the remainder of this chapter, the sums are over all *nonempty* faces. (Alternatively, we could agree that $L_{\varnothing}(t) = 0$.) The constant term of $L_{\mathcal{F}}(t)$ is 1, for any face $\mathcal{F}$ (by Exercise 3.27). Hence the constant terms of the identity above give

$$1 = \sum_{\mathcal{F} \subseteq \mathcal{P}} (-1)^{\dim \mathcal{F}} = \sum_{j=0}^{d} (-1)^j f_j ,$$

which proves our claim. □

There is a natural structure on the faces of a polytope $\mathcal{P}$ induced by the containment relation $\mathcal{F} \subseteq \mathcal{G}$. This relation gives a partial ordering on the set of all faces of $\mathcal{P}$, called the **face lattice** of $\mathcal{P}$.[4] A useful way to illustrate this partially ordered set is through a graph whose nodes correspond to the faces of $\mathcal{P}$, such that two nodes are adjacent if one of their corresponding faces contains the other. In Figure 5.1, we give the face lattice for a triangle. Exercise 2.6 implies that the face lattice of any simplex is a *Boolean lattice*, which is the partially ordered set formed by all subsets of a finite set, where the partial ordering is again subset containment.

We already mentioned that we will unify the Dehn–Sommerville relations (Theorem 5.1) with Ehrhart–Macdonald reciprocity (Theorem 4.1). It is for this reason that we will prove Theorem 5.1 only for *rational* polytopes. To combine the notions of face numbers and lattice-point enumeration, we define

[3] Note that the relative interior of a vertex is the vertex itself.

[4] The usage of the word *lattice* here is disjoint from our previous definition of the word.

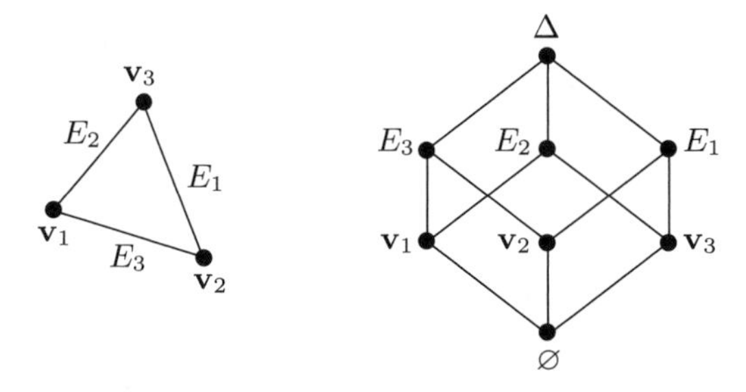

Fig. 5.1. The face lattice of a triangle.

$$F_k(t) := \sum_{\substack{\mathcal{F} \subseteq \mathcal{P} \\ \dim \mathcal{F} = k}} L_{\mathcal{F}}(t) \,,$$

the sum being taken over all k-faces of $\mathcal{P}$. By Ehrhart's theorem (Theorem 3.23), F_k is a quasipolynomial. Since $L_{\mathcal{F}}(0) = 1$ for all $\mathcal{F}$,

$$F_k(0) = f_k \,,$$

the number of k-faces of $\mathcal{P}$. We also remark that the leading coefficient of F_k measures the relative volume of the k-**skeleton** of $\mathcal{P}$, that is, the union of all k-faces; see Section 5.4 for a precise definition of relative volume.

Our common extension of Theorems 5.1 and 4.1 is the subject of the next section.

5.2 Dehn–Sommerville Extended

Theorem 5.3. *If $\mathcal{P}$ is a simple rational d-polytope and $0 \le k \le d$, then*

$$F_k(t) = \sum_{j=0}^{k} (-1)^j \binom{d-j}{d-k} F_j(-t) \,.$$

The classical Dehn–Sommerville equations (Theorem 5.1)—again, only for rational polytopes—are obtained from the constant terms of the counting functions on both sides of the identity. On the other hand, for $k = d$, Theorem 5.3 gives (with t replaced by $-t$)

$$L_{\mathcal{P}}(-t) = F_d(-t) = \sum_{j=0}^{d} (-1)^j F_j(t) = (-1)^d \sum_{j=0}^{d} (-1)^{d-j} F_j(t) \,.$$

The sum on the right-hand side is an inclusion–exclusion formula for the number of lattice points in the interior of $t\mathcal{P}$ (count all the points in $\mathcal{P}$, subtract the ones on the facets, add back what you've overcounted, etc.), so in a sense we recover Ehrhart–Macdonald reciprocity.

Proof. Suppose $\mathcal{F}$ is a k-face of $\mathcal{P}$. Then, again by counting the integer points in $\mathcal{F}$ according to relatively open faces of $\mathcal{F}$,

$$L_{\mathcal{F}}(t) = \sum_{\mathcal{G} \subseteq \mathcal{F}} L_{\mathcal{G}^\circ}(t),$$

or, by the Ehrhart–Macdonald reciprocity (Theorem 4.1),

$$L_{\mathcal{F}}(t) = \sum_{\mathcal{G} \subseteq \mathcal{F}} (-1)^{\dim \mathcal{G}} L_{\mathcal{G}}(-t) = \sum_{j=0}^{k} (-1)^j \sum_{\substack{\mathcal{G} \subseteq \mathcal{F} \\ \dim \mathcal{G} = j}} L_{\mathcal{G}}(-t). \tag{5.1}$$

Now sum both left- and right-hand sides over all k-faces and rearrange the sum on the right-hand side:

$$\begin{aligned}
F_k(t) &= \sum_{\substack{\mathcal{F} \subseteq \mathcal{P} \\ \dim \mathcal{F} = k}} \sum_{j=0}^{k} (-1)^j \sum_{\substack{\mathcal{G} \subseteq \mathcal{F} \\ \dim \mathcal{G} = j}} L_{\mathcal{G}}(-t) \\
&= \sum_{j=0}^{k} (-1)^j \sum_{\substack{\mathcal{F} \subseteq \mathcal{P} \\ \dim \mathcal{F} = k}} \sum_{\substack{\mathcal{G} \subseteq \mathcal{F} \\ \dim \mathcal{G} = j}} L_{\mathcal{G}}(-t) \\
&= \sum_{j=0}^{k} (-1)^j \sum_{\substack{\mathcal{G} \subseteq \mathcal{P} \\ \dim \mathcal{G} = j}} f_k(\mathcal{P}/\mathcal{G}) L_{\mathcal{G}}(-t) \\
&= \sum_{j=0}^{k} (-1)^j \sum_{\substack{\mathcal{G} \subseteq \mathcal{P} \\ \dim \mathcal{G} = j}} \binom{d-j}{d-k} L_{\mathcal{G}}(-t) \\
&= \sum_{j=0}^{k} (-1)^j \binom{d-j}{d-k} F_j(-t).
\end{aligned}$$

Here $f_k(\mathcal{P}/\mathcal{G})$ denotes the number of k-faces of $\mathcal{P}$ containing a given j-face $\mathcal{G}$ of $\mathcal{P}$. Since $\mathcal{P}$ is simple, this number equals $\binom{d-j}{d-k}$ (see Exercise 5.4). □

5.3 Applications to the Coefficients of an Ehrhart Polynomial

We will now apply Theorem 5.3 to the computation of the Ehrhart polynomial of an integral d-polytope $\mathcal{P}$. The only face-lattice-point enumerator involving the face $\mathcal{P}$ is $F_d(t)$, for which Theorem 5.3 specializes to

$$L_{\mathcal{P}}(t) = F_d(t) = \sum_{j=0}^{d} (-1)^j F_j(-t).$$

In fact, we do not have to assume that $\mathcal{P}$ is simple, since this identity simply counts integer points by faces. (Recall that $(-1)^j F_j(-t)$ counts the integer points in the t-dilates of the interior of the j-faces.)[5] The last term on the right-hand side is

$$(-1)^d F_d(-t) = (-1)^d L_{\mathcal{P}}(-t) = L_{\mathcal{P}^\circ}(t)$$

by Ehrhart–Macdonald reciprocity. Shifting this term to the left gives

$$L_{\mathcal{P}}(t) - L_{\mathcal{P}^\circ}(t) = \sum_{j=0}^{d-1} (-1)^j F_j(-t) \, . \tag{5.2}$$

The difference on the left-hand side of this identity has a natural interpretation: it counts the integer points on the *boundary* of $t\mathcal{P}$. (And in fact, the right-hand side is once more an inclusion–exclusion formula for this number.) Let us write $L_{\mathcal{P}}(t) = c_d\, t^d + c_{d-1}\, t^{d-1} + \cdots + c_0$. Then $L_{\mathcal{P}^\circ}(t) = c_d\, t^d - c_{d-1}\, t^{d-1} + \cdots + (-1)^d c_0$, so that

$$L_{\mathcal{P}}(t) - L_{\mathcal{P}^\circ}(t) = 2c_{d-1}\, t^{d-1} + 2c_{d-3}\, t^{d-3} + \cdots \, ,$$

where this sum ends with $2c_0$ if d is odd and $2c_1 t$ if d is even (this should look familiar; see Exercise 4.9). Combining this expression with (5.2) yields the following useful result.

Theorem 5.4. *Suppose $L_{\mathcal{P}}(t) = c_d\, t^d + c_{d-1}\, t^{d-1} + \cdots + c_0$ is the Ehrhart polynomial of $\mathcal{P}$. Then*

$$c_{d-1}\, t^{d-1} + c_{d-3}\, t^{d-3} + \cdots = \frac{1}{2} \sum_{j=0}^{d-1} (-1)^j F_j(-t) \, . \qquad \square$$

We can make the statement of this theorem more precise (but also more messy) by writing

$$F_j(t) = \sum_{\substack{\mathcal{F} \subseteq \mathcal{P} \\ \dim \mathcal{F} = j}} L_{\mathcal{F}}(t) = c_{j,j}\, t^j + c_{j,j-1} t^{j-1} + \cdots + c_{j,0} \, .$$

Then collecting the coefficients of t^k in Theorem 5.4 yields the following relations.

Corollary 5.5. *If k and d are of different parity, then*

$$c_k = \frac{1}{2} \sum_{j=0}^{d-1} (-1)^{j+k} c_{j,k} \, . \qquad \square$$

[5] So one might argue that we did not need the Dehn–Sommerville machinery for the computations in the current section. This argument is correct, although Theorem 5.3 is a strong motivation.

If k and d have the same parity, then the left-hand side has to be replaced by 0.

The first coefficient c_k in the Ehrhart polynomial of a d-polytope $\mathcal{P}$ satisfying the parity condition is c_{d-1}. In this case Corollary 5.5 tells us that c_{d-1} equals $\frac{1}{2}$ times the sum of the leading terms of the Ehrhart polynomials of the facets of $\mathcal{P}$.

The next interesting coefficient is c_{d-3}. For example, if $\dim \mathcal{P} = 4$, we can use Corollary 5.5 to compute c_1 entirely from (the linear coefficients of) the Ehrhart polynomials of the faces of dimension ≤ 3.

5.4 Relative Volume

It's time to return to continuous volume. Recall Lemma 3.19: if $S \subset \mathbb{R}^d$ is d-dimensional, then $\operatorname{vol} S = \lim_{t\to\infty} \frac{1}{t^d} \cdot \#\left(tS \cap \mathbb{Z}^d\right)$. Back in Chapter 3, we stressed the importance of S being d-dimensional, because otherwise (i.e., S is lower-dimensional although living in d-space), by our definition $\operatorname{vol} S = 0$. However, the case that $S \subset \mathbb{R}^d$ is *not* of dimension d is often very interesting; an example is the polytope $\mathcal{P}$ that we encountered in connection with the coin-exchange problem in Chapter 1. We still would like to compute the volume of such objects, in the relative sense. This makes for a slight complication. Let's say $S \subset \mathbb{R}^d$ is of dimension $m < d$, and let $\operatorname{span} S = \{\mathbf{x} + \lambda(\mathbf{y} - \mathbf{x}) : \mathbf{x}, \mathbf{y} \in S,\ \lambda \in \mathbb{R}\}$, the affine span of S. If we follow the same procedure as above (counting boxes or grid points), we compute the volume relative to the sublattice $(\operatorname{span} S) \cap \mathbb{Z}^d$; we call this the **relative volume** of S.

For example, the line segment L from $(0,0)$ to $(4,2)$ in $\mathbb{R}^2$ has the relative volume 2, because in $\operatorname{span} L = \left\{(x,y) \in \mathbb{R}^2 : y = x/2\right\}$, L is covered by two segments of "unit length" in this affine subspace, as pictured in Figure 5.2. A three-dimensional instance that should be reminiscent of Chapter 1 is illustrated in Figure 5.3.

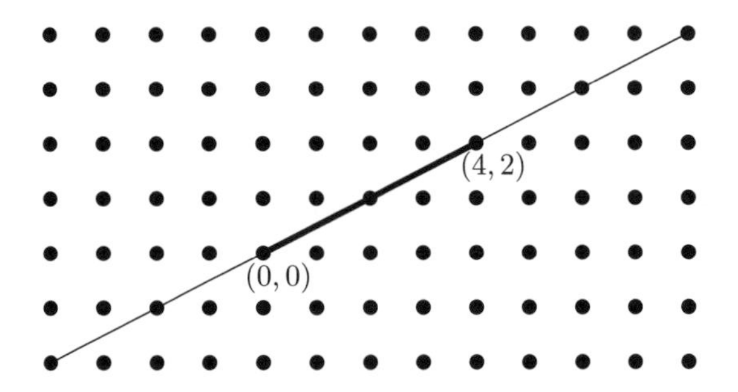

Fig. 5.2. The line segment from $(0,0)$ to $(4,2)$ and its affine sublattice.

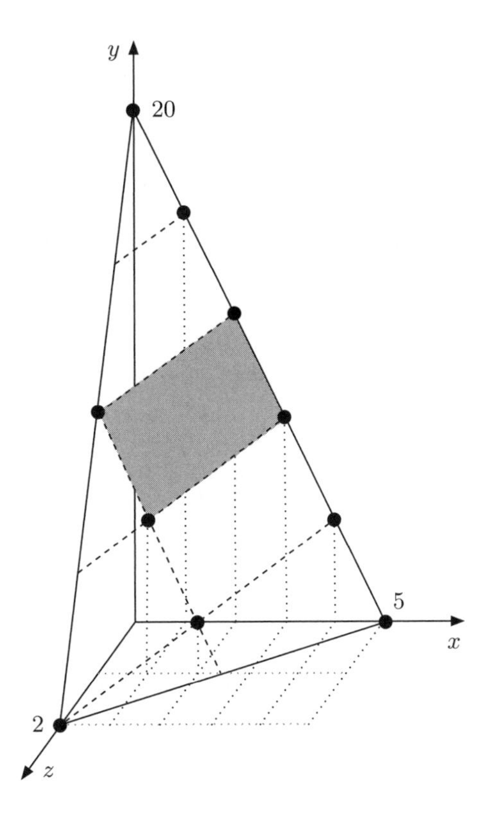

Fig. 5.3. The triangle defined by $\frac{x}{5} + \frac{y}{20} + \frac{z}{2} = 1, x \geq 0, y \geq 0, z \geq 0$. The shaded region is a fundamental domain for the sublattice that lies on the affine span of the triangle.

If $S \subseteq \mathbb{R}^d$ has full dimension d, the relative volume coincides with the "full-dimensional" volume. Henceforth, when we write $\operatorname{vol} S$ we refer to the relative volume of S. With this convention we can rewrite Lemma 3.19 to accommodate a set $S \subset \mathbb{R}^d$ that is m-dimensional: its relative volume can be computed as

$$\operatorname{vol} S = \lim_{t \to \infty} \frac{1}{t^m} \cdot \# \left(tS \cap \mathbb{Z}^d \right).$$

In the case that $\# \left(tS \cap \mathbb{Z}^d \right)$ has the special form of a polynomial—for example, if S is an integral polytope—we can further simplify this theorem. Suppose $\mathcal{P} \subset \mathbb{R}^d$ is an integral m-polytope with Ehrhart polynomial

$$L_{\mathcal{P}}(t) = c_m \, t^m + c_{m-1} \, t^{m-1} + \cdots + c_1 \, t + 1 .$$

Then, according to the above discussion, and much in sync with Lemma 3.19,

$$\operatorname{vol} \mathcal{P} = \lim_{t \to \infty} \frac{1}{t^m} L_{\mathcal{P}}(t) = \lim_{t \to \infty} \frac{c_m t^m + c_{m-1} t^{m-1} + \cdots + c_1 t + 1}{t^m} = c_m \, .$$

The relative volume of $\mathcal{P}$ is the leading term of the corresponding counting function $L_{\mathcal{P}}$.

For example, in the previous section we found that Corollary 5.5 implies that the second leading coefficient c_{d-1} of the Ehrhart polynomial of the d-polytope $\mathcal{P}$ equals $\frac{1}{2}$ times the sum of the leading terms of the Ehrhart polynomials of the facets of $\mathcal{P}$. The leading term for one facet is simply the relative volume of that facet:

Theorem 5.6. *Suppose $L_{\mathcal{P}}(t) = c_d t^d + c_{d-1} t^{d-1} + \cdots + c_0$ is the Ehrhart polynomial of the integral polytope $\mathcal{P}$. Then*

$$c_{d-1} = \frac{1}{2} \sum_{\mathcal{F} \text{ a facet of } \mathcal{P}} \operatorname{vol} \mathcal{F} \, . \qquad \square$$

Notes

1. The Dehn–Sommerville relations (Theorem 5.1) first surfaced in the work of Max Dehn, who proved them in 1905 for dimension five [71]. (The Dehn–Sommerville relations are not that complicated for $d \leq 4$; see Exercise 5.3.) Some two decades later, D. M. Y. Sommerville proved the general case [167]. Theorem 5.1 was neither well known nor much used in the first half of the twentieth century, but only after its rediscovery by Victor Klee [111] and its appearance in Branko Grünbaum's famous and widely read book [90].

2. The Euler relation (Theorem 5.2) is easy to prove directly for $d = 3$ (this case is attributed to Euler), but for higher dimension, one has to be somewhat careful, as we already remarked in the text. The classical proof for general d was found in 1852 by Ludwig Schläfli [158], although it (like numerous later proofs) assumes that the boundary of a convex polytope can be built up inductively in a "good" way. This nontrivial fact—which is called *shelling* of a polytope—was proved by Heinz Bruggesser and Peter Mani in 1971 [50]. Shellability is nicely discussed in [193, Lecture 8]. There are short proofs of the Euler relation that do not use the shelling of a polytope (see, for example, [120, 139, 184]).

3. The reader might suspect that proving Theorems 5.1 and 5.2 for *rational* polytopes suffices for the general case, since it seems that we can transform a polytope with irrational vertices slightly to one with only rational vertices without changing the face structure of the polytope. This is true in our everyday world but fails in dimension ≥ 4 (see [156] for dimension 4 and [193, pp. 172/173] for general dimension).

4. Theorem 5.3 is due to Peter McMullen [127], who, in fact, proved this result in somewhat greater generality. Another generalization of Theorem 5.3 can be found in [59].

Exercises

5.1. Consider a simple 3-polytope with at least five facets. Two players play the following game: Each player, in turn, signs his or her name on a previously unsigned face. The winner is the player who first succeeds in signing three facets that share a common vertex. Show that the player who signs first will always win by playing as well as possible.[6]

5.2. Show that for the d-cube, $f_k = 2^{d-k}\binom{d}{k}$.

5.3. Give an elementary proof of the Dehn–Sommerville relations (Theorem 5.1) for $d \le 4$.

5.4. ♣ Let $\mathcal{P}$ be a simple d-polytope. Prove that the number of k-faces of $\mathcal{P}$ containing a given j-face of $\mathcal{P}$ equals $\binom{d-j}{d-k}$.

5.5. ♣ Show directly, without using Theorem 5.2, that for a d-simplex:

(a) $f_k = \binom{d+1}{k+1}$.

(b) $\displaystyle\sum_{k=0}^{d}(-1)^k f_k = 1$.

5.6. Prove Theorem 5.1 directly (and hence not requiring $\mathcal{P}$ to be an integral polytope). (*Hint:* Orient yourself along the proof of Theorem 5.3, but start with the Euler relation (Theorem 5.2) for a given face $\mathcal{F}$ instead of (5.1).)

5.7. Let $\mathcal{F}$ be a face of a simple polytope $\mathcal{P}$. Prove that

$$\sum_{\mathcal{G} \supseteq \mathcal{F}} (-1)^{\dim \mathcal{G}} \binom{\dim \mathcal{G}}{k} = (-1)^d \binom{\dim \mathcal{F}}{d-k}, \quad k = 0, \dots, d.$$

5.8. Show that the equations in Theorem 5.3 are equivalent to the following identities. If $\mathcal{P}$ is a simple lattice d-polytope and $k \le d$, then

$$\sum_{j=0}^{k} (-1)^{k-j} \binom{d-j}{k-j} F_{d-j}(-n) = \sum_{i=k}^{d} (-1)^{i-k} \binom{i}{k} F_i(n).$$

[6] This was one of the 2002 Putnam contest problems.

5.9. Prove that the equations in the previous exercise imply the following identities which compare the number of lattice points in faces and relative interiors of faces of the simple polytope $\mathcal{P}$:

$$\sum_{j=0}^{k}(-1)^j\binom{d-j}{k-j}\sum_{\substack{\mathcal{F}\subseteq\mathcal{P}\\ \dim\mathcal{F}=d-j}}\#\left(\mathcal{F}\cap\mathbb{Z}^d\right)=\sum_{i=k}^{d}\binom{i}{k}\sum_{\substack{\mathcal{G}\subseteq\mathcal{P}\\ \dim\mathcal{G}=i}}\#\left(\mathcal{G}^\circ\cap\mathbb{Z}^d\right),$$

where $k=0,\dots,d=\dim\mathcal{P}$. For example, for $k=0$, we have

$$\#(\mathcal{P}\cap\mathbb{Z}^d)=\sum_{\mathcal{G}\subseteq\mathcal{P}}\#\left(\mathcal{G}^\circ\cap\mathbb{Z}^d\right),$$

and for $k=d$ we obtain the inclusion–exclusion formula

$$\#(\mathcal{P}^\circ\cap\mathbb{Z}^d)=\sum_{j=0}^{d}(-1)^{d-j}\sum_{\substack{\mathcal{F}\subseteq\mathcal{P}\\ \dim\mathcal{F}=j}}\#\left(\mathcal{F}\cap\mathbb{Z}^d\right).$$

5.10. Another nice reformulation of Theorem 5.3 is the following generalized reciprocity law. For an integral d-polytope $\mathcal{P}$, define the *generalized Ehrhart polynomial*

$$E_k(t):=\sum_{j=0}^{k}(-1)^j\binom{d-j}{k-j}\sum_{\substack{\mathcal{F}\subseteq\mathcal{P}\\ \dim\mathcal{F}=d-j}}L_{\mathcal{F}}(t),\quad k=0,\dots,d.$$

Prove the generalized reciprocity law

$$E_k(-t)=(-1)^d\,E_{d-k}(t),\quad k=0,\dots,d,$$

which implies Ehrhart–Macdonald reciprocity (Theorem 4.1) when $k=0$.

5.11. What happens when $\mathcal{P}$ is *not* simple? Give an example for which Theorem 5.3 fails.

5.12. Give an alternative proof of Theorem 5.6 by considering $L_{\mathcal{P}}(t)-L_{\mathcal{P}^\circ}(t)$ as the lattice-point enumerator of the boundary of $\mathcal{P}$.

6

Magic Squares

The peculiar interest of magic squares and all lusus numerorum *in general lies in the fact that they possess the charm of mystery.*

W. S. Andrews

Fig. 6.1. Magic square at the *Temple de la Sagrada Família* (Barcelona, Spain).

Equipped with a solid base of theoretical results, we are now ready to return to computations. We use Ehrhart theory to assist us in enumerating *magic squares.*

Loosely speaking, a magic square is an $n \times n$ array of integers (usually required to be positive, often restricted to the numbers $1, 2, \ldots, n^2$, usually required to have distinct entries) whose sum along any row, column, and main diagonal is the same number, called the *magic sum.* Magic squares have turned up time and again, some in mathematical contexts, others in philosophical or religious contexts. According to legend, the first magic square (the ancient *Luo Shu square*) was discovered in China sometime before the first century B.C.E. on the back of a turtle emerging from a river. It was the square pictured in Figure 6.2.

4	9	2
3	5	7
8	1	6

Fig. 6.2. The Luo Shu square.

Our task in this chapter is to develop a theory for counting certain classes of magic squares, which we now introduce.

6.1 It's a Kind of Magic

One should notice that the Luo Shu square has the distinct entries $1, 2, \ldots, 9$, so these entries are positive, distinct integers drawn from a particular set. Such requirements are too restrictive for our purposes. We define a **semimagic square** to be a square matrix whose entries are nonnegative integers and whose rows and columns (called *lines* in this setting) sum to the same number. A **magic square** is a semimagic square whose main diagonals also add up to the line sum. Figure 6.3 shows two examples.

3	0	0
0	1	2
0	2	1

1	2	0
0	1	2
2	0	1

Fig. 6.3. A semimagic and a magic square.

We caution the reader about clashing definitions in the literature. For example, some people reserve the term "magic square" for what we will call a **traditional magic square**, a magic square of order n whose entries are the distinct integers $1, 2, \ldots, n^2$. (The Luo Shu square is an example of a traditional magic square.) Others are slightly less restrictive and use the term "magic square" for a magic square with *distinct* entries. We stress that we do not make this requirement in this chapter.

Our goal is to count semimagic and magic squares. In the traditional case, this is in some sense not very interesting:[1] for each order there is a fixed number of traditional magic squares. For example, there are 7040 traditional 4×4 magic squares.

The situation becomes more interesting if we drop the condition of traditionality and study the number of magic squares as a function of the line sum. We denote the total numbers of semimagic and magic squares of order n and line sum t by $H_n(t)$ and $M_n(t)$, respectively.

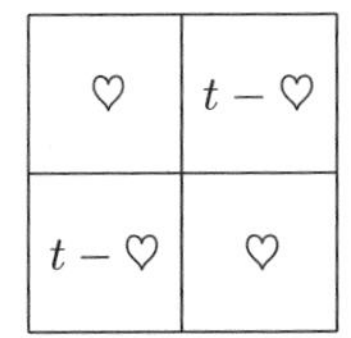

$\frac{t}{2}$	$\frac{t}{2}$
$\frac{t}{2}$	$\frac{t}{2}$

Fig. 6.4. Semimagic and magic squares for $n = 2$.

Example 6.1. We illustrate these notions for the case $n = 2$, which is not very complicated. Here a semimagic square is determined once we know one entry, say the upper left one, denoted by $\heartsuit$ in Figure 6.4. Because of the upper row sum, the upper right entry has to be $t - \heartsuit$, as does the lower left entry (because of the left column sum). But then the lower right entry has to be $t - (t - \heartsuit) = \heartsuit$ (for two reasons: the lower row sum and the right column sum). The entry $\heartsuit$ can be any integer between 0 and t. Since there are $t + 1$ such integers,

$$H_2(t) = t + 1 \,. \tag{6.1}$$

In the magic case, we also have to think of the diagonals. Looking back at our semimagic square in Figure 6.4, the first diagonal gives $2 \cdot \heartsuit = t$, or $\heartsuit = t/2$. In this case, $t - \heartsuit = t/2$, and so a 2×2 magic square has to have identical entries in each position. Because we require the entries to be integers, this is possible only if t is even, in which case we obtain precisely one solution, the square on the right in Figure 6.4. That is,

$$M_2(t) = \begin{cases} 1 & \text{if } t \text{ is even,} \\ 0 & \text{if } t \text{ is odd.} \end{cases}$$

These easy results already hint at something: the counting function H_n is of a different character than the function M_n. □

[1] It is, nevertheless, an incredibly hard problem to count all traditional magic squares of a given size n. At present, these numbers are known only for $n \leq 5$ [165, Sequence A006052].

6.2 Semimagic Squares: Integer Points in the Birkhoff–von Neumann Polytope

Just as the Frobenius problem was intrinsically connected to questions about integer points on line segments, triangles, and higher-dimensional simplices, magic squares and their relatives have a life in the world of geometry. The most famous example is connected to semimagic squares.

A semimagic $n \times n$ square has n^2 nonnegative entries that sum to the same number along any row or column. Consider therefore the polytope

$$\mathcal{B}_n := \left\{ \begin{pmatrix} x_{11} & \cdots & x_{1n} \\ \vdots & & \vdots \\ x_{n1} & \cdots & x_{nn} \end{pmatrix} \in \mathbb{R}^{n^2} : x_{jk} \geq 0, \begin{array}{l} \sum_j x_{jk} = 1 \text{ for all } 1 \leq k \leq n \\ \sum_k x_{jk} = 1 \text{ for all } 1 \leq j \leq n \end{array} \right\}, \tag{6.2}$$

consisting of nonnegative real matrices, in which all rows and columns sum to one. $\mathcal{B}_n$ is called the nth **Birkhoff–von Neumann polytope**, in honor of Garrett Birkhoff (1911–1996)[2] and John von Neumann (1903–1957).[3] Because the matrices contained in the Birkhoff–von Neumann polytope appear frequently in probability and statistics (the line sum 1 representing probability 1), $\mathcal{B}_n$ is often described as the set of all $n \times n$ *doubly stochastic matrices.*

Geometrically, $\mathcal{B}_n$ is a subset of $\mathbb{R}^{n^2}$ and as such difficult to picture once n exceeds 1.[4] However, we can get a glimpse of $\mathcal{B}_2 \subset \mathbb{R}^4$ when we think about what form points in $\mathcal{B}_2$ can possibly attain. Very much in sync with Figure 6.4, such a point is determined by its upper left entry $\heartsuit$, pictured in Figure 6.5. This entry $\heartsuit$ is a real number between 0 and 1, which suggests that $\mathcal{B}_2$ should

$$\begin{pmatrix} \heartsuit & 1-\heartsuit \\ 1-\heartsuit & \heartsuit \end{pmatrix}$$

Fig. 6.5. A point in $\mathcal{B}_2$.

look like a line segment in 4-space. Indeed, the vertices of $\mathcal{B}_2$ should be given by $\heartsuit = 0$ and $\heartsuit = 1$, that is, by the points

$$\begin{pmatrix} 1 & 0 \\ 0 & 1 \end{pmatrix}, \begin{pmatrix} 0 & 1 \\ 1 & 0 \end{pmatrix} \in \mathcal{B}_2 .$$

[2] For more information about Birkhoff, see
http://www-groups.dcs.st-and.ac.uk/~history/Biographies/Birkhoff_Garrett.html.

[3] For more information about von Neumann, see
http://www-groups.dcs.st-and.ac.uk/~history/Biographies/Von_Neumann.html.

[4] The case $n = 1$ is not terribly interesting: $\mathcal{B}_1 = \{1\}$ is a point.

These results generalize: $\mathcal{B}_n$ is an $(n-1)^2$-polytope (see Exercise 6.3), whose vertices (Exercise 6.5) are the **permutation matrices**, namely, those $n \times n$ matrices that have precisely one 1 in each row and column (every other entry being zero). For dimensional reasons, we can talk about the continuous volume of $\mathcal{B}_n$ only in the relative sense, following the definition of Section 5.4.

The connection of the semimagic counting function $H_n(t)$ to the Birkhoff–von Neumann polytope $\mathcal{B}_n$ becomes clear in the light of the lattice-point enumerator for $\mathcal{B}_n$: the counting function $H_n(t)$ enumerates precisely the integer points in $t\mathcal{B}_n$, that is,

$$H_n(t) = \# \left(t\mathcal{B}_n \cap \mathbb{Z}^{n^2} \right) = L_{\mathcal{B}_n}(t) \,.$$

We can say more after noticing that permutation vertices are *integer* points in $\mathcal{B}_n$, and so Ehrhart's theorem (Theorem 3.8) applies:

Theorem 6.2. *$H_n(t)$ is a polynomial in t of degree $(n-1)^2$.* □

The fact that H_n is a polynomial—apart from being mathematically appealing—has the same nice computational consequence that we exploited in Section 3.6: we can calculate this counting function by interpolation. For example, to compute H_2, a linear polynomial, we need to know only two values. In fact, since we know that the constant term of H_2 is 1 (by Corollary 3.15), we need only one value. It is not hard to convince even a lay person that $H_2(1) = 2$ (which two semimagic squares are those?), from which we interpolate

$$H_2(t) = t + 1 \,.$$

To interpolate the polynomial H_3, we need to know 4 values aside from $H_3(0) = 1$. In fact, we do not even have to know that many values, because Ehrhart–Macdonald reciprocity (Theorem 4.1) assists us in computations. To see this, let $H_n^\circ(t)$ denote the number of $n \times n$ square matrices with *positive* integer entries summing up to t along each row and column. A moment's thought (Exercise 6.6) reveals that

$$H_n^\circ(t) = H_n(t-n) \,. \tag{6.3}$$

But there is a second relationship between H_n and H_n°, namely, $H_n^\circ(t)$ counts, by definition, the integer points in the *relative interior* of the Birkhoff–von Neumann polytope $\mathcal{B}_n$, that is, $H_n^\circ(t) = L_{\mathcal{B}_n^\circ}(t)$. Now Ehrhart–Macdonald reciprocity (Theorem 4.1) gives

$$H_n^\circ(-t) = (-1)^{(n-1)^2} H_n(t) \,.$$

Combining this identity with (6.3) gives us a symmetry identity for the counting function for semimagic squares:

Theorem 6.3. *The polynomial* H_n *satisfies*

$$H_n(-n-t) = (-1)^{(n-1)^2} H_n(t)$$

and

$$H_n(-1) = H_n(-2) = \cdots = H_n(-n+1) = 0\,. \qquad \square$$

The roots of H_n at the first $n-1$ negative integers follow from (Exercise 6.7)

$$H_n^\circ(1) = H_n^\circ(2) = \cdots = H_n^\circ(n-1) = 0\,.$$

Theorem 6.3 gives the degree of $\mathcal{B}_n$, and it implies that the numerator of the Ehrhart series of the Birkhoff–von Neumann polytope is palindromic:

Corollary 6.4. *The Ehrhart series of the Birkhoff–von Neumann polytope* $\mathcal{B}_n$ *has the form*

$$\mathrm{Ehr}_{\mathcal{B}_n}(z) = \frac{h_{(n-1)(n-2)}\, z^{(n-1)(n-2)} + \cdots + h_0}{(1-z)^{(n-1)^2+1}}\,,$$

where $h_0, h_1, \dots, h_{(n-1)(n-2)} \in \mathbb{Z}_{\geq 0}$ *satisfy* $h_k = h_{(n-1)(n-2)-k}$ *for* $0 \leq k \leq \frac{(n-1)(n-2)}{2}$.

Proof. Denote the Ehrhart series of $\mathcal{B}_n$ by

$$\mathrm{Ehr}_{\mathcal{B}_n}(z) = \frac{h_{(n-1)^2}\, z^{(n-1)^2} + \cdots + h_0}{(1-z)^{(n-1)^2+1}}\,.$$

The fact that $h_{(n-1)^2} = \cdots = h_{(n-1)^2-(n-2)} = 0$ follows from the second part of Theorem 6.3 and Theorem 4.5. The palindromy of the numerator coefficients follows from the first part of Theorem 6.3 and Exercise 4.6: it implies

$$\mathrm{Ehr}_{\mathcal{B}_n}\left(\frac{1}{z}\right) = (-1)^{(n-1)^2+1} z^n\, \mathrm{Ehr}_{\mathcal{B}_n}(z)\,,$$

which yields $h_k = h_{(n-1)(n-2)-k}$ upon simplifying both sides of the equation. $\square$

Let's return to the interpolation of H_3: Theorem 6.3 gives, in addition to $H_3(0) = 1$, the values

$$H_3(-3) = 1 \qquad \text{and} \qquad H_3(-1) = H_3(-2) = 0\,.$$

These four values together with $H_3(1) = 6$ (see Exercise 6.1) suffice to interpolate the quartic polynomial H_3, and one computes

$$H_3(t) = \frac{1}{8}\,t^4 + \frac{3}{4}\,t^3 + \frac{15}{8}\,t^2 + \frac{9}{4}\,t + 1\,. \tag{6.4}$$

This interpolation example suggests the use of a computer; we let it calculate enough values of H_n and then simply interpolate. As far as computations are concerned, however, we should not get too excited about the fact that we computed H_2 and H_3 so effortlessly. In general, the polynomial H_n has degree $(n-1)^2$, so we need to compute $(n-1)^2+1$ values of H_n to be able to interpolate. Of those, we know n (the constant term and the roots given by Theorem 6.3), so n^2-3n+1 values of H_n remain to be computed. Ehrhart–Macdonald reciprocity reduces the number of values to be computed to $\frac{n^2-3n}{2}+1$. That's still a lot, as anyone can testify who has tried to get a computer to enumerate all semimagic 7×7 squares with line sum 15. Nevertheless, it is a fun fact that we can compute H_n for small n by interpolation. It is amusing to test one's computer against the constant-term computation we will outline below, and we invite the reader to try both. For small n, interpolation is clearly superior to a constant-term computation in the spirit of Chapter 1. The turning point seems to be right around $n=5$: the computer needs more and more time to compute the values $H_n(t)$ as t increases. Methods superior to interpolation are needed.

6.3 Magic Generating Functions and Constant-Term Identities

Now we will construct a generating function for H_n, for which we will use Theorem 2.13. The semimagic counting function H_n is the Ehrhart polynomial of the nth Birkhoff–von Neumann polytope $\mathcal{B}_n$, which, in turn, is defined as a set of matrices by (6.2). First we rewrite the definition of $\mathcal{B}_n$ to fit the general description (2.23) of a polytope. If we consider the points in $\mathcal{B}_n$ as column vectors in $\mathbb{R}^{n^2}$ (rather than as matrices in $\mathbb{R}^{n\times n}$) then

$$\mathcal{B}_n = \left\{ \mathbf{x} \in \mathbb{R}^{n^2}_{\geq 0} : \mathbf{A}\,\mathbf{x} = \mathbf{b} \right\},$$

where

$$\mathbf{A} = \begin{pmatrix} 1 \cdots 1 & & & \\ & 1 \cdots 1 & & \\ & & \ddots & \\ & & & 1 \cdots 1 \\ 1 & 1 & & 1 \\ \ddots & \ddots & \cdots & \ddots \\ 1 & 1 & & 1 \end{pmatrix} \in \mathbb{Z}^{2n \times n^2} \tag{6.5}$$

(here we omit the zero entries) and

$$\mathbf{b} = \begin{pmatrix} 1 \\ 1 \\ \vdots \\ 1 \end{pmatrix} \in \mathbb{Z}^{2n}.$$

From this description of $\mathcal{B}_n$, we can easily build the generating function for H_n. According to Theorem 2.13, for a general rational polytope $\mathcal{P} = \left\{ \mathbf{x} \in \mathbb{R}^d_{\geq 0} : \mathbf{A}\,\mathbf{x} = \mathbf{b} \right\}$, we have

$$L_{\mathcal{P}}(t) = \operatorname{const}\left(\frac{1}{\left(1 - \mathbf{z}^{\mathbf{c}_1}\right)\left(1 - \mathbf{z}^{\mathbf{c}_2}\right)\cdots\left(1 - \mathbf{z}^{\mathbf{c}_d}\right)\mathbf{z}^{t\mathbf{b}}} \right),$$

where $\mathbf{c}_1, \mathbf{c}_2, \ldots, \mathbf{c}_d$ denote the columns of $\mathbf{A}$. In our special case, the columns of $\mathbf{A}$ are of a simple form: they contain exactly two 1's and elsewhere 0's. We need one generating-function variable for each row of $\mathbf{A}$. To keep things as clear as possible, we use $z_1, z_2, \ldots, z_n$ for the first n rows of $\mathbf{A}$ (representing the row constraints of $\mathcal{B}_n$) and $w_1, w_2, \ldots, w_n$ for the last n rows of $\mathbf{A}$ (representing the column constraints of $\mathcal{B}_n$). With this notation, Theorem 2.13 applied to $\mathcal{B}_n$ gives the following starting point for our computations:

Theorem 6.5. *The number $H_n(t)$ of semimagic $n \times n$ squares with line sum t satisfies*

$$H_n(t) = \operatorname{const}\left(\frac{1}{\prod_{1 \le j,k \le n} \left(1 - z_j w_k\right) \left(\prod_{1 \le j \le n} z_j \prod_{1 \le k \le n} w_k \right)^t} \right). \qquad \square$$

This identity is of both theoretical and practical use. One can use it to compute H_3 and even H_4 by hand. For now, we work on refining it further, exemplified by the case $n = 2$.

We first note that in the formula for H_2, the variables w_1 and w_2 are separated, in the sense that we can write this formula as a product of two factors, one involving only w_1 and the other involving only w_2:

$$H_2(t) = \operatorname{const}\left(\frac{1}{z_1^t z_2^t} \, \frac{1}{\left(1 - z_1 w_1\right)\left(1 - z_2 w_1\right) w_1^t} \, \frac{1}{\left(1 - z_1 w_2\right)\left(1 - z_2 w_2\right) w_2^t} \right).$$

Now we put an ordering on the constant-term computation: let us first compute the constant term with respect to w_2, then the one with respect to w_1. (We don't order the computation with respect to z_1 and z_2 just yet.) Since z_1, z_2, and w_1 are considered constants when we do constant-term computations with respect to w_2, we can simplify:

$$\begin{aligned} H_2(t) = \operatorname{const}_{z_1, z_2}\Bigg(\frac{1}{z_1^t z_2^t} \operatorname{const}_{w_1}\bigg(&\frac{1}{\left(1 - z_1 w_1\right)\left(1 - z_2 w_1\right) w_1^t} \\ &\times \operatorname{const}_{w_2}\left(\frac{1}{\left(1 - z_1 w_2\right)\left(1 - z_2 w_2\right) w_2^t} \right) \bigg) \Bigg). \end{aligned}$$

Now we can see the effect of the separate appearance of w_1 and w_2: the constant-term identity *factors*. This is very similar to the factoring that can appear in computations of integrals in several variables. Let us rewrite our identity to emphasize the factoring:

$$H_2(t) = \operatorname{const}_{z_1,z_2}\left(\frac{1}{z_1^t z_2^t}\operatorname{const}_{w_1}\left(\frac{1}{(1-z_1w_1)(1-z_2w_1)\,w_1^t}\right)\right.$$
$$\left.\times\operatorname{const}_{w_2}\left(\frac{1}{(1-z_1w_2)(1-z_2w_2)\,w_2^t}\right)\right).$$

But now the expressions in the last two sets of parentheses are identical, except that in one case the constant-term variable is called w_1 and in the other case w_2. Since these are just "dummy" variables, we can call them w, and combine:

$$H_2(t) = \operatorname{const}_{z_1,z_2}\left(\frac{1}{z_1^t z_2^t}\left(\operatorname{const}_w\frac{1}{(1-z_1w)(1-z_2w)\,w^t}\right)^2\right).$$

(Note the square!) Naturally, all of this factoring works in the general case, and we invite the reader to prove it (Exercise 6.8):

$$H_n(t) = \operatorname{const}_{z_1,\dots,z_n}\left((z_1\cdots z_n)^{-t}\left(\operatorname{const}_w\frac{1}{(1-z_1w)\cdots(1-z_nw)w^t}\right)^n\right). \tag{6.6}$$

We can go further, namely, we can compute the innermost constant term

$$\operatorname{const}_w\frac{1}{(1-z_1w)\cdots(1-z_nw)w^t}.$$

It should come as no surprise that we'll use a partial fraction expansion to do so. The w-poles of the rational function are at $w = 1/z_1$, $w = 1/z_2$, $\dots$, $w = 1/z_n$, and so we expand

$$\frac{1}{(1-z_1w)\cdots(1-z_nw)w^t} = \frac{A_1}{w-\frac{1}{z_1}} + \frac{A_2}{w-\frac{1}{z_2}} + \cdots + \frac{A_n}{w-\frac{1}{z_n}} + \sum_{k=1}^{t}\frac{B_k}{w^k}. \tag{6.7}$$

Just as in Chapter 1, we can forget about the B_k-terms, since they do not contribute to the constant term, that is,

$$\begin{aligned}\operatorname{const}_w&\frac{1}{(1-z_1w)\cdots(1-z_nw)w^t}\\ &= \operatorname{const}_w\left(\frac{A_1}{w-\frac{1}{z_1}} + \frac{A_2}{w-\frac{1}{z_2}} + \cdots + \frac{A_n}{w-\frac{1}{z_n}}\right)\\ &= -A_1z_1 - A_2z_2 - \cdots - A_nz_n.\end{aligned}$$

We invite the reader to show (Exercise 6.9) that

$$\begin{aligned}A_k &= -\frac{z_k^{t-1}}{\left(1-\frac{z_1}{z_k}\right)\cdots\left(1-\frac{z_{k-1}}{z_k}\right)\left(1-\frac{z_{k+1}}{z_k}\right)\cdots\left(1-\frac{z_n}{z_k}\right)}\\ &= -\frac{z_k^{t+n-2}}{\prod_{j\neq k}(z_k-z_j)}.\end{aligned} \tag{6.8}$$

Putting these coefficients back into the partial fraction expansion yields the following identity.

Theorem 6.6. *The number $H_n(t)$ of semimagic $n \times n$ squares with line sum t satisfies*

$$H_n(t) = \operatorname{const}\left((z_1 \cdots z_n)^{-t} \left(\sum_{k=1}^{n} \frac{z_k^{t+n-1}}{\prod_{j \neq k} (z_k - z_j)} \right)^n \right). \qquad \square$$

Amidst all this generality, we almost forgot to compute H_2 with our partial fraction approach. The last theorem says that

$$\begin{aligned} H_2(t) &= \operatorname{const}\left((z_1 z_2)^{-t} \left(\frac{z_1^{t+1}}{z_1 - z_2} + \frac{z_2^{t+1}}{z_2 - z_1} \right)^2 \right) \\ &= \operatorname{const}\left(\frac{z_1^{t+2} z_2^{-t}}{(z_1 - z_2)^2} - 2 \frac{z_1 z_2}{(z_1 - z_2)^2} + \frac{z_1^{-t} z_2^{t+2}}{(z_1 - z_2)^2} \right). \end{aligned} \tag{6.9}$$

Now it's time to put more order on the constant-term computation. Let's say we first compute the constant term with respect to z_1, and after that with respect to z_2. So we have to compute first

$$\operatorname{const}_{z_1}\left(\frac{z_1^{t+2} z_2^{-t}}{(z_1 - z_2)^2} \right), \quad \operatorname{const}_{z_1}\left(\frac{z_1 z_2}{(z_1 - z_2)^2} \right), \quad \text{and} \quad \operatorname{const}_{z_1}\left(\frac{z_1^{-t} z_2^{t+2}}{(z_1 - z_2)^2} \right).$$

To obtain these constant terms, we need to expand the function $\frac{1}{(z_1-z_2)^2}$. As we know from calculus, this expansion depends on the order of the magnitudes of z_1 and z_2. For example, if $|z_1| < |z_2|$ then

$$\frac{1}{z_1 - z_2} = \frac{1}{z_2} \frac{1}{\frac{z_1}{z_2} - 1} = -\frac{1}{z_2} \sum_{k \geq 0} \left(\frac{z_1}{z_2} \right)^k = -\sum_{k \geq 0} \frac{1}{z_2^{k+1}} z_1^k,$$

and hence

$$\frac{1}{(z_1 - z_2)^2} = -\frac{d}{dz_1} \left(\frac{1}{z_1 - z_2} \right) = \sum_{k \geq 1} \frac{k}{z_2^{k+1}} z_1^{k-1} = \sum_{k \geq 0} \frac{k+1}{z_2^{k+2}} z_1^k.$$

So let's assume for the moment that $|z_1| < |z_2|$. This might sound funny, since z_1 and z_2 are *variables*. However, as such they are simply tools that enable us to compute some quantity that is independent of z_1 and z_2. In view of these ideas, we may assume any order of the magnitudes of the variables. In Exercise 6.11 we will check that indeed the order does not matter. Now,

$$\begin{aligned}\operatorname{const}_{z_1}\left(\frac{z_1^{t+2}z_2^{-t}}{(z_1-z_2)^2}\right) &= z_2^{-t}\operatorname{const}_{z_1}\left(\frac{z_1^{t+2}}{(z_1-z_2)^2}\right)\\ &= z_2^{-t}\operatorname{const}_{z_1}\left(z_1^{t+2}\sum_{k\ge0}\frac{k+1}{z_2^{k+2}}\,z_1^k\right)\\ &= z_2^{-t}\operatorname{const}_{z_1}\left(\sum_{k\ge0}\frac{k+1}{z_2^{k+2}}\,z_1^{k+t+2}\right)\\ &= 0\,,\end{aligned} \tag{6.10}$$

since there are only positive powers of z_1 (remember that $t \ge 0$). Analogously (see Exercise 6.10) one checks that

$$\operatorname{const}_{z_1}\left(\frac{z_1z_2}{(z_1-z_2)^2}\right) = 0\,. \tag{6.11}$$

For the last constant term, we compute

$$\begin{aligned}\operatorname{const}_{z_1}\left(\frac{z_1^{-t}z_2^{t+2}}{(z_1-z_2)^2}\right) &= z_2^{t+2}\operatorname{const}_{z_1}\left(z_1^{-t}\sum_{k\ge0}\frac{k+1}{z_2^{k+2}}\,z_1^k\right)\\ &= z_2^{t+2}\operatorname{const}_{z_1}\left(\sum_{k\ge0}\frac{k+1}{z_2^{k+2}}\,z_1^{k-t}\right).\end{aligned}$$

The constant term on the right-hand side is the term with $k = t$, that is,

$$\operatorname{const}_{z_1}\left(\frac{z_1^{-t}z_2^{t+2}}{(z_1-z_2)^2}\right) = z_2^{t+2}\,\frac{t+1}{z_2^{t+2}} = t+1\,.$$

So of the three constant terms only one survives, and with $\operatorname{const}_{z_2}(t+1) = t+1$ we recover what we have known since the beginning of this chapter:

$$H_2(t) = \operatorname{const}\left(\frac{z_1^{t+2}z_2^{-t}}{(z_1-z_2)^2} - 2\frac{z_1z_2}{(z_1-z_2)^2} + \frac{z_1^{-t}z_2^{t+2}}{(z_1-z_2)^2}\right) = t+1\,.$$

This was a lot of work for this seemingly trivial polynomial. Recall, for example, that we can get the same result by an easy interpolation. However, to compute a similar interpolation, e.g., for H_4, we likely would need to use a computer (to obtain the interpolation values). On the other hand, the constant-term computation of H_4 boils down to only five iterated constant terms, which can actually be computed by hand (see Exercise 6.14). The result is

$$\begin{aligned}H_4(t) = {}& \frac{11}{11340}\,t^9 + \frac{11}{630}\,t^8 + \frac{19}{135}\,t^7 + \frac{2}{3}\,t^6 + \frac{1109}{540}\,t^5 + \frac{43}{10}\,t^4 + \frac{35117}{5670}\,t^3\\ &+ \frac{379}{63}\,t^2 + \frac{65}{18}\,t + 1\,.\end{aligned}$$

6.4 The Enumeration of Magic Squares

What happens when we bring the diagonal constraints, which are not present in the semimagic case, into the magic picture? In the introduction of this chapter we have already seen an example, namely the number of 2×2 magic squares,

$$M_2(t) = \begin{cases} 1 & \text{if } t \text{ is even,} \\ 0 & \text{if } t \text{ is odd.} \end{cases}$$

This is a very simple example of a quasipolynomial. In fact, like H_n, the counting function M_n is defined by integral linear equations and inequalities, so it is the lattice-point enumerator of a rational polytope, and Theorem 3.23 gives at once the following result.

Theorem 6.7. *The counting function $M_n(t)$ is a quasipolynomial in t.* $\square$

We invite the reader to prove that the degree of M_n is $n^2 - 2n - 1$ (Exercise 6.16).

Let's see what happens in the first nontrivial case, 3×3 magic squares. We follow our recipe and assign variables $m_1, m_2, \dots, m_9$ to the entries of our 3×3 squares, as in Figure 6.6.

m_1	m_2	m_3
m_4	m_5	m_6
m_7	m_8	m_9

Fig. 6.6. Variables in a magic 3×3 square.

The magic conditions require now that $m_1, m_2, \dots, m_9 \in \mathbb{Z}_{\geq 0}$ and

$$\begin{aligned}
m_1 + m_2 + m_3 &= t\,, \\
m_4 + m_5 + m_6 &= t\,, \\
m_7 + m_8 + m_9 &= t\,, \\
m_1 + m_4 + m_7 &= t\,, \\
m_2 + m_5 + m_8 &= t\,, \\
m_3 + m_6 + m_9 &= t\,, \\
m_1 + m_5 + m_9 &= t\,, \\
m_3 + m_5 + m_7 &= t\,,
\end{aligned}$$

according to the row sums (the first three equations), the column sums (the next three equations), and the diagonal sums (the last two equations). By now, we're experienced in translating this system into a generating function: we need one variable for each equation, so let's take z_1, z_2, z_3 for the first three, w_1, w_2, w_3 for the next three, and y_1, y_2 for the last two equations. The function $M_3(t)$ is the constant term of

$$\frac{1}{(1-z_1w_1y_1)(1-z_1w_2)(1-z_1w_3y_2)(1-z_2w_1)(1-z_2w_2y_1y_2)(1-z_2w_3)} \times \frac{1}{(1-z_3w_1y_2)(1-z_3w_2)(1-z_3w_3y_1)(z_1z_2z_3w_1w_2w_3y_1y_2)^t}. \tag{6.12}$$

It does take some work, but it is instructive to compute this constant term (just try it!). The result is

$$M_3(t) = \begin{cases} \frac{2}{9}t^2 + \frac{2}{3}t + 1 & \text{if } 3|t, \\ 0 & \text{otherwise.} \end{cases} \tag{6.13}$$

As predicted by Theorem 6.7, M_3 is a quasipolynomial. It has degree 2 and period 3. This may be more apparent if we rewrite it as

$$M_3(t) = \begin{cases} \frac{2}{9}t^2 + \frac{2}{3}t + 1 & \text{if } t \equiv 0 \bmod 3, \\ 0 & \text{if } t \equiv 1 \bmod 3, \\ 0 & \text{if } t \equiv 2 \bmod 3, \end{cases}$$

and we can see the three constituents of the quasipolynomial M_3. There is an alternative way to describe M_3; namely, let

$$c_2(t) = \begin{cases} \frac{2}{9} & \text{if } t \equiv 0 \bmod 3, \\ 0 & \text{if } t \equiv 1 \bmod 3, \\ 0 & \text{if } t \equiv 2 \bmod 3, \end{cases}$$

$$c_1(t) = \begin{cases} \frac{2}{3} & \text{if } t \equiv 0 \bmod 3, \\ 0 & \text{if } t \equiv 1 \bmod 3, \\ 0 & \text{if } t \equiv 2 \bmod 3, \end{cases}$$

$$c_0(t) = \begin{cases} 1 & \text{if } t \equiv 0 \bmod 3, \\ 0 & \text{if } t \equiv 1 \bmod 3, \\ 0 & \text{if } t \equiv 2 \bmod 3. \end{cases}$$

Then the quasipolynomial M_3 can be written as

$$M_3(t) = c_2(t)\,t^2 + c_1(t)\,t + c_1(t)\,.$$

Notes

1. Magic squares date back to China in the first millennium B.C.E. [53]; they underwent much further development in the Islamic world late in the first millennium C.E. and in the next millennium (or sooner; the data are lacking) in India [54]. From Islam they passed to Christian Europe in the later Middle Ages, probably initially through the Jewish community [54, Part II, pp. 290 ff.] and later possibly Byzantium [54, Part I, p. 198], and no later than the early eighteenth century (the data are buried in barely tapped archives) to sub-Saharan Africa [191, Chapter 12]. The contents of a magic square have varied with time and writer; usually they have been the first n^2 consecutive integers, but often any arithmetic sequence or arbitrary positive numbers. In the past century, mathematicians have made some simplifications in the interest of obtaining results about the number of squares with a fixed magic sum, in particular, allowing repeated entries as in this chapter.

2. The problem of counting magic squares (other than traditional magic squares) seems to have occurred to anyone only in the twentieth century, no doubt because there was no way to approach the question previously. The first nontrivial formulas addressing the counting problem, namely (6.4) and (6.13) for H_3 and M_3, were established by Percy Macmahon (1854–1929)[5] [124] in 1915. Recently there has grown up a literature on exact formulas (see for example [80, 168] for semimagic squares; for magic squares see [1, 23]; for magic squares with distinct entries see [31, 189]).

3. Another famous kind of square is *latin squares* (see, for example, [72]). Here each row and column has n different numbers, the same n numbers in every row/column (usually taken to be the first n positive integers). There are counting problems associated with latin squares, which can be attacked using Ehrhart theory [31] (see also [165, Sequence A002860]).

4. Recent work includes mathematical-historical research, such as the discovery of unpublished magic squares of Benjamin Franklin [2, 141]. Aside from mathematical research, magic squares and their siblings naturally continue to be an excellent source of topics for popular mathematics books (see, for example, [4] or [144]).

5. The Birkhoff–von Neumann polytope $\mathcal{B}_n$ possesses fascinating combinatorial properties [38, 48, 49, 58, 192] and relates to many mathematical areas [74, 112]. Its name honors Garrett Birkhoff and John von Neumann, who proved that the extremal points of $\mathcal{B}_n$ are the permutation matrices [39, 186] (see Exercise 6.5). A long-standing open problem is the determination of the relative volume of $\mathcal{B}_n$, which is known only for $n \leq 10$ [165, Sequence A037302]. In fact, the last two records ($n = 9$ and 10) for computing $\operatorname{vol} \mathcal{B}_n$ rely on the theory of counting functions that is introduced in this book, more precisely, Theorem 6.6 [28].

[5] For more information about MacMahon, see
`http://www-groups.dcs.st-and.ac.uk/~history/Biographies/MacMahon.html`.

6. An important generalization of the Birkhoff–von Neumann polytope is the *transporation polytopes*, which consist of *contingency tables*. They have applications to statistics and in particular to disclosure limitation procedures [68]. The Birkhoff–von Neumann polytopes are special transportation polytopes that consist of two-way contingency tables with given 1-marginals.

7. The polynomiality of H_n (Theorem 6.2) and its symmetry (Theorem 6.3) were conjectured in 1966 by Harsh Anand, Vishwa Dumir, and Hansraj Gupta [3] and proved seven years later independently by Eugène Ehrhart [80] and Richard Stanley [168]. Stanley also conjectured that the numerator coefficients in Corollary 6.4 are unimodal, a fact that was proved only in 2005, by Christos Athanasiadis [8]. The quasipolynomiality of M_n (Theorem 6.7) and its degree are discussed in [23]. The period of M_n is in general not known. In [23] it is conjectured that it is always nontrivial for $n > 1$. The work in [1] gives some credence to this conjecture by proving that the polytope of magic $n \times n$ squares is *not* integral for $n \geq 2$.

8. We close with a story about Cornelius Agrippa's *De Occulta Philosophia*, written in 1510. In it he describes the spiritual powers of magic squares and produces some squares of orders from three up to nine. His work, although influential in the mathematical community, enjoyed only brief success, for the Counter-Reformation and the witch hunts of the Inquisition began soon thereafter: Agrippa himself was accused of being allied with the devil.

Exercises

6.1. ♣ Find and prove a formula for $H_n(1)$.

6.2. Let $(x_{ij})_{1 \leq i,j \leq 3}$ be a magic 3×3 square.

(a) Show that the center term x_{22} is the average over all x_{ij}.
(b) Show that $M_3(t) = 0$ if 3 does not divide t.

6.3. ♣ Prove that $\dim \mathcal{B}_n = (n-1)^2$.

6.4. Prove the following characterization of a vertex of a convex polytope $\mathcal{P}$: A point $\mathbf{v} \in \mathcal{P}$ is a vertex of $\mathcal{P}$ if for any line L through $\mathbf{v}$ and any neighborhood N of $\mathbf{v}$ there exists a point in $L \cap N$ that is not in $\mathcal{P}$.

6.5. ♣ Prove that the vertices of $\mathcal{B}_n$ are the $n \times n$-permutation matrices.

6.6. ♣ Let $H_n^\circ(t)$ denote the number of $n \times n$ matrices with *positive* integer entries summing up to t along each row and column. Show that $H_n^\circ(t) = H_n(t-n)$ for $t > n$.

6.7. ♣ Show that $H_n^\circ(1) = H_n^\circ(2) = \cdots = H_n^\circ(n-1) = 0$.

6.8. ♣ Prove (6.6):

$$H_n(t) = \text{const}_{z_1,\dots,z_n} \left((z_1 \cdots z_n)^{-t} \left(\text{const}_w \frac{1}{(1 - z_1 w) \cdots (1 - z_n w) w^t} \right)^n \right).$$

6.9. ♣ Compute the partial fraction coefficients (6.8).

6.10. ♣ Verify (6.11).

6.11. Repeat the constant-term computation of H_2 starting from (6.9), but now by first computing the constant term with respect to z_2, and after that with respect to z_1.

6.12. Use your favorite computer program to calculate the formula for $H_3(t)$, $H_4(t), \dots$ by interpolation.

6.13. Compute H_3 using Theorem 6.6.

6.14. Compute H_4 using Theorem 6.6.

6.15. Show that

$$\sum_{k=1}^{n} \frac{z_k^{t+n-1}}{\prod_{j \neq k} (z_k - z_j)} = \sum_{m_1 + \cdots + m_n = t} z_1^{m_1} \cdots z_n^{m_n},$$

and use this identity to give an alternative proof of Theorem 6.6.

6.16. ♣ Prove that for $n \geq 3$, the degree of M_n is $n^2 - 2n - 1$.

6.17. Compute the vertices of the polytope of 3×3 magic squares.

6.18. ♣ Verify (6.12) and use it to compute M_3.

6.19. Compute M_3 by interpolation. (*Hint:* Use Exercises 6.2 and 6.17.)

6.20. A **symmetric semimagic square** is a semimagic square that is a symmetric matrix. Show that the number of *symmetric* semimagic $n \times n$ squares with line sum t is a quasipolynomial in t. Determine its degree and period.

Open Problems

6.21. Compute the number of *traditional* magic $n \times n$ squares for $n > 5$.

6.22. Compute $\text{vol}\,\mathcal{B}_n$ for $n > 10$. Compute H_n for $n > 9$.

6.23. Prove that the period of M_n is nontrivial for $n > 1$.

6.24. The vertices of the Birkhoff–von Neumann polytope are in one-to-one correspondence with the elements of the symmetric group S_n. Consider a subgroup of S_n and take the convex hull of the corresponding permutation matrices. Compute the Ehrhart polynomials of this polytope. (The face numbers of the polytope corresponding to the subgroup A_n, the even permutations, were studied in [102].)

6.25. Prove that the graph formed by the vertices and edges of any 2-way transportation polytope is Hamiltonian.

Part II

Beyond the Basics

7

Finite Fourier Analysis

God created infinity, and man, unable to understand infinity, created finite sets.

Gian-Carlo Rota (1932–1999)

We now consider the vector space of all complex-valued periodic functions on the integers with period b. It turns out that every such function $a(n)$ on the integers can be written as a polynomial in the bth root of unity $\xi^n := e^{2\pi i n/b}$. Such a representation for $a(n)$ is called a **finite Fourier series.** Here we develop the finite Fourier theory using rational functions and their partial fraction decomposition. We then define the Fourier transform and the convolution of finite Fourier series, and show how one can use these ideas to prove identities on trigonometric functions, as well as find connections to the classical Dedekind sums.

The more we know about roots of unity and their various sums, the deeper are the results that we can prove (see Exercise 7.19); in fact, certain statements about sums of roots of unity even imply the Riemann hypothesis! However, this chapter is elementary and draws connections to the sawtooth functions and Dedekind sums, two basic sums over roots of unity. The general philosophy here is that finite sums of rational functions of roots of unity are basic ingredients in many mathematical structures.

7.1 A Motivating Example

To ease the reader into the general theory, let's work out the finite Fourier series for a simple example first, an arithmetic function with a period of 3.

Example 7.1. Consider the following arithmetic function, of period 3:

$$n : 0, 1, 2, 3, 4, 5, \ldots$$
$$a(n) : 1, 5, 2, 1, 5, 2, \ldots$$

We first embed this sequence into a generating function as follows:

$$F(z) := 1 + 5z + 2z^2 + z^3 + 5z^4 + 2z^5 + \cdots = \sum_{n \geq 0} a(n)\, z^n .$$

Since the sequence is periodic, we can simplify $F(z)$ using a geometric series argument:

$$\begin{aligned} F(z) &= \sum_{n \geq 0} a(n)\, z^n \\ &= 1 + 5z + 2z^2 + z^3 \left(1 + 5z + 2z^2\right) + z^6 \left(1 + 5z + 2z^2\right) + \cdots \\ &= \left(1 + 5z + 2z^2\right) \sum_{k \geq 0} z^{3k} \\ &= \frac{1 + 5z + 2z^2}{1 - z^3} . \end{aligned}$$

We now use the same technique that was employed in Chapter 1, namely the technique of expanding a rational function into its partial fraction decomposition. Here all the poles are simple, and located at the three cube roots of unity, so that

$$F(z) = \frac{\hat{a}(0)}{1 - z} + \frac{\hat{a}(1)}{1 - \rho z} + \frac{\hat{a}(2)}{1 - \rho^2 z} , \tag{7.1}$$

where the constants $\hat{a}(0), \hat{a}(1), \hat{a}(2)$ remain to be found, and where $\rho := e^{2\pi i/3}$, a third root of unity. Using the geometric series for each of these terms separately, we arrive at

$$F(z) = \sum_{n \geq 0} \left(\hat{a}(0) + \hat{a}(1)\rho^n + \hat{a}(2)\rho^{2n}\right) z^n ,$$

so that we've derived the finite Fourier series of our sequence $a(n)$! The only remaining piece of information that we need is the computation of the constants $\hat{a}(j)$, for $j = 0, 1, 2$. It turns out that this is also quite easy to do. We have, from (7.1) above, the identity

$$\begin{aligned} &\hat{a}(0) (1 - \rho z) \left(1 - \rho^2 z\right) + \hat{a}(1) (1 - z) \left(1 - \rho^2 z\right) + \hat{a}(2) (1 - z) (1 - \rho z) \\ &\quad = 1 + 5z + 2z^2 , \end{aligned}$$

valid for all $z \in \mathbb{C}$. Upon letting $z = 1$, ρ^2, and ρ, respectively, we obtain

$$\begin{aligned} 3\, \hat{a}(0) &= 1 + 5 + 2 , \\ 3\, \hat{a}(1) &= 1 + 5\rho^2 + 2\rho^4 , \\ 3\, \hat{a}(2) &= 1 + 5\rho + 2\rho^2 , \end{aligned}$$

where we've used the identity $(1 - \rho)(1 - \rho^2) = 3$ (see Exercise 7.2). We can simplify a bit to get $\hat{a}(0) = \frac{8}{3}$, $\hat{a}(1) = \frac{-4-3\rho}{3}$, and $\hat{a}(2) = \frac{-1+3\rho}{3}$. Thus the finite Fourier series for our sequence is

$$a(n) = \frac{8}{3} + \left(-\frac{4}{3} - \rho\right)\rho^n + \left(-\frac{1}{3} + \rho\right)\rho^{2n}. \qquad \square$$

The object of the next section is to show that this simple process follows just as easily for any periodic function on $\mathbb{Z}$. The ensuing sections contain some applications of the finite Fourier series of periodic functions.

7.2 Finite Fourier Series for Periodic Functions on $\mathbb{Z}$

The general theory is just as easy conceptually as the example above, and we now develop it. Consider any periodic sequence on $\mathbb{Z}$, defined by $\{a(n)\}_{n=0}^{\infty}$, of period b. Throughout the chapter, we fix the bth root of unity $\xi := e^{2\pi i/b}$. As before, we embed our periodic sequence $\{a(n)\}_{n=0}^{\infty}$ into a generating function,

$$F(z) := \sum_{n \geq 0} a(n)\, z^n,$$

and use the periodicity of the sequence to immediately get

$$\begin{aligned} F(z) &= \left(\sum_{k=0}^{b-1} a(k)\, z^k\right) + \left(\sum_{k=0}^{b-1} a(k)\, z^k\right) z^b + \left(\sum_{k=0}^{b-1} a(k)\, z^k\right) z^{2b} + \cdots \\ &= \frac{\sum_{k=0}^{b-1} a(k)\, z^k}{1 - z^b} = \frac{P(z)}{1 - z^b}, \end{aligned}$$

where the last step simply defines the polynomial $P(z) = \sum_{k=0}^{b-1} a(k)\, z^k$. Now we expand the rational generating function $F(z)$ into partial fractions, as before:

$$F(z) = \frac{P(z)}{1 - z^b} = \sum_{m=0}^{b-1} \frac{\hat{a}(m)}{1 - \xi^m z}.$$

As in the example of the previous section, we expand each of the terms $\frac{1}{1-\xi^m z}$ as a geometric series, and substitute into the sum above to get

$$\begin{aligned} F(z) = \sum_{n \geq 0} a(n)\, z^n &= \sum_{m=0}^{b-1} \frac{\hat{a}(m)}{1 - \xi^m z} \\ &= \sum_{m=0}^{b-1} \hat{a}(m) \sum_{n \geq 0} \xi^{mn} z^n = \sum_{n \geq 0} \left(\sum_{m=0}^{b-1} \hat{a}(m)\, \xi^{mn}\right) z^n. \end{aligned}$$

Comparing the coefficients of any fixed z^n gives us the finite Fourier series for $a(n)$, namely

$$a(n) = \sum_{m=0}^{b-1} \hat{a}(m)\, \xi^{mn}.$$

We now find a formula for the Fourier coefficients $\hat{a}(n)$, as in the example. To recapitulate, we have

$$P(z) = \sum_{m=0}^{b-1} \hat{a}(m) \frac{1 - z^b}{1 - \xi^m z} = \sum_{m=0}^{b-1} \hat{a}(m) \prod_{1 \le k \le b, k \ne m} \left(1 - \xi^k z\right),$$

where we have used the factorization $1 - z^b = \prod_{k=1}^{b} (1 - \xi^k z)$ of Exercise 7.1. To solve for $P(\xi^{-n})$, we note that

$$\lim_{z \to \xi^{-n}} \frac{1 - z^b}{1 - \xi^m z} = 0 \qquad \text{if } m - n \not\equiv 0 \bmod b,$$

and

$$\lim_{z \to \xi^{-n}} \frac{1 - z^b}{1 - \xi^m z} = \lim_{z \to \xi^{-n}} \frac{b z^{b-1}}{\xi^m} = b\, \xi^{n-m} = b \qquad \text{if } m - n \equiv 0 \bmod b.$$

Thus $P(\xi^{-n}) = b\, \hat{a}(n)$ and so

$$\hat{a}(n) = \frac{1}{b} P(\xi^{-n}) = \frac{1}{b} \sum_{k=0}^{b-1} a(k)\, \xi^{-nk}.$$

We have just proved the main result of finite Fourier series, using only elementary properties of rational functions:

Theorem 7.2 (Finite Fourier series expansion and Fourier inversion). *Let $a(n)$ be any periodic function on $\mathbb{Z}$, with period b. Then we have the following finite Fourier series expansion:*

$$a(n) = \sum_{k=0}^{b-1} \hat{a}(k)\, \xi^{nk},$$

where the Fourier coefficients are

$$\hat{a}(n) = \frac{1}{b} \sum_{k=0}^{b-1} a(k)\, \xi^{-nk}, \tag{7.2}$$

with $\xi = e^{2\pi i/b}$. □

The coefficients $\hat{a}(m)$ are known as the **Fourier coefficients** of the function $a(n)$, and if $\hat{a}(m) \neq 0$ we sometimes say that the function **has frequency** m. The finite Fourier series of a periodic function provides us with surprising power and insight into its structure. We are able to analyze the function using its frequencies (only finitely many), and this window into the frequency domain becomes indispensable for computations and simplifications.

We note that the Fourier coefficients $\hat{a}(n)$ and the original sequence elements $a(n)$ are related by a linear transformation given by the matrix

$$L = \left(\xi^{(i-1)(j-1)} \right), \tag{7.3}$$

where $1 \le i, j \le b$, as is evident from (7.2) in the proof above. We further note that the second half of the proof, namely solving for the Fourier coefficients $\hat{a}(n)$, is just tantamount to inverting this matrix L.

One of the main building blocks of our lattice-point enumeration formulas in polytopes is the **sawtooth function**, defined by

$$((x)) := \begin{cases} \{x\} - \frac{1}{2} & \text{if } x \notin \mathbb{Z}, \\ 0 & \text{if } x \in \mathbb{Z}. \end{cases} \tag{7.4}$$

(As a reminder, $\{x\} = x - \lfloor x \rfloor$ is the fractional part of x.) The graph of this function is displayed in Figure 7.1. We have seen a closely related function before, in Chapter 1, in our study of the coin-exchange problem. Equation (1.8) gave us the finite Fourier series for essentially this function from the discrete-geometry perspective of the coin-exchange problem; however, we now compute the finite Fourier series for this periodic function directly, pretending that we do not know about its other life as a counting function.

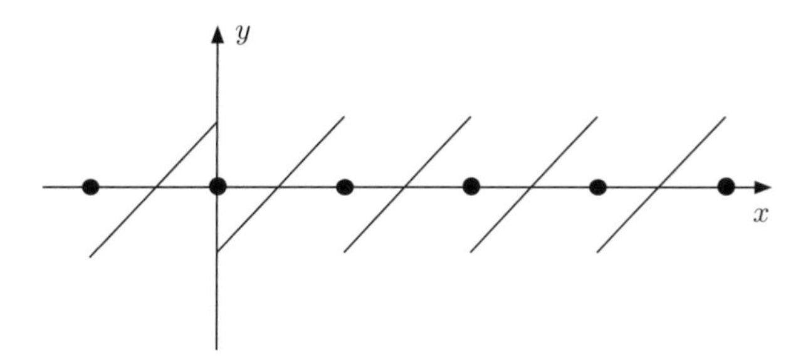

Fig. 7.1. The sawtooth function $y = ((x))$.

Lemma 7.3. *The finite Fourier series for the discrete sawtooth function $\left(\left(\frac{a}{b}\right)\right)$, a periodic function of $a \in \mathbb{Z}$ with period b, is given by*

$$\left(\left(\frac{a}{b}\right)\right) = \frac{1}{2b} \sum_{k=1}^{b-1} \frac{1+\xi^k}{1-\xi^k} \, \xi^{ak} = \frac{i}{2b} \sum_{k=1}^{b-1} \cot \frac{\pi k}{b} \, \xi^{ak}.$$

Here the second equality follows from $\frac{1+e^{2\pi i x}}{1-e^{2\pi i x}} = i \cot(\pi x)$, by the definition of the cotangent.

Proof. Using Theorem 7.2, we know that our periodic function has a finite Fourier series $\left(\left(\frac{a}{b}\right)\right) = \sum_{k=0}^{b-1} \hat{a}(k)\, \xi^{ak}$, where

$$\hat{a}(k) = \frac{1}{b} \sum_{m=0}^{b-1} \left(\left(\frac{m}{b}\right)\right) \xi^{-mk}.$$

We first compute $\hat{a}(0) = \frac{1}{b} \sum_{m=0}^{b-1} \left(\left(\frac{m}{b}\right)\right) = 0$, by Exercise 7.14. For $k \neq 0$, we have

$$\begin{aligned}\hat{a}(k) &= \frac{1}{b} \sum_{m=1}^{b-1} \left(\frac{m}{b} - \frac{1}{2}\right) \xi^{-mk} = \frac{1}{b^2} \sum_{m=1}^{b-1} m\, \xi^{-mk} + \frac{1}{2b} \\ &= \frac{1}{b} \left(\frac{\xi^k}{1-\xi^k} + \frac{1}{2}\right) = \frac{1}{2b} \frac{1+\xi^k}{1-\xi^k},\end{aligned}$$

where we used Exercise 7.5 in the penultimate equality above. □

We define the **Dedekind sum** by

$$s(a,b) = \sum_{k=0}^{b-1} \left(\left(\frac{ka}{b}\right)\right) \left(\left(\frac{k}{b}\right)\right),$$

for any two relatively prime integers a and $b > 0$. Note that the Dedekind sum is a periodic function of the variable a, with period b, by the periodicity of the sawtooth function. That is,

$$s(a + jb, b) = s(a, b) \qquad \text{for all } \; j \in \mathbb{Z}. \tag{7.5}$$

Using the finite Fourier series for the sawtooth function, we can now easily reformulate the Dedekind sums as a finite sum over the bth roots of unity or cotangents:

Lemma 7.4.

$$s(a,b) = \frac{1}{4b} \sum_{\mu=1}^{b-1} \frac{1+\xi^\mu}{1-\xi^\mu} \frac{1+\xi^{-\mu a}}{1-\xi^{-\mu a}} = \frac{1}{4b} \sum_{\mu=1}^{b-1} \cot\frac{\pi\mu}{b} \cot\frac{\pi\mu a}{b}.$$

Proof.

$$\begin{aligned}s(a,b) &= \sum_{k=0}^{b-1} \left(\left(\frac{ka}{b}\right)\right) \left(\left(\frac{k}{b}\right)\right) \\ &= \frac{1}{4b^2} \sum_{k=0}^{b-1} \left(\left(\sum_{\mu=1}^{b-1} \frac{1+\xi^\mu}{1-\xi^\mu}\, \xi^{\mu k a}\right) \left(\sum_{\nu=1}^{b-1} \frac{1+\xi^\nu}{1-\xi^\nu}\, \xi^{\nu k}\right)\right) \\ &= \frac{1}{4b^2} \sum_{\mu=1}^{b-1} \sum_{\nu=1}^{b-1} \frac{1+\xi^\mu}{1-\xi^\mu} \frac{1+\xi^\nu}{1-\xi^\nu} \left(\sum_{k=0}^{b-1} \xi^{k(\nu+\mu a)}\right).\end{aligned}$$

We note that the last sum $\sum_{k=0}^{b-1} \xi^{k(\nu+\mu a)}$ vanishes, unless $\nu \equiv -\mu a \bmod b$ (Exercise 7.6), in which case the sum equals b, and we obtain

$$s(a,b) = \frac{1}{4b} \sum_{\mu=1}^{b-1} \frac{1+\xi^{\mu}}{1-\xi^{\mu}} \frac{1+\xi^{-\mu a}}{1-\xi^{-\mu a}} .$$

Rewriting the right-hand side in terms of cotangents gives

$$s(a,b) = \frac{i^2}{4b} \sum_{\mu=1}^{b-1} \cot \frac{\pi \mu}{b} \cot \frac{-\pi \mu a}{b} = \frac{1}{4b} \sum_{\mu=1}^{b-1} \cot \frac{\pi \mu}{b} \cot \frac{\pi \mu a}{b} ,$$

because the cotangent is an odd function. □

7.3 The Finite Fourier Transform and Its Properties

Given a periodic function f on $\mathbb{Z}$, we have seen that f possesses a finite Fourier series, with the finite collection of Fourier coefficients that we called $\hat{f}(0), \hat{f}(1), \dots, \hat{f}(b-1)$. We now regard f as a function on the finite set $G = \{0, 1, 2, \dots, b-1\}$, and let V_G be the vector space of all complex-valued functions on G. Equivalently, V_G is the vector space of all complex-valued, *periodic functions on* $\mathbb{Z}$ with period b.

We define the **Fourier transform** of f, denoted by $\mathbf{F}(f)$, to be the periodic function on $\mathbb{Z}$ defined by the sequence of uniquely determined values

$$\hat{f}(0),\ \hat{f}(1),\ \dots,\ \hat{f}(b-1) .$$

Thus

$$\mathbf{F}(f)(m) = \hat{f}(m) .$$

Theorem 7.2 above gave us these coefficients as a linear combination of the values $f(k)$, with $k = 0, 1, 2, \dots, b-1$. Thus $\mathbf{F}(f)$ is a linear transformation of the function f, thought of as a vector in V_G. In other words, we've shown that $\mathbf{F}(f)$ is a one-to-one and onto linear transformation of V_G.

The vector space V_G is a vector space of dimension b; indeed, an explicit basis can easily be given for V_G using the "delta functions" (see Exercise 7.7) defined by

$$\delta_m(x) := \begin{cases} 1 & \text{if } x = m + kb, \text{ for some integer } k, \\ 0 & \text{otherwise.} \end{cases}$$

In other words, $\delta_m(x)$ is the periodic function on $\mathbb{Z}$ that picks out the arithmetic progression $\{m + kb : \ k \in \mathbb{Z}\}$.

But there is another natural basis for V_G. For any fixed integer a the roots of unity $\{\mathbf{e}_a(x) := e^{2\pi i a x/b} : \ x \in \mathbb{Z}\}$ can be thought of as a single function

$\mathbf{e}_a(x) \in V_G$ because of its periodicity on $\mathbb{Z}$. As we saw in Theorem 7.2, the functions $\{\mathbf{e}_1(x), \dots, \mathbf{e}_b(x)\}$ give a basis for the vector space of functions V_G. A natural question now arises: how are the two bases related to each other? An initial observation is

$$\widehat{\delta_a}(n) = \frac{1}{b}\, e^{-2\pi i a n/b},$$

which simply follows from the computation

$$\widehat{\delta_a}(n) = \frac{1}{b}\sum_{k=0}^{b-1} \delta_a(k)\, \xi^{-kn} = \frac{1}{b}\,\xi^{-an} = \frac{1}{b}\, e^{-2\pi i a n/b}.$$

So to get from the first basis to the second basis, we need precisely the finite Fourier transform!

It is extremely useful to define the following **inner product** on this vector space:

$$\langle f, g\rangle = \sum_{k=0}^{b-1} f(k)\,\overline{g(k)}\,, \tag{7.6}$$

for any two functions $f, g \in V_G$. Here the bar denotes complex conjugation. The following elementary properties show that $\langle f, g\rangle$ is an inner product (see Exercise 7.8):

1. $\langle f, f\rangle \geq 0$, with equality if and only if $f = 0$, the zero function.
2. $\langle f, g\rangle = \overline{\langle g, f\rangle}$.

Equipped with this inner product, V_G can now be regarded as a metric space. We can now measure *distances* between any two functions, and in particular between any two basis elements $\mathbf{e}_a(x) := e^{2\pi i a x/b}$ and $\mathbf{e}_c(x) := e^{2\pi i c x/b}$. Any positive definite inner product gives rise to the distance function $d(f, g) = \langle f - g, f - g\rangle$.

Lemma 7.5 (Orthogonality relations).

$$\frac{1}{b}\langle \mathbf{e}_a, \mathbf{e}_c\rangle = \delta_a(c) = \begin{cases} 1 & \text{if } b \mid (a-c), \\ 0 & \text{otherwise.} \end{cases}$$

Proof. We compute the inner product

$$\langle \mathbf{e}_a, \mathbf{e}_c\rangle = \sum_{m=0}^{b-1} \mathbf{e}_a(m)\,\overline{\mathbf{e}_c(m)} = \sum_{m=0}^{b-1} e^{2\pi i (a-c)m/b}.$$

If $b \mid (a-c)$, then each term equals 1 in the latter sum, and hence the sum equals b. This verifies the first case of the lemma.

If $b \nmid (a-c)$, then $\mathbf{e}_{a-c}(m) = e^{2\pi i m(a-c)/b}$ is a nontrivial root of unity, and we have the finite geometric series

$$\sum_{m=0}^{b-1} e^{\frac{2\pi i(a-c)m}{b}} = \frac{e^{b\frac{2\pi i m(a-c)}{b}} - 1}{e^{\frac{2\pi i m(a-c)}{b}} - 1} = 0\,,$$

verifying the second case of the lemma. □

Example 7.6. We recall the sawtooth function again, since it is one of the building blocks of lattice point enumeration, and compute its Fourier transform. Namely, we define

$$B(k) := \left(\left(\frac{k}{b}\right)\right) = \begin{cases} \left\{\frac{k}{b}\right\} - \frac{1}{2} & \text{if } \frac{k}{b} \notin \mathbb{Z}, \\ 0 & \text{if } \frac{k}{b} \in \mathbb{Z}, \end{cases}$$

a periodic function on the integers with period b. What is its finite Fourier transform? We have already seen the answer, in the course of the proof of Lemma 7.3:

$$\widehat{B}(n) = \frac{1}{2b}\frac{1+\xi^n}{1-\xi^n} = \frac{i}{2b}\cot\frac{\pi n}{b}$$

for $n \neq 0$, and $\widehat{B}(0) = 0$. As always, $\xi = e^{2\pi i/b}$. □

In the next section we delve more deeply into the behavior of this inner product, where the Parseval identity is proved.

7.4 The Parseval Identity

A nontrivial property of the inner product defined above is the following identity, linking the "norm of a function" to the "norm of its Fourier transform." It is known as the Parseval identity, and also goes by the name of the Plancherel theorem.

Theorem 7.7 (Parseval identity). *For all $f \in V_G$,*

$$\langle f, f \rangle = b \,\langle \hat{f}, \hat{f} \rangle\,.$$

Proof. Using the definition $\mathbf{e}_m(x) = \xi^{mx}$ and the relation

$$\hat{f}(x) = \frac{1}{b}\sum_{m=0}^{b-1} f(m)\,\overline{\mathbf{e}_m(x)}$$

from Theorem 7.2, we have

$$\begin{aligned}
\langle \hat{f}, \hat{f} \rangle &= \left\langle \frac{1}{b} \sum_{m=0}^{b-1} f(m) \, \overline{\mathbf{e}_m} \, , \, \frac{1}{b} \sum_{n=0}^{b-1} f(n) \, \overline{\mathbf{e}_n} \right\rangle \\
&= \frac{1}{b^2} \sum_{k=0}^{b-1} \sum_{m=0}^{b-1} f(m) \, \overline{\mathbf{e}_m(k)} \sum_{n=0}^{b-1} \overline{f(n)} \, \mathbf{e}_n(k) \\
&= \frac{1}{b^2} \sum_{m=0}^{b-1} \sum_{n=0}^{b-1} f(m) \, \overline{f(n)} \, \langle \mathbf{e}_m, \mathbf{e}_n \rangle \\
&= \frac{1}{b} \sum_{m=0}^{b-1} \sum_{n=0}^{b-1} f(m) \, \overline{f(n)} \, \delta_m(n) \\
&= \frac{1}{b} \langle f, f \rangle \, ,
\end{aligned}$$

where the essential step in the proof was using the orthogonality relations (Lemma 7.5) in the fourth equality above. □

A basically identical proof yields the following stronger result, showing that the "distance between any two functions" is essentially equal to the "distance between their Fourier transforms."

Theorem 7.8. *For all $f, g \in V_G$, we have*

$$\langle f, g \rangle = b \, \langle \hat{f}, \hat{g} \rangle \, . \qquad \square$$

Example 7.9. A nice application of the generalized Parseval identity above now gives us Lemma 7.4 very quickly, the reformulation of the Dedekind sum as a sum over roots of unity. Namely, we first fix an integer a relatively prime to b and define $f(k) = \left(\left(\frac{k}{b}\right)\right)$, and $g(k) = \left(\left(\frac{ka}{b}\right)\right)$. Then, using Example 7.6, we have $\hat{f}(n) = \frac{i}{2b} \cot \frac{\pi n}{b}$. To find the Fourier transform of g we need an extra twist. Since

$$\left(\left(\frac{ka}{b}\right)\right) = \frac{i}{2b} \sum_{m=1}^{b-1} \cot \frac{\pi m}{b} \, \xi^{mka} ,$$

we can multiply each index m by a^{-1}, the multiplicative inverse of a modulo b (recall that we require a and b to be relatively prime for the reformulation of the Dedekind sum). Since a^{-1} is relatively prime to b, this multiplication just permutes $m = 1, 2, \ldots, b-1$ modulo b, but the sum stays invariant (see Exercise 1.9):

$$\sum_{m=1}^{b-1} \cot \frac{\pi m}{b} \, \xi^{mka} = \sum_{m=1}^{b-1} \cot \frac{\pi m a^{-1}}{b} \, \xi^{m a^{-1} k a} = \sum_{m=1}^{b-1} \cot \frac{\pi m a^{-1}}{b} \, \xi^{mk} ,$$

that is,

$$\hat{g}(n) = \frac{i}{2b} \sum_{m=1}^{b-1} \cot \frac{\pi m a^{-1}}{b} \, \xi^{mk} .$$

Hence Theorem 7.8 immediately gives us the reformulation of the Dedekind sum:

$$\begin{aligned}
s(a,b) &:= \sum_{k=0}^{b-1} \left(\left(\frac{k}{b}\right)\right)\left(\left(\frac{ka}{b}\right)\right) \\
&= b \sum_{m=1}^{b-1} \left(\frac{i}{2b}\cot\frac{\pi m}{b}\right)\overline{\left(\frac{i}{2b}\cot\frac{\pi m a^{-1}}{b}\right)} \\
&= \frac{1}{4b}\sum_{m=1}^{b-1} \cot\frac{\pi m}{b}\cot\frac{\pi m a^{-1}}{b} \\
&= \frac{1}{4b}\sum_{m=1}^{b-1} \cot\frac{\pi m a}{b}\cot\frac{\pi m}{b}\,.
\end{aligned}$$

For the last equality, we again used the trick of replacing m by ma. □

7.5 The Convolution of Finite Fourier Series

Another basic tool in finite Fourier analysis is the convolution of two finite Fourier series. Namely, let $f(t) = \frac{1}{b}\sum_{k=0}^{b-1} a_k\, \xi^{kt}$ and $g(t) = \frac{1}{b}\sum_{k=0}^{b-1} c_k \xi^{kt}$, where $\xi = e^{2\pi i/b}$. We define the **convolution** of f and g by

$$(f * g)(t) = \sum_{m=0}^{b-1} f(t-m)g(m)\,.$$

Indeed, it is this convolution tool (the proof of the convolution theorem below is almost trivial!) that is responsible for the fastest known algorithm for multiplying two polynomials of degree b in $O(b\log(b))$ steps (see the Notes at the end of this chapter).

Theorem 7.10 (Convolution theorem for finite Fourier series). *Let $f(t) = \frac{1}{b}\sum_{k=0}^{b-1} a_k\, \xi^{kt}$ and $g(t) = \frac{1}{b}\sum_{k=0}^{b-1} c_k\, \xi^{kt}$, where $\xi = e^{2\pi i/b}$. Then their convolution satisfies*

$$(f * g)(t) = \frac{1}{b}\sum_{k=0}^{b-1} a_k c_k\, \xi^{kt}.$$

Proof. The proof is straightforward: we just compute the left-hand side, and obtain

$$\begin{aligned}\sum_{m=0}^{b-1} f(t-m)g(m) &= \frac{1}{b^2}\sum_{m=0}^{b-1}\left(\sum_{k=0}^{b-1} a_k\,\xi^{k(t-m)}\right)\left(\sum_{l=0}^{b-1} c_l\,\xi^{lm}\right)\\ &= \frac{1}{b^2}\sum_{k=0}^{b-1}\sum_{l=0}^{b-1} a_k c_l\left(\sum_{m=0}^{b-1}\xi^{kt+(l-k)m}\right)\\ &= \frac{1}{b}\sum_{k=0}^{b-1} a_k c_k\,\xi^{kt},\end{aligned}$$

because the sum $\sum_{m=0}^{b-1}\xi^{(l-k)m}$ vanishes, unless $l=k$ (see Exercise 7.6). In the case that $l=k$, we have $\sum_{m=0}^{b-1}\xi^{(l-k)m}=b$. □

It is an easy exercise (Exercise 7.22) to show that this convolution theorem is equivalent to the following statement:

$$\mathbf{F}(f*g) = b\,\mathbf{F}(f)\mathbf{F}(g)\,.$$

Note that the proof of Theorem 7.10 is essentially identical with the proof of Lemma 7.4 above; we could have proved the lemma, in fact, by applying the convolution theorem. We now show how Theorem 7.10 can be used to derive identities on trigonometric functions.

Example 7.11. We claim that

$$\sum_{k=1}^{b-1}\cot^2\left(\frac{\pi k}{b}\right) = \frac{(b-1)(b-2)}{3}\,.$$

The sum suggests the use of the convolution theorem, with a function whose Fourier coefficients are $a_k = c_k = \cot\frac{\pi k}{b}$. But we already know such a function! It is just the sawtooth function $\frac{2b}{i}\left(\left(\frac{m}{b}\right)\right)$. Therefore

$$-\frac{1}{4b}\sum_{k=1}^{b-1}\cot^2\left(\frac{\pi k}{b}\right)\xi^{kt} = \sum_{m=1}^{b-1}\left(\left(\frac{t-m}{b}\right)\right)\left(\left(\frac{m}{b}\right)\right),$$

where the equality follows from Theorem 7.10. On setting $t=0$, we obtain

$$\begin{aligned}\sum_{m=1}^{b-1}\left(\left(\frac{-m}{b}\right)\right)\left(\left(\frac{m}{b}\right)\right) &= -\sum_{m=1}^{b-1}\left(\left(\frac{m}{b}\right)\right)\left(\left(\frac{m}{b}\right)\right)\\ &= -\frac{1}{b^2}\sum_{m=1}^{b-1} m^2 + \frac{1}{b}\sum_{m=1}^{b-1} m - \frac{1}{4}(b-1)\\ &= -\frac{(b-1)(b-2)}{12b},\end{aligned}$$

as desired. We used the identity $\left(\left(\frac{-m}{b}\right)\right) = -\left(\left(\frac{m}{b}\right)\right)$ in the first equality above, and some algebra was used in the last equality. Notice, moreover, that the convolution theorem gave us more than we asked for, namely an identity for every value of t. □

Notes

1. Finite Fourier analysis offers a wealth of applications and is, for example, one of the main tools in quantum information theory. For the reader interested in going further than the humble beginnings outlined in this chapter, we heartily recommend Audrey Terras's monograph [180].

2. The Dedekind sum is our main motivation for studying finite Fourier series, and in fact, Chapter 8 is devoted to a detailed investigation of these sums, in which the Fourier–Dedekind sums of Chapter 1 also finally reappear.

3. The reader may consult [116, p. 501] for a proof that two polynomials of degree N can be multiplied in $O(N \log N)$ steps. The actual proof of this fact runs along the following conceptual lines. First, let the two given polynomials of degree N be $f(x) = \sum_{n=0}^{N} a(n) x^n$ and $g(x) = \sum_{n=0}^{N} b(n) x^n$. Then we know that $f(\xi)$ and $g(\xi)$ are two finite Fourier series, and we abbreviate them by f and g, respectively. We now note that $fg = \mathbf{F}(\mathbf{F}^{-1}(f) * \mathbf{F}^{-1}(g))$. If we can compute the finite Fourier transform (and its inverse) quickly, then this argument shows that we can multiply polynomials quickly. It is a fact of life that we *can* compute the Fourier transform of a periodic function of period N in $O(N \log N)$ steps, by an algorithm known as the *fast Fourier transform* (again, see [116] for a complete description).

4. The continuous Fourier transform, defined by $\int_{-\infty}^{\infty} f(t) e^{-2\pi i t x} dt$, can be related to the finite Fourier transform in the following way. We approximate the continuous integral by discretizing a large interval $[0, a]$. Precisely, we let $\Delta := \frac{a}{b}$, and we let $t_k := k\Delta = \frac{ka}{b}$. Then

$$\int_0^a f(t) e^{-2\pi i t x} dt \approx \sum_{k=1}^{b} f(t_k) e^{-2\pi i t_k x} t_k \,,$$

a finite Fourier series for the function $f\left(\frac{a}{b} x\right)$ as a function of $x \in \mathbb{Z}$. Hence finite Fourier series find an application to continuous Fourier analysis as an approximation tool.

Exercises

Throughout the exercises, we fix an integer $b > 1$ and let $\xi = e^{2\pi i/b}$.

7.1. ♣ Show that $1 - x^b = \prod_{k=1}^{b} (1 - \xi^k x)$.

7.2. ♣ Show that $\prod_{k=1}^{b-1} (1 - \xi^k) = b$.

7.3. Consider the matrix that came up in the proof of Theorem 7.2, namely $L = (a_{ij})$, with $a_{ij} := \xi^{(i-1)(j-1)}$ and with $1 \le i, j \le b$. Show that the matrix $\frac{1}{\sqrt{b}} L$ is a unitary matrix (recall that a matrix U is unitary if $U^* U = I$, where U^* is the conjugate transpose of U). Thus, this exercise shows that the Fourier transform of a periodic function is always given by a unitary transformation.

7.4. Show that $\left| \det \left(\frac{1}{\sqrt{b}} L \right) \right| = 1$, where $|z|$ denotes the norm of the complex number z. (It turns out the $\det(L)$ can sometimes be a complex number, but we will not use this fact here.)

7.5. ♣ For any integer a relatively prime to b, show that

$$\frac{1}{b} \sum_{k=1}^{b-1} k \, \xi^{-ak} = \frac{\xi^a}{1 - \xi^a} \, .$$

7.6. ♣ Let n be an integer. Show that the sum $\sum_{k=0}^{b-1} \xi^{kn}$ vanishes, unless $n \equiv 0 \pmod{b}$, in which case it is equal to b.

7.7. ♣ For an integer m, define the delta function $\delta_m(x)$ by

$$\delta_m(x) = \begin{cases} 1 & \text{if } x = m + ab, \text{ for some integer } a, \\ 0 & \text{otherwise.} \end{cases}$$

The b functions $\delta_1(x), \ldots, \delta_b(x)$ are clearly in the vector space V_G, since they are periodic on $\mathbb{Z}$ with period b. Show that they form a basis for V_G.

7.8. ♣ Prove that for all $f, g \in V_G$:

(a) $\langle f, f \rangle \ge 0$, with equality if and only if $f = 0$, the zero function.
(b) $\langle f, g \rangle = \overline{\langle g, f \rangle}$.

7.9. Show that $\sum_{k=1}^{b-1} \frac{1}{1-\xi^k} = \frac{b-1}{2}$.

7.10. Show that $\langle \delta_a, \delta_c \rangle = \delta_a(c)$.

7.11. Prove that $(f * \delta_a)(x) = f(x - a)$.

7.12. Prove that $\delta_a * \delta_c = \delta_{a+c(\text{mod } b)}$.

7.13. Prove that $\widehat{f(x-a)} = \hat{f}(x) e^{\frac{2\pi i a x}{b}}$.

7.14. ♣ For any real number x, prove that $((x)) = \sum_{k=0}^{b-1} \left(\left(\frac{x+k}{b} \right) \right)$.

7.15. If x is not an integer, show that $\sum_{n=0}^{b-1} \cot \left(\pi \frac{n+x}{b} \right) = b \cot (\pi x)$.

7.16. For any integer a relatively prime to b, show that

$$\sum_{\xi} \frac{\xi^{a+1}-1}{(\xi^a-1)(\xi-1)} = 0\,,$$

where the sum is taken over all b^{th} roots of unity ξ except $\xi = 1$.

7.17. We call a root of unity $e^{2\pi i a/b}$ a **primitive** b^{th} root of unity if a is relatively prime to b. Let $\Phi_b(x)$ denote the polynomial with leading coefficient 1 and of degree $\phi(b)$[1] whose roots are the $\phi(b)$ distinct primitive b^{th} roots of unity. This polynomial is known as the *cyclotomic polynomial of order* b. Show that

$$\prod_{d|b} \Phi_d(x) = x^b - 1\,,$$

where the product is taken over all positive divisors d of b.

7.18. We define the Möbius μ-function for positive integers n by

$$\mu(n) = \begin{cases} 1 & \text{if } n = 1, \\ 0 & \text{if } n \text{ is divisible by a square,} \\ (-1)^k & \text{if } n \text{ is square-free and has } k \text{ prime divisors.} \end{cases}$$

Deduce from the previous exercise that

$$\Phi_b(x) = \prod_{d|b} \left(x^d - 1\right)^{\mu(b/d)}\,.$$

7.19. Prove that for any positive integer b,

$$\sum_{1\le a\le b,(a,b)=1} e^{2\pi i a/b} = \mu(b)\,,$$

the Möbius μ-function.

7.20. Show that for any positive integer k, $s(1,k) = -\frac{1}{4} + \frac{1}{6k} + \frac{k}{12}$.

7.21. Show that $\sum_{k=1}^{b-1} \tan^2\left(\frac{\pi k}{b}\right) = b(b-1)$.

7.22. ♣ Show that Theorem 7.10 is equivalent to the following statement:

$$\mathbf{F}(f * g) = b\,\mathbf{F}(f)\mathbf{F}(g)\,.$$

7.23. Consider the trace of the linear transformation $L = \left(\xi^{(i-1)(j-1)}\right)$, defined in (7.3). The trace of L is $G(b) := \sum_{m=0}^{b-1} \xi^{m^2}$, known as a *Gauß sum.* Show that $|G(b)| = \sqrt{b}$ if b is an odd prime.

---

[1] $\phi(b) := \#\{k \in [1, b-1] : (k,b) = 1\}$ is the Euler ϕ-function.

8

Dedekind Sums, the Building Blocks of Lattice-point Enumeration

If things are nice there is probably a good reason why they are nice: and if you don't know at least one reason for this good fortune, then you still have work to do.

Richard Askey

We've encountered Dedekind sums in our study of finite Fourier analysis and we became intimately acquainted with their siblings in our study of the coin-exchange problem in Chapter 1. They have one shortcoming, however (which we'll remove): the definition of $s(a,b)$ requires us to sum over b terms, which is rather slow when $b = 2^{100}$, for example. Luckily, there is a magical *reciprocity law* for the Dedekind sum $s(a,b)$ that allows us to compute it in roughly $\log_2(b) = 100$ steps. This is the kind of magic that saves the day when we try to enumerate lattice points in integral polytopes of dimensions $d \leq 4$. There is an ongoing effort to extend these ideas to higher dimensions, but there is much room for improvement. In this chapter we focus on the computational-complexity issues that arise when we try to compute Dedekind sums explicitely.

8.1 Fourier–Dedekind Sums and the Coin-Exchange Problem Revisited

Recall from Chapter 1 the Fourier–Dedekind sum (defined in (1.13))

$$s_n\left(a_1, a_2, \ldots, a_d; b\right) = \frac{1}{b} \sum_{k=1}^{b-1} \frac{\xi_b^{kn}}{\left(1 - \xi_b^{ka_1}\right)\left(1 - \xi_b^{ka_2}\right) \cdots \left(1 - \xi_b^{ka_d}\right)},$$

which appeared as a main player in our analysis of the Frobenius coin-exchange problem. We can now recognize the Fourier–Dedekind sums as honest finite Fourier series with period b. The Fourier–Dedekind sums unify many variations of the Dedekind sum that have appeared in the literature, and form the building blocks of Ehrhart quasipolynomials. For example, we showed in Chapter 1 that $s_n(a_1, a_2, \dots, a_d; b)$ appears in the Ehrhart quasipolynomial of the d-simplex

$$\left\{ (x_1, \dots, x_{d+1}) \in \mathbb{R}_{\geq 0}^{d+1} : \, a_1 x_1 + \cdots + a_d x_d + b x_{d+1} = 1 \right\}.$$

Example 8.1. We first notice that when $n = 0$ and $d = 2$, the Fourier–Dedekind sum reduces to a classical Dedekind sum (which—finally—explains the name): for relatively prime positive integers a and b,

$$\begin{aligned}
s_0(a, 1; b) &= \frac{1}{b} \sum_{k=1}^{b-1} \frac{1}{\left(1 - \xi_b^{ka}\right)\left(1 - \xi_b^{k}\right)} \\
&= \frac{1}{b} \sum_{k=1}^{b-1} \left(\frac{1}{1 - \xi_b^{ka}} - \frac{1}{2} \right) \left(\frac{1}{1 - \xi_b^{k}} - \frac{1}{2} \right) \\
&\quad + \frac{1}{2b} \sum_{k=1}^{b-1} \frac{1}{1 - \xi_b^{k}} + \frac{1}{2b} \sum_{k=1}^{b-1} \frac{1}{1 - \xi_b^{ka}} - \frac{1}{b} \sum_{k=1}^{b-1} \frac{1}{4} \\
&= \frac{1}{4b} \sum_{k=1}^{b-1} \left(\frac{1 + \xi_b^{ka}}{1 - \xi_b^{ka}} \right) \left(\frac{1 + \xi_b^{k}}{1 - \xi_b^{k}} \right) + \frac{1}{b} \sum_{k=1}^{b-1} \frac{1}{1 - \xi_b^{k}} - \frac{b-1}{4b} .
\end{aligned}$$

In the last step we used the fact that multiplying the index k by a does not change the middle sum. This middle sum can be further simplified by recalling (1.8):

$$\frac{1}{b} \sum_{k=1}^{b-1} \frac{1}{\left(1 - \xi_b^{k}\right) \xi_b^{kn}} = - \left\{ \frac{n}{b} \right\} + \frac{1}{2} - \frac{1}{2b} ,$$

whence

$$\begin{aligned}
s_0(a, 1; b) &= \frac{1}{4b} \sum_{k=1}^{b-1} \left(\frac{1 + \xi_b^{ka}}{1 - \xi_b^{ka}} \right) \left(\frac{1 + \xi_b^{k}}{1 - \xi_b^{k}} \right) + \frac{1}{2} - \frac{1}{2b} - \frac{b-1}{4b} \\
&= - \frac{1}{4b} \sum_{k=1}^{b-1} \cot \left(\frac{\pi k a}{b} \right) \cot \left(\frac{\pi k}{b} \right) + \frac{b-1}{4b} \qquad (8.1) \\
&= -s(a, b) + \frac{b-1}{4b} .
\end{aligned}$$

□

Example 8.2. The next special evaluation of a Fourier–Dedekind sum is very similar to the computation above, so that we leave it to the reader to prove (Exercise 8.5) that for a_1, a_2 relatively prime to b,

$$s_0\left(a_1, a_2; b\right) = -s\left(a_1 a_2^{-1}, b\right) + \frac{b-1}{4b} , \qquad (8.2)$$

where $a_2^{-1} a_2 \equiv 1 \bmod b$. □

Returning to the general Fourier–Dedekind sum, we now prove the first of a series of *reciprocity laws*: identities for certain sums of Fourier–Dedekind sums. We first recall how these sums came up in Chapter 1, namely, from the partial fraction expansion of the function

$$\begin{aligned} f(z) &= \frac{1}{(1-z^{a_1})\cdots(1-z^{a_d})\,z^n} \\ &= \frac{A_1}{z} + \frac{A_2}{z^2} + \cdots + \frac{A_n}{z^n} + \frac{B_1}{z-1} + \frac{B_2}{(z-1)^2} + \cdots + \frac{B_d}{(z-1)^d} \\ &\quad + \sum_{k=1}^{a_1-1} \frac{C_{1k}}{z-\xi_{a_1}^k} + \sum_{k=1}^{a_2-1} \frac{C_{2k}}{z-\xi_{a_2}^k} + \cdots + \sum_{k=1}^{a_d-1} \frac{C_{dk}}{z-\xi_{a_d}^k} . \end{aligned} \tag{8.3}$$

(Here we assume that $a_1, a_2, \ldots, a_d$ are pairwise relatively prime.) Theorem 1.7 states that with the help of the partial fraction coefficients $B_1, \ldots, B_d$ and Fourier–Dedekind sums, we can compute the restricted partition function for $A = \{a_1, a_2, \ldots, a_d\}$:

$$\begin{aligned} p_A(n) &= -B_1 + B_2 - \cdots + (-1)^d B_d + s_{-n}(a_2, a_3, \ldots, a_d; a_1) \\ &\quad + s_{-n}(a_1, a_3, a_4, \ldots, a_d; a_2) + \cdots + s_{-n}(a_1, a_2, \ldots, a_{d-1}; a_d) . \end{aligned}$$

We note that $B_1, B_2, \ldots, B_d$ are polynomials in n (Exercise 8.6), whence we call

$$\operatorname{poly}_A(n) := -B_1 + B_2 - \cdots + (-1)^d B_d$$

the **polynomial part** of the restricted partition function $p_A(n)$.

Example 8.3. The first few expressions for $\operatorname{poly}_{\{a_1,\ldots,a_d\}}(n)$ are

$$\begin{aligned} \operatorname{poly}_{\{a_1\}}(n) &= \frac{1}{a_1} , \\ \operatorname{poly}_{\{a_1,a_2\}}(n) &= \frac{n}{a_1 a_2} + \frac{1}{2}\left(\frac{1}{a_1} + \frac{1}{a_2}\right) , \\ \operatorname{poly}_{\{a_1,a_2,a_3\}}(n) &= \frac{n^2}{2a_1a_2a_3} + \frac{n}{2}\left(\frac{1}{a_1a_2} + \frac{1}{a_1a_3} + \frac{1}{a_2a_3}\right) \\ &\quad + \frac{1}{12}\left(\frac{3}{a_1} + \frac{3}{a_2} + \frac{3}{a_3} + \frac{a_1}{a_2a_3} + \frac{a_2}{a_1a_3} + \frac{a_3}{a_1a_2}\right) , \end{aligned} \tag{8.4}$$

$$
\begin{aligned}
\operatorname{poly}_{\{a_1,a_2,a_3,a_4\}}(n) = {} & \frac{n^3}{6a_1a_2a_3a_4} \\
& + \frac{n^2}{4}\left(\frac{1}{a_1a_2a_3}+\frac{1}{a_1a_2a_4}+\frac{1}{a_1a_3a_4}+\frac{1}{a_2a_3a_4}\right) \\
& + \frac{n}{4}\left(\frac{1}{a_1a_2}+\frac{1}{a_1a_3}+\frac{1}{a_1a_4}+\frac{1}{a_2a_3}+\frac{1}{a_2a_4}+\frac{1}{a_3a_4}\right) \\
& + \frac{n}{12}\left(\frac{a_1}{a_2a_3a_4}+\frac{a_2}{a_1a_3a_4}+\frac{a_3}{a_1a_2a_4}+\frac{a_4}{a_1a_2a_3}\right) \\
& + \frac{1}{24}\left(\frac{a_1}{a_2a_3}+\frac{a_1}{a_2a_4}+\frac{a_1}{a_3a_4}+\frac{a_2}{a_1a_3}+\frac{a_2}{a_1a_4}+\frac{a_2}{a_3a_4}\right. \\
& \quad \left.+\frac{a_3}{a_1a_2}+\frac{a_3}{a_1a_4}+\frac{a_3}{a_2a_4}+\frac{a_4}{a_1a_2}+\frac{a_4}{a_1a_3}+\frac{a_4}{a_2a_3}\right) \\
& + \frac{1}{8}\left(\frac{1}{a_1}+\frac{1}{a_2}+\frac{1}{a_3}+\frac{1}{a_4}\right). \qquad \square
\end{aligned}
$$

We are about to combine the Ehrhart results of Chapter 3 with the partial fraction expansion of Chapter 1 that gave rise to the Fourier–Dedekind sums.

Theorem 8.4 (Zagier reciprocity). *For any pairwise relatively prime positive integers* $a_1, a_2, \ldots, a_d$,

$$
\begin{aligned}
& s_0(a_2, a_3, \ldots, a_d; a_1) + s_0(a_1, a_3, a_4, \ldots, a_d; a_2) + \cdots \\
& \quad + s_0(a_1, a_2, \ldots, a_{d-1}; a_d) \\
& = 1 - \operatorname{poly}_{\{a_1,a_2,\ldots,a_d\}}(0) .
\end{aligned}
$$

At first sight, this reciprocity law should come as a surprise. The Fourier–Dedekind sums can be complicated, long sums, yet when combined in this fashion, they add up to a trivial rational function in $a_1, a_2, \ldots, a_d$.

Proof. We compute the constant term of the quasipolynomial $p_A(n)$:

$$
\begin{aligned}
p_A(0) = {} & \operatorname{poly}_A(0) + s_0(a_2, a_3, \ldots, a_d; a_1) \\
& + s_0(a_1, a_3, a_4, \ldots, a_d; a_2) + \cdots + s_0(a_1, a_2, \ldots, a_{d-1}; a_d) .
\end{aligned}
$$

On the other hand, Exercise 3.27 (the extension of Corollary 3.15 to Ehrhart quasipolynomials) states that $p_A(0) = 1$, whence

$$
\begin{aligned}
1 = {} & \operatorname{poly}_A(0) + s_0(a_2, a_3, \ldots, a_d; a_1) \\
& + s_0(a_1, a_3, a_4, \ldots, a_d; a_2) + \cdots + s_0(a_1, a_2, \ldots, a_{d-1}; a_d) . \qquad \square
\end{aligned}
$$

8.2 The Dedekind Sum and Its Reciprocity and Computational Complexity

We derived in (8.1) the classical Dedekind sum $s(a,b)$ as a special evaluation of the Fourier–Dedekind sum. Naturally, Theorem 8.4 takes on a particular form when we specialize this reciprocity law to the classical Dedekind sum.

Corollary 8.5 (Dedekind's reciprocity law). *For any relatively prime positive integers a and b,*

$$s(a,b) + s(b,a) = \frac{1}{12}\left(\frac{a}{b} + \frac{b}{a} + \frac{1}{ab}\right) - \frac{1}{4}.$$

Proof. A special case of Theorem 8.4 is

$$\begin{aligned} s_0(a,1;b) + s_0(b,a;1) + s_0(1,b;a) &= 1 - \operatorname{poly}_{\{a,1,b\}}(0) \\ &= 1 - \frac{1}{12}\left(\frac{3}{a} + 3 + \frac{3}{b} + \frac{a}{b} + \frac{1}{ab} + \frac{b}{a}\right) \\ &= \frac{3}{4} - \frac{1}{12}\left(\frac{a}{b} + \frac{b}{a} + \frac{1}{ab}\right) - \frac{1}{4a} - \frac{1}{4b}. \end{aligned}$$

Now we use the fact that $s_0(b,a;1) = 0$ and the identity (8.1):

$$s_0(a,1;b) = -s(a,b) + \frac{1}{4} - \frac{1}{4b}. \qquad \square$$

Dedekind's reciprocity law allows us to compute the Dedekind sum $s(a,b)$ as quickly as the gcd algorithm for a and b. Let's get a better feeling for the way we can compute the Dedekind sum by working out an example. We remind the reader of another crucial property of the Dedekind sums that we already pointed out in (7.5): $s(a,b)$ remains invariant when we replace a by its residue modulo b, that is,

$$s(a,b) = s(a \bmod b, b). \tag{8.5}$$

Example 8.6. Let $a = 100$ and $b = 147$. Now we alternately use Corollary 8.5 and the reduction identity (8.5):

$$
\begin{aligned}
s(100,147) &= \frac{1}{12}\left(\frac{100}{147}+\frac{147}{100}+\frac{1}{14700}\right)-\frac{1}{4}-s(147,100) \\
&= -\frac{1249}{17640}-s(47,100) \\
&= -\frac{1249}{17640}-\left(\frac{1}{12}\left(\frac{47}{100}+\frac{100}{47}+\frac{1}{4700}\right)-\frac{1}{4}-s(100,47)\right) \\
&= -\frac{773}{20727}+s(6,47) \\
&= -\frac{773}{20727}+\frac{1}{12}\left(\frac{6}{47}+\frac{47}{6}+\frac{1}{282}\right)-\frac{1}{4}-s(47,6) \\
&= \frac{166}{441}-s(5,6) \\
&= \frac{166}{441}-\left(\frac{1}{12}\left(\frac{5}{6}+\frac{6}{5}+\frac{1}{30}\right)-\frac{1}{4}-s(6,5)\right) \\
&= \frac{2003}{4410}+s(1,5) \\
&= \frac{2003}{4410}-\frac{1}{4}+\frac{1}{30}+\frac{5}{12} \\
&= \frac{577}{882}.
\end{aligned}
$$

In the last step we used Exercise 7.20: $s(1,k) = -\frac{1}{4}+\frac{1}{6k}+\frac{k}{12}$. A priori, $s(100,147)$ takes 147 steps to compute, whereas we were able to compute it in nine steps using Dedekind's reciprocity law and (8.5). □

As a second corollary to Theorem 8.4, we mention the following three-term reciprocity law for the special Fourier–Dedekind sum $s_0(a,b;c)$. This reciprocity law could be restated in terms of the classical Dedekind sum via the identity (8.2).

Corollary 8.7. *For pairwise relatively prime positive integers a, b, and c,*

$$
s_0(a,b;c)+s_0(c,a;b)+s_0(b,c;a) = 1-\frac{1}{12}\left(\frac{3}{a}+\frac{3}{b}+\frac{3}{c}+\frac{a}{bc}+\frac{b}{ca}+\frac{c}{ab}\right).
$$

□

8.3 Rademacher Reciprocity for the Fourier–Dedekind Sum

The next reciprocity law will be again for the general Fourier–Dedekind sums. It extends Theorem 8.4 beyond $n=0$.

Theorem 8.8 (Rademacher reciprocity). *Let $a_1, a_2, \ldots, a_d$ be pairwise relatively prime positive integers. Then for $n = 1, 2, \ldots, (a_1 + \cdots + a_d - 1)$,*

$$s_n(a_2, a_3, \dots, a_d; a_1) + s_n(a_1, a_3, a_4, \dots, a_d; a_2) + \cdots \\ + s_n(a_1, a_2, \dots, a_{d-1}; a_d) = -\operatorname{poly}_{\{a_1, a_2, \dots, a_d\}}(-n).$$

Proof. We recall the definition

$$p_A^\circ(n) = \#\left\{(m_1, \dots, m_d) \in \mathbb{Z}^d : \text{ all } m_j > 0, \ m_1 a_1 + \cdots + m_d a_d = n\right\}$$

of Exercise 1.31, that is, $p_A^\circ(n)$ counts the number of partitions of n using only the elements of A as parts, *where each part is used at least once.* This counting function is, naturally, connected to p_A through Ehrhart–Macdonald reciprocity (Theorem 4.1):

$$p_A^\circ(n) = (-1)^{d-1} p_A(-n),$$

that is,

$$(-1)^{d-1} p_A^\circ(n) = \operatorname{poly}_A(-n) + s_n(a_2, a_3, \dots, a_d; a_1) \\ + s_n(a_1, a_3, a_4, \dots, a_d; a_2) + \cdots + s_n(a_1, a_2, \dots, a_{d-1}; a_d).$$

On the other hand, by its very definition,

$$p_A^\circ(n) = 0 \qquad \text{for } n = 1, 2, \dots, (a_1 + \cdots + a_d - 1),$$

so that for those n,

$$0 = \operatorname{poly}_A(-n) + s_n(a_2, a_3, \dots, a_d; a_1) \\ + s_n(a_1, a_3, a_4, \dots, a_d; a_2) + \cdots + s_n(a_1, a_2, \dots, a_{d-1}; a_d).$$

□

Just as Zagier reciprocity takes on a special form for the classical Dedekind sum, Rademacher reciprocity specializes for $d = 2$ to a reciprocity identity for the **Dedekind–Rademacher sum**

$$r_n(a, b) := \sum_{k=0}^{b-1} \left(\left(\frac{ka + n}{b}\right)\right) \left(\left(\frac{k}{b}\right)\right).$$

The classical Dedekind sum is, naturally, the specialization $r_0(a, b) = s(a, b)$. To be able to state the reciprocity law for the Dedekind–Rademacher sums, we define the function

$$\chi_a(n) := \begin{cases} 1 & \text{if } a \mid n, \\ 0 & \text{otherwise,} \end{cases}$$

which will come in handy as a book keeping device.

Corollary 8.9 (Reciprocity law for Dedekind–Rademacher sums). *Let a and b be relatively prime positive integers. Then for $n = 1, 2, \dots, a + b$,*

$$\begin{aligned} r_n(a,b) + r_n(b,a) = & \frac{n^2}{2ab} - \frac{n}{2}\left(\frac{1}{ab} + \frac{1}{a} + \frac{1}{b}\right) + \frac{1}{12}\left(\frac{a}{b} + \frac{b}{a} + \frac{1}{ab}\right) \\ & + \frac{1}{2}\left(\left(\left(\frac{a^{-1}n}{b}\right)\right) + \left(\left(\frac{b^{-1}n}{a}\right)\right) + \left(\left(\frac{n}{a}\right)\right) + \left(\left(\frac{n}{b}\right)\right)\right) \\ & + \frac{1}{4}\left(1 + \chi_a(n) + \chi_b(n)\right), \end{aligned}$$

where $a^{-1}a \equiv 1 \bmod b$ *and* $b^{-1}b \equiv 1 \bmod a$.

This identity follows almost instantly once we are able to express the Dedekind–Rademacher sum in terms of Fourier–Dedekind sums.

Lemma 8.10. *Suppose a and b are relatively prime positive integers and $n \in \mathbb{Z}$. Then*

$$r_n(a,b) = -s_n(a,1;b) + \frac{1}{2}\left(\left(\frac{n}{b}\right)\right) + \frac{1}{2}\left(\left(\frac{na^{-1}}{b}\right)\right) - \frac{1}{4b} + \frac{1}{4}\chi_b(n),$$

where $a^{-1}a \equiv 1 \bmod b$.

Proof. We start by rewriting the finite Fourier series (1.8) for the sawtooth function $((x))$:

$$\begin{aligned} \frac{1}{b}\sum_{k=1}^{b-1} \frac{\xi_b^{kn}}{1-\xi_b^k} &= -\left\{\frac{-n}{b}\right\} + \frac{1}{2} - \frac{1}{2b} \\ &= -\left(\left(\frac{-n}{b}\right)\right) + \frac{1}{2}\chi_b(n) - \frac{1}{2b} \\ &= \left(\left(\frac{n}{b}\right)\right) + \frac{1}{2}\chi_b(n) - \frac{1}{2b}. \end{aligned}$$

Hence we also have

$$\begin{aligned} \frac{1}{b}\sum_{k=1}^{b-1} \frac{\xi_b^{kn}}{1-\xi_b^{ka}} = \frac{1}{b}\sum_{k=1}^{b-1} \frac{\xi_b^{ka^{-1}n}}{1-\xi_b^k} &= \left(\left(\frac{a^{-1}n}{b}\right)\right) + \frac{1}{2}\chi_b\left(a^{-1}n\right) - \frac{1}{2b} \\ &= \left(\left(\frac{a^{-1}n}{b}\right)\right) + \frac{1}{2}\chi_b\left(n\right) - \frac{1}{2b}. \end{aligned}$$

Now we use the convolution theorem for finite Fourier series (Theorem 7.10) for the functions

$$f(n) := \frac{1}{b}\sum_{k=1}^{b-1} \frac{\xi_b^{kn}}{1-\xi_b^k} \qquad \text{and} \qquad g(n) := \frac{1}{b}\sum_{k=1}^{b-1} \frac{\xi_b^{kn}}{1-\xi_b^{ka}}.$$

It gives

$$\frac{1}{b}\sum_{k=1}^{b-1} \frac{\xi_b^{kn}}{\left(1-\xi_b^k\right)\left(1-\xi_b^{ka}\right)} = \sum_{m=0}^{b-1} f(n-m)\, g(m) =$$

$$\sum_{m=0}^{b-1} \left(\left(\left(\frac{n-m}{b} \right)\right) + \frac{1}{2}\chi_b(n-m) - \frac{1}{2b} \right) \left(\left(\left(\frac{a^{-1}m}{b} \right)\right) + \frac{1}{2}\chi_b(m) - \frac{1}{2b} \right).$$

We invite the reader to check (Exercise 8.9) that the sum on the right-hand side simplifies to

$$-\sum_{m=0}^{b-1} \left(\left(\frac{am+n}{b} \right)\right) \left(\left(\frac{m}{b} \right)\right) + \frac{1}{2}\left(\left(\frac{a^{-1}n}{b} \right)\right) + \frac{1}{2}\left(\left(\frac{n}{b} \right)\right) - \frac{1}{4b} + \frac{1}{4}\chi_b(n),$$

whence

$$s_n(a,1;b) = -r_n(a,b) + \frac{1}{2}\left(\left(\frac{a^{-1}n}{b} \right)\right) + \frac{1}{2}\left(\left(\frac{n}{b} \right)\right) - \frac{1}{4b} + \frac{1}{4}\chi_b(n). \qquad \square$$

Proof of Corollary 8.9. We use the special case of Theorem 8.8

$$\begin{aligned} s_n(a,1;b) + s_n(1,a;b) + s_n(a,b;1) &= -\operatorname{poly}_{\{a,1,b\}}(-n) \\ &= -\frac{n^2}{2ab} + \frac{n}{2}\left(\frac{1}{ab} + \frac{1}{a} + \frac{1}{b} \right) - \frac{1}{12}\left(\frac{3}{a} + \frac{3}{b} + 3 + \frac{a}{b} + \frac{b}{a} + \frac{1}{ab} \right), \end{aligned}$$

which holds for $n = 1, 2, \dots, a+b$. Lemma 8.10 allows us to translate this identity into one for Dedekind–Rademacher sums:

$$\begin{aligned} r_n(a,b) + r_n(b,a) =& \frac{n^2}{2ab} - \frac{n}{2}\left(\frac{1}{ab} + \frac{1}{a} + \frac{1}{b} \right) + \frac{1}{12}\left(\frac{a}{b} + \frac{b}{a} + \frac{1}{ab} \right) \\ &+ \frac{1}{2}\left(\left(\left(\frac{a^{-1}n}{b} \right)\right) + \left(\left(\frac{b^{-1}n}{a} \right)\right) + \left(\left(\frac{n}{a} \right)\right) + \left(\left(\frac{n}{b} \right)\right) \right) \\ &+ \frac{1}{4}\left(1 + \chi_a(n) + \chi_b(n)\right). \qquad \square \end{aligned}$$

The two-term reciprocity law allows us to compute the Dedekind–Rademacher sum as quickly as the gcd algorithm, just as obtained for the classical Dedekind sum. This fact has an interesting consequence: In Theorem 2.10 and Exercise 2.34 we showed implicitly (see Exercise 8.10) that Dedekind–Rademacher sums are the only nontrivial ingredients of the Ehrhart quasipolynomials of rational polygons. Corollary 8.9 ensures that these Ehrhart quasipolynomials can be computed almost instantly.

8.4 The Mordell–Pommersheim Tetrahedron

In this section we return to Ehrhart polynomials and illustrate how Dedekind sums appear naturally in generating-function computations. We will study the tetrahedron that historically first gave rise to the connection of Dedekind sums and lattice-point enumeration in polytopes. It is given by

$$\mathcal{P} = \left\{ (x, y, z) \in \mathbb{R}^3 : \, x, y, z \geq 0, \ \frac{x}{a} + \frac{y}{b} + \frac{z}{c} \leq 1 \right\}, \tag{8.6}$$

a tetrahedron with vertices $(0,0,0)$, $(a,0,0)$, $(0,b,0)$, and $(0,0,c)$, where a, b, c are positive integers. We insert the slack variable n and interpret

$$\begin{aligned} L_{\mathcal{P}}(t) &= \# \left\{ (k, l, m) \in \mathbb{Z}^3 : \, k, l, m \geq 0, \ \frac{k}{a} + \frac{l}{b} + \frac{m}{c} \leq t \right\} \\ &= \# \left\{ (k, l, m, n) \in \mathbb{Z}^4 : \, k, l, m, n \geq 0, \ bck + acl + abm + n = abct \right\} \end{aligned}$$

as the Taylor coefficient of z^{abct} for the function

$$\begin{aligned} &\left(\sum_{k \geq 0} z^{bck} \right) \left(\sum_{l \geq 0} z^{acl} \right) \left(\sum_{m \geq 0} z^{abm} \right) \left(\sum_{n \geq 0} z^{n} \right) \\ &= \frac{1}{(1 - z^{bc})(1 - z^{ac})(1 - z^{ab})(1 - z)} \, . \end{aligned}$$

As we have done numerous times before, we shift this coefficient to the constant term:

$$L_{\mathcal{P}}(t) = \operatorname{const} \left(\frac{1}{(1 - z^{bc})(1 - z^{ac})(1 - z^{ab})(1 - z) \, z^{abct}} \right) .$$

To reduce the number of poles, it is convenient to change this function slightly; the constant term of $1/\left(1 - z^{bc}\right)\left(1 - z^{ac}\right)\left(1 - z^{ab}\right)(1 - z)$ is 1, so that

$$L_{\mathcal{P}}(t) = \operatorname{const} \left(\frac{z^{-abct} - 1}{(1 - z^{bc})(1 - z^{ac})(1 - z^{ab})(1 - z)} \right) + 1 \, .$$

This trick becomes useful in the next step, namely expanding the function into partial fractions. Strictly speaking, we cannot do that, since the numerator is not a polynomial in z. However, we can think of this rational function as a sum of two functions. The higher-order poles of both summands that we will not include in our computation below cancel each other, so we can ignore them at this stage. The only poles of

$$\frac{z^{-abct} - 1}{(1 - z^{bc})(1 - z^{ac})(1 - z^{ab})(1 - z)} \tag{8.7}$$

are at the ath, b^{th}, c^{th} roots of unity and at 0. (As before, we don't have to bother with the coefficients of $z = 0$ of the partial fraction expansion.) To make life momentarily easier (the general case is the subject of Exercise 8.12), let's assume that a, b, and c are pairwise relatively prime; then all the poles besides 0 and 1 are simple. The computation of the coefficients for $z = 1$ is very similar to what we did with the restricted partition function in Chapter 1. The coefficient in the partial fraction expansion of a nontrivial root of unity, say ξ_a^k, is also computed practically as easily as in earlier examples: it is

$$-\frac{t}{a\left(1-\xi_a^{kbc}\right)\left(1-\xi_a^k\right)} \tag{8.8}$$

(see Exercise 8.11). Summing this fraction over $k = 1, 2, \dots, a-1$ gives rise to the Fourier–Dedekind sum

$$-\frac{t}{a}\sum_{k=1}^{a-1}\frac{1}{\left(1-\xi_a^{kbc}\right)\left(1-\xi_a^k\right)} = -t\, s_0\left(bc, 1; a\right).$$

Putting this coefficient and its siblings for the other roots of unity into the partial fraction expansion and computing the constant term yields (Exercise 8.11)

$$\begin{aligned} L_{\mathcal{P}}(t) = {} & \frac{abc}{6}\,t^3 + \frac{ab+ac+bc+1}{4}\,t^2 \\ & + \left(\frac{a+b+c}{4} + \frac{1}{4}\left(\frac{1}{a}+\frac{1}{b}+\frac{1}{c}\right) + \frac{1}{12}\left(\frac{bc}{a}+\frac{ca}{b}+\frac{ab}{c}+\frac{1}{abc}\right)\right)t \\ & + \left(s_0\left(bc,1;a\right) + s_0\left(ca,1;b\right) + s_0\left(ab,1;c\right)\right)t \\ & + 1\,. \end{aligned}$$

We recognize instantly that the Fourier–Dedekind sums in this Ehrhart polynomial are in fact classical Dedekind sums by (8.1), and so we arrive at the following celebrated result.

Theorem 8.11. *Let $\mathcal{P}$ be given by* (8.6) *with a, b, and c pairwise relatively prime. Then*

$$\begin{aligned} L_{\mathcal{P}}(t) = {} & \frac{abc}{6}\,t^3 + \frac{ab+ac+bc+1}{4}\,t^2 + \left(\frac{3}{4} + \frac{a+b+c}{4}\right. \\ & \left. + \frac{1}{12}\left(\frac{bc}{a}+\frac{ca}{b}+\frac{ab}{c}+\frac{1}{abc}\right) - s\left(bc,a\right) - s\left(ca,b\right) - s\left(ab,c\right)\right)t + 1\,. \end{aligned}$$

□

We finish this chapter by giving the Ehrhart series of the Mordell–Pommersheim tetrahedron $\mathcal{P}$. It follows simply from the transformation formulas (computing the Ehrhart numerator coefficients from the Ehrhart polynomial coefficients) of Corollary 3.16 and Exercise 3.10, and hence the Ehrhart series of $\mathcal{P}$ naturally contains Dedekind sums.

Corollary 8.12. *Let $\mathcal{P}$ be given by* (8.6) *with a, b, and c pairwise relatively prime. Then*

$$\operatorname{Ehr}_{\mathcal{P}}(z) = \frac{h_3\,z^3 + h_2\,z^2 + h_1\,z + 1}{(1-z)^4}\,,$$

where

$$\begin{aligned}
h_3 &= \frac{abc}{6} - \frac{ab+ac+bc+a+b+c}{4} - \frac{1}{2} + \frac{1}{12}\left(\frac{bc}{a} + \frac{ca}{b} + \frac{ab}{c} + \frac{1}{abc}\right) \\
&\quad - s(bc,a) - s(ca,b) - s(ab,c) \\
h_2 &= \frac{2abc}{3} + \frac{a+b+c}{2} + \frac{3}{2} + \frac{1}{6}\left(\frac{bc}{a} + \frac{ca}{b} + \frac{ab}{c} + \frac{1}{abc}\right) \\
&\quad - 2\left(s(bc,a) + s(ca,b) + s(ab,c)\right) \\
h_1 &= \frac{abc}{6} + \frac{ab+ac+bc+a+b+c}{4} - 2 + \frac{1}{12}\left(\frac{bc}{a} + \frac{ca}{b} + \frac{ab}{c} + \frac{1}{abc}\right) \\
&\quad - s(bc,a) - s(ca,b) - s(ab,c). \qquad \square
\end{aligned}$$

It is a curious fact that the above expressions for h_1, h_2, and h_3 are nonnegative integers due to Corollary 3.11.

Notes

1. The classical Dedekind sums came to life in the 1880s when Richard Dedekind (1831–1916)[1] studied the transformation properties of the *Dedekind η-function* [70]

$$\eta(z) := e^{\pi i z/12} \prod_{n \geq 1} \left(1 - e^{2\pi i n z}\right),$$

a useful computational gadget in the land of modular forms in number theory. Dedekind's reciprocity law (Corollary 8.5) follows from one of the functional transformation identities for η. Dedekind also proved that

$$12k\, s(h,k) \equiv k + 1 - 2\left(\frac{h}{k}\right) \pmod 8,$$

establishing a beautiful connection between the Dedekind sum and the Jacobi symbol $\left(\frac{h}{k}\right)$ (the reader may want to consult the lovely Carus monograph entitled *Dedekind Sums*, by Emil Grosswald and Hans Rademacher, where the above result is proved [151, p. 34]), and then used this identity to show that the reciprocity law for the Dedekind sums (for which [151] contains several different proofs) is equivalent to the reciprocity law for the Jacobi symbol.

2. The Dedekind sums and their generalizations appear in various contexts besides analytic number theory and discrete geometry. Other mathematical areas in which Dedekind sums show up include topology [101, 131, 190], algebraic number theory [129, 166], and algebraic geometry [85]. They also have connections to algorithmic complexity [114] and continued fractions [10, 99, 138].

[1] For more information about Dedekind, see
`http://www-groups.dcs.st-and.ac.uk/~history/Biographies/Dedekind.html`.

3. The reciprocity laws (Theorems 8.4 and 8.8) for the Fourier–Dedekind sums were proved in [25]. Theorem 8.4 is equivalent to the reciprocity law for Don Zagier's *higher-dimensional Dedekind sums* [190]. Corollary 8.7 (stated in terms of the classical Dedekind sum) is originally due to Hans Rademacher [149]. Theorem 8.8 generalizes reciprocity laws by Rademacher [150] (essentially Corollary 8.9) and Ira Gessel [87].

4. The Fourier–Dedekind sums form only one set of generalizations of the classical Dedekind sums. A long, but by no means complete, list of other generalizations is [5, 6, 21, 34, 35, 36, 56, 77, 78, 87, 92, 93, 113, 129, 131, 130, 150, 179, 190].

5. The connection of Dedekind sums and lattice-point problems, namely Theorem 8.11 for $t = 1$, was first established by Louis Mordell in 1951 [136]. Some 42 years later, James Pommersheim established a proof of Theorem 8.11 as part of a much more general machinery [146]. In fact, Pommersheim's work implies that the classical Dedekind sum is the only nontrivial ingredient one needs for Ehrhart polynomials in dimensions three and four.

6. We touched the question of efficient computability of Ehrhart (quasi-)polynomials in this chapter. Unfortunately, our current state of knowledge on generalized Dedekind sums does not suffice to make any general statement. However, Alexander Barvinok proved in 1994 [15] that in fixed dimension, the rational generating function of the Ehrhart quasipolynomial of a rational polytope can be efficiently computed. Barvinok's proof did not employ Dedekind sums but rather used a decomposition theorem of Brion, which is the subject of Chapter 9.

Exercises

8.1. Show that $s(a,b) = 0$ if and only if $a^2 \equiv -1 \bmod b$.

8.2. Prove that $6b\, s(a,b) \in \mathbb{Z}$. (*Hint:* Start with rewriting the definition of the Dedekind sum as $6b\, s(a,b) = \frac{6a}{b} \sum_{k=1}^{b-1} k^2 - 6 \sum_{k=1}^{b-1} k \left\lfloor \frac{ka}{b} \right\rfloor - 3 \sum_{k=1}^{b-1} k$.)

8.3. Let a and b be any two relatively prime positive integers. Show that the reciprocity law for the Dedekind sums implies that for $b \equiv r \bmod a$,

$$12ab\, s(a,b) = -12ab\, s(r,a) + a^2 + b^2 - 3ab + 1 \, .$$

Deduce the following identities:

(a) For $b \equiv 1 \bmod a$,

$$12ab\, s(a,b) = -a^2 b + b^2 + a^2 - 2b + 1 \, .$$

(b) For $b \equiv 2 \mod a$,

$$12ab\, s(a,b) = -\frac{1}{2}a^2b + a^2 + b^2 - \frac{5}{2}b + 1\,.$$

(c) For $b \equiv -1 \mod a$,

$$12ab\, s(a,b) = a^2b + a^2 + b^2 - 6ab + 2b + 1\,.$$

8.4. Denote by f_n the sequence of Fibonacci numbers, defined by

$$f_1 = f_2 = 1 \qquad \text{and} \qquad f_{n+2} = f_{n+1} + f_n \quad \text{for } n \geq 1\,.$$

Prove that

$$s\left(f_{2k}, f_{2k+1}\right) = 0$$

and

$$12 f_{2k-1} f_{2k}\, s\left(f_{2k-1}, f_{2k}\right) = f_{2k-1}^2 + f_{2k}^2 - 3 f_{2k-1} f_{2k} + 1\,.$$

8.5. ♣ Prove (8.2):

$$s_0(a_1, a_2; b) = -s\left(a_1 a_2^{-1}, b\right) + \frac{b-1}{4b}\,,$$

where $a_2^{-1} a_2 \equiv 1 \mod b$.

8.6. Prove that $B_1, B_2, \ldots, B_d$ in the partial fraction expansion (8.3),

$$\begin{aligned} f(z) &= \frac{1}{\left(1 - z^{a_1}\right) \cdots \left(1 - z^{a_d}\right) z^n} \\ &= \frac{A_1}{z} + \frac{A_2}{z^2} + \cdots + \frac{A_n}{z^n} + \frac{B_1}{z-1} + \frac{B_2}{(z-1)^2} + \cdots + \frac{B_d}{(z-1)^d} \\ &\quad + \sum_{k=1}^{a_1 - 1} \frac{C_{1k}}{z - \xi_{a_1}^k} + \sum_{k=1}^{a_2 - 1} \frac{C_{2k}}{z - \xi_{a_2}^k} + \cdots + \sum_{k=1}^{a_d - 1} \frac{C_{dk}}{z - \xi_{a_d}^k}\,, \end{aligned}$$

are polynomials in n (of degree less than d) and rational functions in $a_1, \ldots, a_d$.

8.7. ♣ Verify the first few expressions for $\operatorname{poly}_{\{a_1, \ldots, a_d\}}(n)$ in (8.4).

8.8. Show that the Dedekind–Rademacher sum satisfies $r_{-n}(a,b) = r_n(a,b)$.

8.9. ♣ Show that

$$\begin{aligned} &\sum_{m=0}^{b-1} \left(\left(\left(\frac{n-m}{b} \right) \right) + \frac{1}{2} \chi_b(n-m) - \frac{1}{2b} \right) \left(\left(\left(\frac{a^{-1} m}{b} \right) \right) + \frac{1}{2} \chi_b(m) - \frac{1}{2b} \right) \\ &\quad = \sum_{m=0}^{b-1} \left(\left(\frac{am - n}{b} \right) \right) \left(\left(\frac{m}{b} \right) \right) + \frac{1}{2} \left(\left(\frac{a^{-1} n}{b} \right) \right) + \frac{1}{2} \left(\left(\frac{-n}{b} \right) \right) - \frac{1}{4b} \\ &\qquad + \frac{1}{4} \chi_b(n)\,. \end{aligned}$$

8.10. Rephrase the Ehrhart quasipolynomials for rational triangles given in Theorem 2.10 and Exercise 2.34 in terms of Dedekind–Rademacher sums.

8.11. ♣ Prove Theorem 8.11 by verifying (8.8) and computing the coefficients for $z = 1$ in the partial fraction expansion of (8.7).

8.12. Generalize the Ehrhart polynomial of the Mordell–Pommersheim tetrahedron to the case that a, b, and c are not necessarily pairwise relatively prime.

8.13. Compute the Ehrhart polynomial of the 4-simplex

$$\left\{ (x_1, x_2, x_3, x_4) \in \mathbb{R}^4_{\geq 0} : \frac{x_1}{a} + \frac{x_2}{b} + \frac{x_3}{c} + \frac{x_4}{d} \leq 1 \right\},$$

where a, b, c, d are pairwise relatively prime positive integers. (*Hint:* You may use Corollary 5.5 to compute the linear term.)

Open Problems

8.14. Find new relations between various Dedekind sums.

8.15. It is known [21] that the Fourier–Dedekind sums are efficiently computable. Find a fast algorithm that can be implemented in practice.

8.16. For any fixed integers b and k, find a nice characterization for the set of all $a \in \mathbb{Z}$ such that $s(a, b) = k$.

9

The Decomposition of a Polytope into Its Cones

Mathematics compares the most diverse phenomena and discovers the secret analogies that unite them.

Jean Baptiste Joseph Fourier (1768–1830)

In this chapter, we return to integer-point transforms of rational cones and polytopes and connect them in a magical way that was first discovered by Michel Brion. The power of Brion's theorem has been applied to various domains, such as Barvinok's algorithm in integer linear programming, and to higher-dimensional Euler–Maclaurin summation formulas, which we study in Chapter 10. In a sense, Brion's theorem is the natural extension of the familiar finite geometric series identity $\sum_{m=a}^{b} z^m = \frac{z^{b+1}-z^a}{z-1}$ to higher dimensions.

9.1 The Identity "$\sum_{m\in\mathbb{Z}} z^m = 0$"... ...or "Much Ado About Nothing"

We start gently by illustrating Brion's theorem in dimension one. To this end, let's consider the line segment $\mathcal{I} := [20, 34]$. We recall that its integer-point transform lists the lattice points in $\mathcal{I}$ in the form of monomials:

$$\sigma_{\mathcal{I}}(z) = \sum_{m\in\mathcal{I}\cap\mathbb{Z}} z^m = z^{20} + z^{21} + \cdots + z^{34}.$$

Already in this simple example, we are too lazy to list all integers in $\mathcal{I}$ and use $\cdots$ to write the polynomial $\sigma_{\mathcal{I}}$. Is there a more compact way to write $\sigma_{\mathcal{I}}$? The reader might have guessed it even before we asked the question: this integer-point transform equals the rational function

$$\sigma_{\mathcal{I}}(z) = \frac{z^{20} - z^{35}}{1 - z}.$$

This last sentence is not quite correct: the definition of $\sigma_{\mathcal{I}}(z)$ yielded a polynomial in z, whereas the rational function above is not defined at $z = 1$. We can overcome this deficiency by noticing that the limit of this rational function as $z \to 1$ equals the evaluation of the polynomials $\sigma_{\mathcal{I}}(1) = 15$, by L'Hôpital's rule. Notice that the rational-function representation of $\sigma_{\mathcal{I}}$ has the unquestionable advantage of being much more compact than the original polynomial representation. The reader who is not convinced of this advantage should replace the right vertex 34 of $\mathcal{I}$ by 3400.

Now let's rewrite the rational form of the integer-point transform of $\mathcal{I}$ slightly:

$$\sigma_{\mathcal{I}}(z) = \frac{z^{20} - z^{35}}{1 - z} = \frac{z^{20}}{1 - z} + \frac{z^{34}}{1 - \frac{1}{z}} . \tag{9.1}$$

There is a natural geometric interpretation of the two summands on the right-hand side. The first term represents the integer-point transform of the interval $[20, \infty)$:

$$\sigma_{[20,\infty)}(z) = \sum_{m \geq 20} z^m = \frac{z^{20}}{1 - z} .$$

The second term in (9.1) corresponds to the integer-point transform of the interval $(-\infty, 34]$:

$$\sigma_{(-\infty,34]}(z) = \sum_{m \leq 34} z^m = \frac{z^{34}}{1 - \frac{1}{z}} .$$

So (9.1) says that on a rational-function level,

$$\sigma_{[20,\infty)}(z) + \sigma_{(-\infty,34]}(z) = \sigma_{[20,34]}(z) . \tag{9.2}$$

This identity, which we illustrate graphically in Figure 9.1, should come

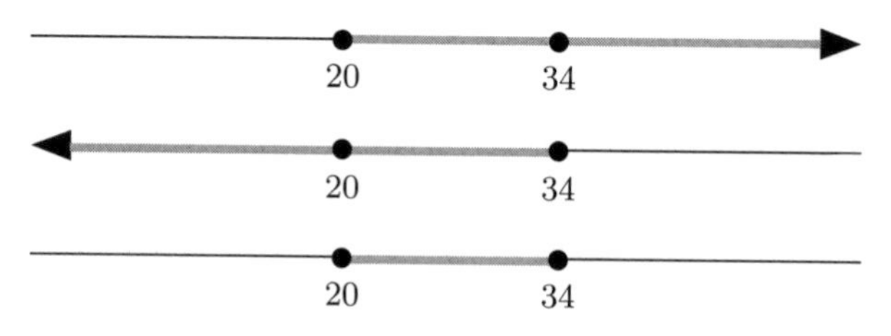

Fig. 9.1. Decomposing a line segment into two infinite rays.

as a mild surprise. Two rational functions that represent infinite sequences somehow collapse, when being summed up, to a polynomial with a finite number of terms. We emphasize that (9.2) does not make sense on the level of infinite series; in fact, the two infinite series involved here have disjoint regions of convergence.

Even more magical is the geometry behind this identity: on the right-hand side we have a polynomial that lists the integer points in a *finite* interval $\mathcal{P}$, while on the left-hand side each of the rational generating functions represents the integer points in an *infinite* ray that begins at a vertex of $\mathcal{P}$. The two half-lines will be called *vertex cones* below, and indeed the remainder of this chapter is devoted to proving that an identity similar to (9.2) holds in general dimension.

We now extend the definition of integer-point transforms $\sigma_{\mathcal{A}}(\mathbf{z})$ from the case of cones $\mathcal{A}$ to the case of affine spaces $\mathcal{A}$. Any such affine space $\mathcal{A} \subseteq \mathbb{R}^d$ equals $\mathbf{w} + \mathcal{V}$ for some $\mathbf{w} \in \mathbb{R}^d$ and some n-dimensional vector subspace $\mathcal{V} \subseteq \mathbb{R}^d$, and if $\mathcal{A}$ contains integer points (which is the only interesting case for our purposes), we may choose $\mathbf{w} \in \mathbb{Z}^d$. The integer points $\mathcal{V} \cap \mathbb{Z}^d$ in $\mathcal{V}$ form a vector space, and hence there exists a basis $\mathbf{v}_1, \mathbf{v}_2, \dots, \mathbf{v}_n$ for $\mathcal{V} \cap \mathbb{Z}^d$. This implies that any integer point $\mathbf{m} \in \mathcal{A} \cap \mathbb{Z}^d$ can be uniquely written as

$$\mathbf{m} = \mathbf{w} + k_1\mathbf{v}_1 + k_2\mathbf{v}_2 + \cdots + k_n\mathbf{v}_n \qquad \text{for some} \qquad k_1, k_2, \dots, k_n \in \mathbb{Z}\,.$$

Using this fixed lattice basis for $\mathcal{V}$, we define the **skewed orthants** of $\mathcal{A}$ as the sets of the form $\{\mathbf{w} + \lambda_1\mathbf{v}_1 + \lambda_2\mathbf{v}_2 + \cdots + \lambda_n\mathbf{v}_n\}$, where for each $1 \le j \le n$, we require either $\lambda_j \ge 0$ or $\lambda_j < 0$. So there are 2^n such skewed orthants, and their disjoint union equals $\mathcal{A}$. We denote them by $\mathcal{O}_1, \mathcal{O}_2, \dots, \mathcal{O}_{2^n}$. All of them are (half-open) pointed cones, and so their integer-point transforms are rational.

Lemma 9.1. *Suppose $\mathcal{A}$ is an n-dimensional affine space with skewed orthants $\mathcal{O}_1, \mathcal{O}_2, \dots, \mathcal{O}_{2^n}$. Then as rational functions,*

$$\sigma_{\mathcal{O}_1}(\mathbf{z}) + \sigma_{\mathcal{O}_2}(\mathbf{z}) + \cdots + \sigma_{\mathcal{O}_{2^n}}(\mathbf{z}) = 0\,.$$

Proof. Suppose

$$\mathcal{A} = \{\mathbf{w} + \lambda_1\mathbf{v}_1 + \lambda_2\mathbf{v}_2 + \cdots + \lambda_n\mathbf{v}_n : \lambda_1, \lambda_2, \dots, \lambda_n \in \mathbb{R}\}\,.$$

Then a typical skewed orthant $\mathcal{O}$ looks like

$$\mathcal{O} = \{\mathbf{w} + \lambda_1\mathbf{v}_1 + \lambda_2\mathbf{v}_2 + \cdots + \lambda_n\mathbf{v}_n : \lambda_1, \dots, \lambda_k \ge 0,\ \lambda_{k+1}, \dots, \lambda_n < 0\}\,,$$

and its integer-point transform is

$$\begin{aligned}
&\sigma_{\mathcal{O}}(\mathbf{z}) \\
&= \mathbf{z}^{\mathbf{w}} \left(\sum_{j_1 \ge 0} \mathbf{z}^{j_1\mathbf{v}_1}\right) \cdots \left(\sum_{j_k \ge 0} \mathbf{z}^{j_k\mathbf{v}_k}\right) \left(\sum_{j_{k+1} < 0} \mathbf{z}^{j_{k+1}\mathbf{v}_{k+1}}\right) \cdots \left(\sum_{j_n < 0} \mathbf{z}^{j_n\mathbf{v}_n}\right) \\
&= \mathbf{z}^{\mathbf{w}} \frac{1}{1 - \mathbf{z}^{\mathbf{v}_1}} \cdots \frac{1}{1 - \mathbf{z}^{\mathbf{v}_k}} \frac{1}{\mathbf{z}^{\mathbf{v}_{k+1}} - 1} \cdots \frac{1}{\mathbf{z}^{\mathbf{v}_n} - 1}\,.
\end{aligned}$$

Now consider the skewed orthant $\mathcal{O}'$ with the same conditions on the λ's as in $\mathcal{O}$ except that we switch $\lambda_1 \ge 0$ to $\lambda_1 < 0$. Then the integer-point transform of $\mathcal{O}'$ is

$$\sigma_{\mathcal{O}'}(\mathbf{z}) = \mathbf{z}^{\mathbf{w}} \frac{1}{\mathbf{z}^{\mathbf{v}_1} - 1} \frac{1}{1 - \mathbf{z}^{\mathbf{v}_2}} \cdots \frac{1}{1 - \mathbf{z}^{\mathbf{v}_k}} \frac{1}{\mathbf{z}^{\mathbf{v}_{k+1}} - 1} \cdots \frac{1}{\mathbf{z}^{\mathbf{v}_n} - 1},$$

so that $\sigma_{\mathcal{O}}(\mathbf{z}) + \sigma_{\mathcal{O}'}(\mathbf{z}) = 0$. Since we can pair up all skewed orthants in this fashion, the sum of all their rational generating functions is zero. □

Since $\mathcal{O}_1 \cup \mathcal{O}_2 \cup \cdots \cup \mathcal{O}_{2^n}$ is equal to $\mathcal{A}$ as a disjoint union, it now makes sense to set

$$\sigma_{\mathcal{A}}(\mathbf{z}) := 0 \tag{9.3}$$

for any n-dimensional affine space with $n > 0$. Lemma 9.1 says that this definition is not as arbitrary as it might seem, and the following result strengthens our motivation for the definition (9.3).

Theorem 9.2. *Given half-open pointed cones $\mathcal{K}_1, \mathcal{K}_2, \ldots, \mathcal{K}_m \subseteq \mathbb{R}^d$ with a common apex in $\mathbb{Z}^d$ such that the disjoint union of $\mathcal{K}_1, \mathcal{K}_2, \ldots, \mathcal{K}_m$ is an affine space, then as rational functions,*

$$\sigma_{\mathcal{K}_1}(\mathbf{z}) + \sigma_{\mathcal{K}_2}(\mathbf{z}) + \cdots + \sigma_{\mathcal{K}_m}(\mathbf{z}) = 0.$$

Proof. Suppose the disjoint union of $\mathcal{K}_1, \mathcal{K}_2, \ldots, \mathcal{K}_m$ is the n-dimensional affine space $\mathcal{A}$, and $\mathbf{w} \in \mathbb{Z}^d$ is the common apex of $\mathcal{K}_1, \mathcal{K}_2, \ldots, \mathcal{K}_m$. Now we decompose $\mathcal{A}$ into the skewed orthants $\mathcal{O}_1, \mathcal{O}_2, \ldots, \mathcal{O}_{2^n}$, which are also pointed cones with common apex $\mathbf{w}$. The intersection of one of the $\mathcal{K}_j$'s with one of the $\mathcal{O}_k$'s is again a half-open pointed cone, and all these cones form yet another disjoint union of $\mathcal{A}$, which is a common refinement of the dissection of $\mathcal{A}$ in terms of the $\mathcal{K}_j$'s and in terms of the $\mathcal{O}_k$'s:

$$\mathcal{A} = \bigcup_{\substack{1 \le j \le m \\ 1 \le k \le 2^n}} (\mathcal{K}_j \cap \mathcal{O}_k).$$

For each $1 \le j \le m$, $\mathcal{K}_j = \bigcup_{k=1}^{2^n} (\mathcal{K}_j \cap \mathcal{O}_k)$ as a disjoint union, and so we can write the integer-point transform of $\mathcal{K}_j$ as

$$\sigma_{\mathcal{K}_j}(\mathbf{z}) = \sum_{k=1}^{2^n} \sigma_{\mathcal{K}_j \cap \mathcal{O}_k}(\mathbf{z}),$$

as a rational-function identity. Similarly, we obtain for each $1 \le k \le 2^n$,

$$\sigma_{\mathcal{O}_k}(\mathbf{z}) = \sum_{j=1}^{m} \sigma_{\mathcal{K}_j \cap \mathcal{O}_k}(\mathbf{z}).$$

Thus

$$\sum_{j=1}^{m} \sigma_{\mathcal{K}_j}(\mathbf{z}) = \sum_{j=1}^{m} \sum_{k=1}^{2^n} \sigma_{\mathcal{K}_j \cap \mathcal{O}_k}(\mathbf{z}) = \sum_{k=1}^{2^n} \sum_{j=1}^{m} \sigma_{\mathcal{K}_j \cap \mathcal{O}_k}(\mathbf{z}) = \sum_{k=1}^{2^n} \sigma_{\mathcal{O}_k}(\mathbf{z}) = 0,$$

by Lemma 9.1. □

9.2 Tangent Cones and Their Rational Generating Functions

The goal of this section, apart from setting the language that allows us to prove Brion's theorem, is to prove a sort of analogue of (9.2) in general dimension.

We recall a definition that was touched on only briefly, in Exercise 3.14: A **hyperplane arrangement** $\mathcal{H}$ is a finite collection of hyperplanes. An arrangement $\mathcal{H}$ is **rational** if all its hyperplanes are, that is, if each hyperplane in $\mathcal{H}$ is of the form $\left\{\mathbf{x} \in \mathbb{R}^d : a_1x_1 + a_2x_2 + \cdots + a_dx_d = b\right\}$ for some $a_1, a_2, \ldots, a_d, b \in \mathbb{Z}$. An arrangement $\mathcal{H}$ is called a **central** hyperplane arrangement if its hyperplanes meet in (at least) one point.

Our next definition generalizes (finally) the notion of a pointed cone, defined in Chapter 3. A **convex cone** is the intersection of finitely many half-spaces of the form $\left\{\mathbf{x} \in \mathbb{R}^d : a_1x_1 + a_2x_2 + \cdots + a_dx_d \le b\right\}$ for which the corresponding hyperplanes $\left\{\mathbf{x} \in \mathbb{R}^d : a_1x_1 + a_2x_2 + \cdots + a_dx_d = b\right\}$ form a central arrangement. This definition extends that of a pointed cone: a cone is pointed if the defining hyperplanes meet in *exactly* one point. A cone is **rational** if all of its defining hyperplanes are rational. Cones and polytopes are special cases of **polyhedra**, which are convex bodies defined as the intersection of finitely many half-spaces.

We now attach a cone to each face $\mathcal{F}$ of $\mathcal{P}$, namely its **tangent cone**, defined by

$$\mathcal{K}_{\mathcal{F}} := \left\{\mathbf{x} + \lambda\left(\mathbf{y} - \mathbf{x}\right) : \mathbf{x} \in \mathcal{F},\, \mathbf{y} \in \mathcal{P},\, \lambda \in \mathbb{R}_{\ge 0}\right\}.$$

It turns out that $\mathcal{K}_{\mathcal{F}}$ is the smallest convex cone containing both $\operatorname{span}\mathcal{F}$ and $\mathcal{P}$. We note that $\mathcal{K}_{\mathcal{P}} = \operatorname{span}\mathcal{P}$. For a vertex $\mathbf{v}$ of $\mathcal{P}$, the tangent cone $\mathcal{K}_{\mathbf{v}}$ is often called a **vertex cone**; it is pointed. For a k-face $\mathcal{F}$ of $\mathcal{P}$ with $k > 0$, the tangent cone $\mathcal{K}_{\mathcal{F}}$ is not pointed. For example, the tangent cone of an edge of a 3-polytope is a wedge.

Lemma 9.3. *For any face $\mathcal{F}$ of $\mathcal{P}$,* $\operatorname{span}\mathcal{F} \subseteq \mathcal{K}_{\mathcal{F}}$.

Proof. As $\mathbf{x}$ and $\mathbf{y}$ vary over all points of $\mathcal{F}$, $\mathbf{x} + \lambda\left(\mathbf{y} - \mathbf{x}\right)$ varies over $\operatorname{span}\mathcal{F}$. □

We note that this lemma implies that $\mathcal{K}_{\mathcal{F}}$ contains a line, unless $\mathcal{F}$ is a vertex. More precisely, if $\mathcal{K}_{\mathcal{F}}$ is not pointed, it contains the affine space $\operatorname{span}\mathcal{F}$, which is called the **apex** of the tangent cone. (A pointed cone has a point as apex.)

An affine space $\mathcal{A} \subseteq \mathbb{R}^d$ equals $\mathbf{w} + \mathcal{V}$ for some $\mathbf{w} \in \mathbb{R}^d$ and some vector subspace $\mathcal{V} \subseteq \mathbb{R}^d$. The **orthogonal complement** $\mathcal{A}^\perp$ of this affine space $\mathcal{A}$ is defined by

$$\mathcal{A}^\perp := \left\{\mathbf{x} \in \mathbb{R}^d : \mathbf{x} \cdot \mathbf{v} = 0 \text{ for all } \mathbf{v} \in \mathcal{V}\right\}.$$

We note that $\mathcal{A} \oplus \mathcal{A}^\perp = \mathbb{R}^d$, which gives us the following result.

Lemma 9.4. *For any face $\mathcal{F}$ of $\mathcal{P}$, the tangent cone $\mathcal{K}_{\mathcal{F}}$ has the decomposition*

$$\mathcal{K}_{\mathcal{F}} = \operatorname{span} \mathcal{F} \oplus \left((\operatorname{span} \mathcal{F})^{\perp} \cap \mathcal{K}_{\mathcal{F}} \right).$$

Consequently, unless $\mathcal{F}$ is a vertex,

$$\sigma_{\mathcal{K}_{\mathcal{F}}}(\mathbf{z}) = 0 .$$

Proof. Since $\operatorname{span} \mathcal{F} \oplus (\operatorname{span} \mathcal{F})^{\perp} = \mathbb{R}^d$,

$$\begin{aligned} \mathcal{K}_{\mathcal{F}} &= \left(\operatorname{span} \mathcal{F} \oplus (\operatorname{span} \mathcal{F})^{\perp} \right) \cap \mathcal{K}_{\mathcal{F}} \\ &= (\operatorname{span} \mathcal{F} \cap \mathcal{K}_{\mathcal{F}}) \oplus \left((\operatorname{span} \mathcal{F})^{\perp} \cap \mathcal{K}_{\mathcal{F}} \right) \\ &= \operatorname{span} \mathcal{F} \oplus \left((\operatorname{span} \mathcal{F})^{\perp} \cap \mathcal{K}_{\mathcal{F}} \right), \end{aligned}$$

where the last step follows from Lemma 9.3. The second part of the lemma is immediate since $\sigma_{\operatorname{span} \mathcal{F} \oplus ((\operatorname{span} \mathcal{F})^{\perp} \cap \mathcal{K}_{\mathcal{F}})}(\mathbf{z}) = \sigma_{\operatorname{span} \mathcal{F}}(\mathbf{z}) \, \sigma_{(\operatorname{span} \mathcal{F})^{\perp} \cap \mathcal{K}_{\mathcal{F}}}(\mathbf{z})$ and $\sigma_{\operatorname{span} \mathcal{F}}(\mathbf{z}) = 0$. □

Although we do not need this fact in the sequel, it's nice to know that $(\operatorname{span} \mathcal{F})^{\perp} \cap \mathcal{K}_{\mathcal{F}}$ is a pointed cone (see Exercise 9.1).

9.3 Brion's Theorem

The following theorem is a classical identity of convex geometry named after Charles Julien Brianchon (1783–1864)[1] and Jørgen Pedersen Gram (1850–1916).[2] It holds for any convex polytope. However, its proof for *simplices* is considerably simpler than that for the general case. We need only the Brianchon–Gram identity for simplices, so we restrict ourselves to this special case. (One could prove the general case along similar lines as below; however, we would need some additional machinery not covered in this book.) The **indicator function** 1_S of a set $S \subset \mathbb{R}^d$ is defined by

$$1_S(\mathbf{x}) := \begin{cases} 1 & \text{if } \mathbf{x} \in S, \\ 0 & \text{if } \mathbf{x} \notin S. \end{cases}$$

Theorem 9.5 (Brianchon–Gram identity for simplices). *Let Δ be a d-simplex. Then*

$$1_{\Delta}(\mathbf{x}) = \sum_{\mathcal{F} \subseteq \Delta} (-1)^{\dim \mathcal{F}} 1_{\mathcal{K}_{\mathcal{F}}}(\mathbf{x}) ,$$

where the sum is taken over all nonempty faces $\mathcal{F}$ of Δ.

[1] For more information about Brainchon, see
`http://www-groups.dcs.st-and.ac.uk/~history/Biographies/Brainchon.html`.

[2] For more information about Gram, see
`http://www-groups.dcs.st-and.ac.uk/~history/Biographies/Gram.html`.

Proof. We distinguish between two disjoint cases: whether or not $\mathbf{x}$ is in the simplex.

Case 1: $\mathbf{x} \in \Delta$. Then $\mathbf{x} \in \mathcal{K}_\mathcal{F}$ for all $\mathcal{F} \subseteq \Delta$, and the identity becomes

$$1 = \sum_{\mathcal{F} \subseteq \Delta} (-1)^{\dim \mathcal{F}} = \sum_{k=0}^{\dim \Delta} (-1)^k f_k \, .$$

This is the Euler relation for simplices, which we proved in Exercise 5.5.

Case 2: $\mathbf{x} \notin \Delta$. Then there is a unique minimal face $\mathcal{F} \subseteq \Delta$ (minimal with respect to dimension) such that $\mathbf{x} \in \mathcal{K}_\mathcal{F}$ and $\mathbf{x} \in \mathcal{K}_\mathcal{G}$ for all faces $\mathcal{G} \subseteq \Delta$ that contain $\mathcal{F}$ (Exercise 9.2). The identity to be proved is now

$$0 = \sum_{\mathcal{G} \supseteq \mathcal{F}} (-1)^{\dim \mathcal{G}} . \tag{9.4}$$

The validity of this identity again follows from the logic of Exercise 5.5; the proof of (9.4) is the subject of Exercise 9.4. □

Corollary 9.6 (Brion's theorem for simplices). *Suppose Δ is a rational simplex. Then we have the following identity of rational functions:*

$$\sigma_\Delta(\mathbf{z}) = \sum_{\mathbf{v} \text{ a vertex of } \Delta} \sigma_{\mathcal{K}_\mathbf{v}}(\mathbf{z}) \, .$$

Proof. We translate the Brianchon–Gram theorem into the language of integer-point transforms: we sum both sides of the identity in Theorem 9.5 for all $\mathbf{m} \in \mathbb{Z}^d$,

$$\sum_{\mathbf{m} \in \mathbb{Z}^d} 1_\Delta(\mathbf{m})\, \mathbf{z}^\mathbf{m} = \sum_{\mathbf{m} \in \mathbb{Z}^d} \sum_{\mathcal{F} \subseteq \Delta} (-1)^{\dim \mathcal{F}} 1_{\mathcal{K}_\mathcal{F}}(\mathbf{m})\, \mathbf{z}^\mathbf{m} ,$$

which is equivalent to

$$\sigma_\Delta(\mathbf{z}) = \sum_{\mathcal{F} \subseteq \Delta} (-1)^{\dim \mathcal{F}} \sigma_{\mathcal{K}_\mathcal{F}}(\mathbf{z}) \, .$$

But Lemma 9.4 implies that $\sigma_{\mathcal{K}_\mathcal{F}}(\mathbf{z}) = 0$ except when $\mathcal{F}$ is a vertex. Hence

$$\sigma_\Delta(\mathbf{z}) = \sum_{\mathbf{v} \text{ a vertex of } \Delta} \sigma_{\mathcal{K}_\mathbf{v}}(\mathbf{z}) \, .$$

□

Now we extend Corollary 9.6 to any convex rational polytope:

Theorem 9.7 (Brion's theorem). *Suppose $\mathcal{P}$ is a rational convex polytope. Then we have the following identity of rational functions:*

$$\sigma_\mathcal{P}(\mathbf{z}) = \sum_{\mathbf{v} \text{ a vertex of } \mathcal{P}} \sigma_{\mathcal{K}_\mathbf{v}}(\mathbf{z}) \, . \tag{9.5}$$

Proof. We use the same irrational trick as in the proofs of Theorems 3.12 and 4.3. Namely we start by triangulating $\mathcal{P}$ into the simplices $\Delta_1, \Delta_2, \dots, \Delta_m$ (using no new vertices). Consider the hyperplane arrangement

$$\mathcal{H} := \{\operatorname{span} \mathcal{F} : \mathcal{F} \text{ is a facet of } \Delta_1, \Delta_2, \dots, \text{ or } \Delta_m\} .$$

We will now shift the hyperplanes in $\mathcal{H}$, obtaining a new hyperplane arrangement $\mathcal{H}^{\text{shift}}$. Those hyperplanes of $\mathcal{H}$ that defined $\mathcal{P}$ now define, after shifting, a new polytope that we will call $\mathcal{P}^{\text{shift}}$. Exercise 9.6 ensures that we can shift $\mathcal{H}$ in such a way that:

- no hyperplane in $\mathcal{H}^{\text{shift}}$ contains any lattice point;
- $\mathcal{H}^{\text{shift}}$ yields a triangulation of $\mathcal{P}^{\text{shift}}$;
- the lattice points contained in a vertex cone of $\mathcal{P}$ are precisely the lattice points contained in the corresponding vertex cone of $\mathcal{P}^{\text{shift}}$.

This setup implies that

- the lattice points in $\mathcal{P}$ are precisely the lattice points in $\mathcal{P}^{\text{shift}}$;
- the lattice points in a vertex cone of $\mathcal{P}^{\text{shift}}$ can be written as a *disjoint* union of lattice points in vertex cones of simplices of the triangulation that $\mathcal{H}^{\text{shift}}$ induces on $\mathcal{P}^{\text{shift}}$.

The latter two conditions, in turn, mean that Brion's identity (9.5) follows from Brion's theorem for simplices: the integer-point transforms on both sides of the identity can be written as a sum of integer-point transforms of simplices and their vertex cones. □

9.4 Brion Implies Ehrhart

We conclude this chapter by showing that Ehrhart's theorem (Theorem 3.23) for rational polytopes (which includes the integral case, Theorem 3.8) follows from Brion's theorem (Theorem 9.7) in a relatively straightforward manner.

Second proof of Theorem 3.23. As in our first proof of Ehrhart's theorem, it suffices to prove Theorem 3.23 for simplices, because we can triangulate any polytope (using only the vertices). So suppose Δ is a rational d-simplex whose vertices have coordinates with denominator p. Our goal is to show that, for a fixed $0 \le r < p$, the function $L_\Delta(r + pt)$ is a polynomial in t; this means that L_Δ is a quasipolynomial with period dividing p.

By Theorem 9.7,

$$\begin{aligned} L_\Delta(r+pt) &= \sum_{\mathbf{m} \in (r+pt)\Delta \cap \mathbb{Z}^d} 1 \\ &= \lim_{\mathbf{z} \to \mathbf{1}} \sigma_{(r+pt)\Delta}(\mathbf{z}) \\ &= \lim_{\mathbf{z} \to \mathbf{1}} \sum_{\mathbf{v} \text{ vertex of } \Delta} \sigma_{(r+pt)\mathcal{K}_\mathbf{v}}(\mathbf{z}) . \end{aligned} \tag{9.6}$$

We used the limit computation for the integer-point transform $\sigma_{(r+pt)\Delta}$ rather than the evaluation $\sigma_{(r+pt)\Delta}(\mathbf{1})$, because this evaluation would have yielded singularities in the rational generating functions of the vertex cones. Note that the vertex cones $\mathcal{K}_{\mathbf{v}}$ are all simplicial, because Δ is a simplex. So suppose

$$\mathcal{K}_{\mathbf{v}} = \{\mathbf{v} + \lambda_1 \mathbf{w}_1 + \lambda_2 \mathbf{w}_2 + \cdots + \lambda_d \mathbf{w}_d : \lambda_1, \lambda_2, \ldots, \lambda_d \geq 0\};$$

then

$$\begin{aligned}(r+pt)\mathcal{K}_{\mathbf{v}} &= \{(r+pt)\mathbf{v} + \lambda_1 \mathbf{w}_1 + \lambda_2 \mathbf{w}_2 + \cdots + \lambda_d \mathbf{w}_d : \lambda_1, \lambda_2, \ldots, \lambda_d \geq 0\} \\ &= tp\mathbf{v} + \{r\mathbf{v} + \lambda_1 \mathbf{w}_1 + \lambda_2 \mathbf{w}_2 + \cdots + \lambda_d \mathbf{w}_d : \lambda_1, \lambda_2, \ldots, \lambda_d \geq 0\} \\ &= tp\mathbf{v} + r\mathcal{K}_{\mathbf{v}}.\end{aligned}$$

What's important to note here is that $p\mathbf{v}$ is an integer vector. In particular, we can safely write

$$\sigma_{(r+pt)\mathcal{K}_{\mathbf{v}}}(\mathbf{z}) = \mathbf{z}^{tp\mathbf{v}} \sigma_{r\mathcal{K}_{\mathbf{v}}}(\mathbf{z})$$

(we say "safely" because $tp\mathbf{v} \in \mathbb{Z}^d$, so $\mathbf{z}^{tp\mathbf{v}}$ is indeed a monomial). Now we can rewrite (9.6) as

$$L_\Delta(r+pt) = \lim_{\mathbf{z} \to \mathbf{1}} \sum_{\mathbf{v} \text{ vertex of } \Delta} \mathbf{z}^{tp\mathbf{v}} \sigma_{r\mathcal{K}_{\mathbf{v}}}(\mathbf{z}). \tag{9.7}$$

The exact forms of the rational functions $\sigma_{r\mathcal{K}_{\mathbf{v}}}(\mathbf{z})$ is not important, except for the fact that they do not depend on t. We know that the sum of the generating functions of all vertex cones is a polynomial in $\mathbf{z}$; that is, the singularities of the rational functions cancel. To compute $L_\Delta(r+pt)$ from (9.7), we write all the rational functions on the right-hand side over one denominator and use L'Hôpital's rule to compute the limit of this one huge rational function. The variable t appears only in the simple monomials $\mathbf{z}^{tp\mathbf{v}}$, so the effect of L'Hôpital's rule is that we obtain linear factors of t every time we differentiate the numerator of this rational function. At the end we evaluate the remaining rational function at $\mathbf{z} = 1$. The result is a polynomial in t. □

Notes

1. Theorem 9.5 (in its general form for convex polytopes) has an interesting history. In 1837 Charles Brianchon proved a version of this theorem involving volumes of polytopes in $\mathbb{R}^3$ [44]. In 1874 Jørgen Gram gave a proof of the same result [88]; apparently he was unaware of Brianchon's paper. In 1927 Duncan Sommerville published a proof for general d [167], which was corrected in the 1960s by Victor Klee [111], Branko Grünbaum [90, Section 14.1], and many others.

2. Michel Brion discovered Theorem 9.7 in 1988 [45]. His proof involved the Baum–Fulton–Quart Riemann–Roch formula for equivariant K-theory of toric varieties. A more elementary proof of Theorem 9.7 was found by Masa-Nori Ishida a few years later [103]. Our approach in this chapter follows [27].

3. As we have already remarked earlier, Brion's theorem led to an efficient algorithm by Alexander Barvinok to compute Ehrhart quasipolynomials [15]. More precisely, Barvinok proved that in fixed dimension, one can efficiently[3] compute the Ehrhart series $\sum_{t \geq 0} L_{\mathcal{P}}(t)\, z^t$ as a short sum of rational functions.[4] Brion's theorem essentially reduces the problem to computing the integer-point transforms of the rational tangent cones of the polytope. Barvinok's ingenious idea was to use a *signed decomposition* of a rational cone to compute its integer-point transform: the cone is written as a sum and difference of *unimodular* cones, which we will encounter in Section 10.4 and which have a trivial integer-point transform. Finding a signed decomposition involves triangulations, Minkowski's theorem on lattice points in convex bodies (see, for example, [57, 133, 140, 163]), and the LLL algorithm which finds a short vector in a lattice [121]. At any rate, Barvinok proved that one can find a signed decomposition quickly, which is the main step towards computing the Ehrhart series of the polytope. Barvinok's algorithm has been implemented in the software packages `barvinok` [185] and `LattE` [66, 67, 115]. Barvinok's algorithm is described in detail in [13].

Exercises

9.1. ♣ Prove that for any face $\mathcal{F}$ of a polytope, $(\operatorname{span} \mathcal{F})^{\perp} \cap \mathcal{K}_{\mathcal{F}}$ is a pointed cone. (*Hint:* Show that if H is a defining hyperplane for $\mathcal{F}$, then $H \cap (\operatorname{span} \mathcal{F})^{\perp}$ is a hyperplane in the vector space $(\operatorname{span} \mathcal{F})^{\perp}$.)

9.2. ♣ Suppose Δ is a simplex and $\mathbf{x} \notin \Delta$. Prove that there is a unique minimal face $\mathcal{F} \subseteq \Delta$ (minimal with respect to dimension) such that the corresponding tangent cone $\mathcal{K}_{\mathcal{F}}$ contains $\mathbf{x}$. Show that $\mathbf{x} \in \mathcal{K}_{\mathcal{G}}$ for all faces $\mathcal{G} \subseteq \Delta$ that contain $\mathcal{F}$, and $x \notin \mathcal{K}_{\mathcal{G}}$ for all other faces $\mathcal{G}$.

9.3. Show that Exercise 9.2 fails to be true if Δ is a quadrilateral (for example). Show that the Brianchon–Gram identity holds for your quadrilateral.

9.4. ♣ Prove (9.4): for a face $\mathcal{F}$ of a simplex Δ,

$$\sum_{\mathcal{G} \supseteq \mathcal{F}} (-1)^{\dim \mathcal{G}} = 0 \, ,$$

where the sum is taken over all faces of Δ that contain $\mathcal{F}$.

[3] "Efficiently" here means that for every dimension, there exists a polynomial that gives an upper bound on the running time of the algorithm, when evaluated at the logarithm of the input data of the polytope (e.g., its vertices).

[4] "Short" means that the set of data needed to output this sum of rational functions is also of polynomial size in the logarithm of the input data of the polytope.

9.5. Give a direct proof of Brion's theorem for the 1-dimensional case.

9.6. ♣ Provide the details of the irrational-shift argument in the proof of Theorem 9.7: Given a rational polytope $\mathcal{P}$, triangulate it into the simplices $\Delta_1, \Delta_2, \dots, \Delta_m$ (using no new vertices). Consider the hyperplane arrangement

$$\mathcal{H} := \{\operatorname{span} \mathcal{F} : \mathcal{F} \text{ is a facet of } \Delta_1, \Delta_2, \dots, \Delta_m\}.$$

We will now shift the hyperplanes in $\mathcal{H}$, obtaining a new hyperplane arrangement $\mathcal{H}^{\text{shift}}$. Those hyperplanes of $\mathcal{H}$ that defined $\mathcal{P}$ now define, after shifting, a new polytope that we will call $\mathcal{P}^{\text{shift}}$. Prove that we can shift $\mathcal{H}$ in such a way that:

- no hyperplane in $\mathcal{H}^{\text{shift}}$ contains any lattice point;
- $\mathcal{H}^{\text{shift}}$ yields a triangulation of $\mathcal{P}^{\text{shift}}$;
- the lattice points contained in a vertex cone of $\mathcal{P}$ are precisely the lattice points contained in the corresponding vertex cone of $\mathcal{P}^{\text{shift}}$.

9.7. ♣ Prove the following "open polytope" analogue for Brion's theorem: If $\mathcal{P}$ is a rational convex polytope, then we have the identity of rational functions

$$\sigma_{\mathcal{P}^\circ}(\mathbf{z}) = \sum_{\mathbf{v} \text{ a vertex of } \mathcal{P}} \sigma_{\mathcal{K}_\mathbf{v}^\circ}(\mathbf{z}).$$

10

Euler–Maclaurin Summation in $\mathbb{R}^d$

All means (even continuous) sanctify the discrete end.

Doron Zeilberger

Thus far we have often been concerned with the difference between the discrete volume of a polytope $\mathcal{P}$ and its continuous volume. In other words, the quantity

$$\sum_{\mathbf{m} \in \mathcal{P} \cap \mathbb{Z}^d} 1 - \int_{\mathcal{P}} d\mathbf{y} \,, \tag{10.1}$$

which is by definition $L_{\mathcal{P}}(1) - \mathrm{vol}(\mathcal{P})$, has been on our minds for a long time and has arisen naturally in many different contexts. An important extension is the difference between the discrete integer-point transform and its continuous sibling:

$$\sum_{\mathbf{m} \in \mathcal{P} \cap \mathbb{Z}^d} e^{\mathbf{m} \cdot \mathbf{x}} - \int_{\mathcal{P}} e^{\mathbf{y} \cdot \mathbf{x}} d\mathbf{y} \,, \tag{10.2}$$

where we have replaced the variable $\mathbf{z}$ that we have commonly used in generating functions by the exponential variable $(z_1, z_2, \dots, z_d) = (e^{x_1}, e^{x_2}, \dots, e^{x_d})$. Note that upon setting $\mathbf{x} = 0$ in (10.2) we get the former quantity (10.1). Relations between the two quantities $\sum_{\mathbf{m} \in \mathcal{P} \cap \mathbb{Z}^d} e^{\mathbf{m} \cdot \mathbf{x}}$ and $\int_{\mathcal{P}} e^{\mathbf{y} \cdot \mathbf{x}} d\mathbf{y}$ are known as Euler–Maclaurin summation formulas for polytopes. The "behind-the-scenes" operators that are responsible for affording us with such connections are the differential operators known as Todd operators, whose definition utilizes the Bernoulli numbers in a surprising way.

10.1 Todd Operators and Bernoulli Numbers

Recall the Bernoulli numbers B_k from Section 2.4, defined by the generating function

$$\frac{z}{e^z - 1} = \sum_{k \geq 0} \frac{B_k}{k!} z^k .$$

We now introduce a differential operator via essentially the same generating function, namely

$$\mathrm{Todd}_h := 1 + \sum_{k \geq 1} (-1)^k \frac{B_k}{k!} \left(\frac{d}{dh} \right)^k . \tag{10.3}$$

This **Todd operator** is often abbreviated as

$$\mathrm{Todd}_h = \frac{\frac{d}{dh}}{1 - e^{-\frac{d}{dh}}} ,$$

but we should keep in mind that this is only a shorthand notation for the infinite series (10.3). We first show that the exponential function is an eigenfunction of the Todd operator.

Lemma 10.1. *For $z \in \mathbb{C} \setminus \{0\}$ with $|z| < 2\pi$,*

$$\mathrm{Todd}_h \, e^{zh} = \frac{z \, e^{zh}}{1 - e^{-z}} .$$

Proof.

$$\begin{aligned} \mathrm{Todd}_h \, e^{zh} &= \sum_{k \geq 0} (-1)^k \frac{B_k}{k!} \left(\frac{d}{dh} \right)^k e^{zh} \\ &= \sum_{k \geq 0} (-1)^k \frac{B_k}{k!} z^k e^{zh} \\ &= e^{zh} \sum_{k \geq 0} (-z)^k \frac{B_k}{k!} \\ &= e^{zh} \frac{-z}{e^{-z} - 1} . \end{aligned}$$

The condition $|z| < 2\pi$ is needed in the last step, by Exercise 2.14. □

The Todd operator is a discretizing operator, in the sense that it transforms a continuous integral into a discrete sum, as the following theorem shows.

Theorem 10.2 (Euler–Maclaurin in dimension 1). *For all $a < b \in \mathbb{Z}$ and $z \in \mathbb{C}$ with $|z| < 2\pi$,*

$$\mathrm{Todd}_{h_1} \mathrm{Todd}_{h_2} \int_{a-h_2}^{b+h_1} e^{zx} \, dx \Bigg|_{h_1 = h_2 = 0} = \sum_{k=a}^{b} e^{kz} .$$

Proof. Case 1: $z = 0$. Then $e^{zx} = 1$, and so

$$\begin{aligned}
&\mathrm{Todd}_{h_1}\,\mathrm{Todd}_{h_2} \int_{a-h_2}^{b+h_1} e^{zx}\,dx \Big|_{h_1=h_2=0} \\
&= \mathrm{Todd}_{h_1}\,\mathrm{Todd}_{h_2} \int_{a-h_2}^{b+h_1} dx \Big|_{h_1=h_2=0} \\
&= b - a + \mathrm{Todd}_{h_1} h_1 + \mathrm{Todd}_{h_2} h_2 \big|_{h_1=h_2=0} \\
&= b - a + h_1 + \frac{1}{2} + h_2 + \frac{1}{2} \Big|_{h_1=h_2=0} \\
&= b - a + 1
\end{aligned}$$

by Exercise 10.1. Since $\sum_{k=a}^{b} e^{k \cdot 0} = b - a + 1$, we've verified the theorem in this case.

Case 2: $z \neq 0$. Then

$$\begin{aligned}
\mathrm{Todd}_{h_1}\,\mathrm{Todd}_{h_2} \int_{a-h_2}^{b+h_1} e^{zx}\,dx &= \mathrm{Todd}_{h_1}\,\mathrm{Todd}_{h_2} \frac{1}{z}\left(e^{z(b+h_1)} - e^{z(a-h_2)}\right) \\
&= \frac{1}{z}\left(\mathrm{Todd}_{h_1} e^{zb+zh_1} - \mathrm{Todd}_{h_2} e^{za-zh_2}\right) \\
&= \frac{e^{zb}}{z}\,\mathrm{Todd}_{h_1} e^{zh_1} - \frac{e^{za}}{z}\,\mathrm{Todd}_{h_2} e^{-zh_2} \\
&= \frac{e^{zb}}{z}\,\frac{z\,e^{zh_1}}{1-e^{-z}} - \frac{e^{za}}{z}\,\frac{-z\,e^{-zh_2}}{1-e^{z}},
\end{aligned}$$

where the last step follows from Lemma 10.1. Hence

$$\begin{aligned}
\mathrm{Todd}_{h_1}\,\mathrm{Todd}_{h_2} \int_{a-h_2}^{b+h_1} e^{zx}\,dx \Big|_{h_1=h_2=0} &= e^{zb}\frac{1}{1-e^{-z}} + e^{za}\frac{1}{1-e^{z}} \\
&= \frac{e^{z(b+1)} - e^{za}}{e^{z}-1} \\
&= \sum_{k=a}^{b} e^{kz}. \qquad \square
\end{aligned}$$

We will need a similar multivariate version of the Todd operator later, so that we define for $\mathbf{h} = (h_1, h_2, \dots, h_m)$,

$$\mathrm{Todd}_{\mathbf{h}} := \prod_{j=1}^{m} \left(\frac{\frac{\partial}{\partial h_j}}{1 - \exp\left(-\frac{\partial}{\partial h_j}\right)} \right),$$

keeping in mind that this is a product over infinite series of the form (10.3).

10.2 A Continuous Version of Brion's Theorem

In the following two sections, we develop the tools that, once fused with the Todd operator, will enable us to extend Euler–Maclaurin summation to higher dimensions. A lemma, which is of independent interest, but will be used in the proof of the continous version of Brion's theorem, now follows.

Lemma 10.3. *Suppose $\mathbf{w}_1, \mathbf{w}_2, \ldots, \mathbf{w}_d \in \mathbb{Z}^d$ are linearly independent, and let*

$$\Pi = \{\lambda_1 \mathbf{w}_1 + \lambda_2 \mathbf{w}_2 + \cdots + \lambda_d \mathbf{w}_d : 0 \le \lambda_1, \lambda_2, \ldots, \lambda_d < 1\} .$$

Then

$$\# \left(\Pi \cap \mathbb{Z}^d\right) = \operatorname{vol} \Pi = \left|\det \left(\mathbf{w}_1, \ldots, \mathbf{w}_d\right)\right|$$

and for any positive integer t,

$$\# \left(t\Pi \cap \mathbb{Z}^d\right) = \left(\operatorname{vol} \Pi\right) t^d .$$

In other words, for the half-open parallelepiped Π, the discrete volume $\# \left(t\Pi \cap \mathbb{Z}^d\right)$ coincides with the continuous volume $\left(\operatorname{vol} \Pi\right) t^d$.

Proof. Because Π is half open, we can tile the tth dilate $t\Pi$ by t^d translates of Π, and hence

$$L_\Pi(t) = \# \left(t\Pi \cap \mathbb{Z}^d\right) = \# \left(\Pi \cap \mathbb{Z}^d\right) t^d .$$

On the other hand, by the results of Chapter 3, $L_\Pi(t)$ is a polynomial with leading coefficient $\operatorname{vol} \Pi = \left|\det \left(\mathbf{w}_1, \ldots, \mathbf{w}_d\right)\right|$. Since we have equality of these polynomials for all positive integers t,

$$\# \left(\Pi \cap \mathbb{Z}^d\right) = \operatorname{vol} \Pi . \qquad \square$$

We now give an integral analogue of Theorem 9.7 for simple rational polytopes. We start by translating Brion's integer-point transforms

$$\sigma_{\mathcal{P}}(\mathbf{z}) = \sum_{\mathbf{v} \text{ a vertex of } \mathcal{P}} \sigma_{\mathcal{K}_\mathbf{v}}(\mathbf{z})$$

into an exponential form:

$$\sigma_{\mathcal{P}}(\exp \mathbf{z}) = \sum_{\mathbf{v} \text{ a vertex of } \mathcal{P}} \sigma_{\mathcal{K}_\mathbf{v}}(\exp \mathbf{z}) .$$

For the continuous analogue of Brion's theorem, we replace the sum on the left-hand side,

$$\sigma_{\mathcal{P}}(\exp \mathbf{z}) = \sum_{\mathbf{m} \in \mathcal{P} \cap \mathbb{Z}^d} (\exp \mathbf{z})^{\mathbf{m}} = \sum_{\mathbf{m} \in \mathcal{P} \cap \mathbb{Z}^d} \exp(\mathbf{m} \cdot \mathbf{z}) ,$$

by an integral.

Theorem 10.4 (Brion's theorem: continuous form). *Suppose $\mathcal{P}$ is a simple rational convex d-polytope. For a vertex cone $\mathcal{K}_{\mathbf{v}}$ of $\mathcal{P}$, fix a set of generators $\mathbf{w}_1(\mathbf{v}), \mathbf{w}_2(\mathbf{v}), \ldots, \mathbf{w}_d(\mathbf{v}) \in \mathbb{Z}^d$. Then*

$$\int_{\mathcal{P}} \exp(\mathbf{x} \cdot \mathbf{z})\, d\mathbf{x} \;=\; (-1)^d \sum_{\mathbf{v} \text{ a vertex of } \mathcal{P}} \frac{\exp(\mathbf{v} \cdot \mathbf{z})\, |\det(\mathbf{w}_1(\mathbf{v}), \ldots, \mathbf{w}_d(\mathbf{v}))|}{\prod_{k=1}^{d} (\mathbf{w}_k(\mathbf{v}) \cdot \mathbf{z})}$$

for all $\mathbf{z}$ such that the denominators on the right-hand side do not vanish.

Proof. We start with the assumption that $\mathcal{P}$ is an *integral* polytope; we will see in the process of the proof that this assumption can be relaxed. Let's write out the exponential form of Brion's theorem (Theorem 9.7), using the assumption that the vertex cones are simplicial (because $\mathcal{P}$ is simple). By Theorem 3.5,

$$\sum_{\mathbf{m} \in \mathcal{P} \cap \mathbb{Z}^d} \exp(\mathbf{m} \cdot \mathbf{z}) \;=\; \sum_{\mathbf{v} \text{ a vertex of } \mathcal{P}} \frac{\exp(\mathbf{v} \cdot \mathbf{z})\, \sigma_{\Pi_{\mathbf{v}}}(\exp \mathbf{z})}{\prod_{k=1}^{d} (1 - \exp(\mathbf{w}_k(\mathbf{v}) \cdot \mathbf{z}))}, \tag{10.4}$$

where

$$\Pi_{\mathbf{v}} = \{\lambda_1 \mathbf{w}_1(\mathbf{v}) + \lambda_2 \mathbf{w}_2(\mathbf{v}) + \cdots + \lambda_d \mathbf{w}_d(\mathbf{v}) : 0 \le \lambda_1, \lambda_2, \ldots, \lambda_d < 1\}$$

is the fundamental parallelepiped of the vertex cone $\mathcal{K}_{\mathbf{v}}$. We would like to rewrite (10.4) with the lattice $\mathbb{Z}^d$ replaced by the refined lattice $\left(\frac{1}{n}\mathbb{Z}\right)^d$, because then the left-hand side of (10.4) will give rise to the sought-after integral by letting n approach infinity. The right-hand side of (10.4) changes accordingly; now every integral point has to be scaled down by $\frac{1}{n}$:

$$\sum_{\mathbf{m} \in \mathcal{P} \cap \left(\frac{1}{n}\mathbb{Z}\right)^d} \exp(\mathbf{m} \cdot \mathbf{z}) \;=\; \sum_{\mathbf{v} \text{ a vertex of } \mathcal{P}} \frac{\exp(\mathbf{v} \cdot \mathbf{z}) \sum_{\mathbf{m} \in \Pi_{\mathbf{v}} \cap \mathbb{Z}^d} \exp\left(\frac{\mathbf{m}}{n} \cdot \mathbf{z}\right)}{\prod_{k=1}^{d} \left(1 - \exp\left(\frac{\mathbf{w}_k(\mathbf{v})}{n} \cdot \mathbf{z}\right)\right)}. \tag{10.5}$$

The proof of this identity is in essence the same as that of Theorem 3.5; we leave it as Exercise 10.2. Now our sought-after integral is

$$\begin{aligned} \int_{\mathcal{P}} \exp(\mathbf{x} \cdot \mathbf{z})\, d\mathbf{x} &= \lim_{n \to \infty} \frac{1}{n^d} \sum_{\mathbf{m} \in \mathcal{P} \cap \left(\frac{1}{n}\mathbb{Z}\right)^d} \exp(\mathbf{m} \cdot \mathbf{z}) \\ &= \lim_{n \to \infty} \frac{1}{n^d} \sum_{\mathbf{v} \text{ a vertex of } \mathcal{P}} \frac{\exp(\mathbf{v} \cdot \mathbf{z}) \sum_{\mathbf{m} \in \Pi_{\mathbf{v}} \cap \mathbb{Z}^d} \exp\left(\frac{\mathbf{m}}{n} \cdot \mathbf{z}\right)}{\prod_{k=1}^{d} \left(1 - \exp\left(\frac{\mathbf{w}_k(\mathbf{v})}{n} \cdot \mathbf{z}\right)\right)}. \end{aligned} \tag{10.6}$$

At this point we can see that our assumption that $\mathcal{P}$ has integral vertices can be relaxed to the rational case, since we may compute the limit only for n's that are multiples of the denominator of $\mathcal{P}$. The numerators of the terms on the right-hand side have a simple limit:

$$\lim_{n\to\infty} \exp(\mathbf{v}\cdot\mathbf{z}) \sum_{\mathbf{m}\in\Pi_{\mathbf{v}}\cap\mathbb{Z}^d} \exp\left(\frac{\mathbf{m}}{n}\cdot\mathbf{z}\right) = \exp(\mathbf{v}\cdot\mathbf{z}) \sum_{\mathbf{m}\in\Pi_{\mathbf{v}}\cap\mathbb{Z}^d} 1$$
$$= \exp(\mathbf{v}\cdot\mathbf{z}) \left|\det\left(\mathbf{w}_1(\mathbf{v}),\dots,\mathbf{w}_d(\mathbf{v})\right)\right| ,$$

where the last identity follows from Lemma 10.3. Hence (10.6) simplifies to

$$\int_{\mathcal{P}} \exp(\mathbf{x}\cdot\mathbf{z})\, d\mathbf{x} = \sum_{\mathbf{v}\text{ a vertex of }\mathcal{P}} \frac{\exp(\mathbf{v}\cdot\mathbf{z}) \left|\det\left(\mathbf{w}_1(\mathbf{v}),\dots,\mathbf{w}_d(\mathbf{v})\right)\right|}{\prod_{k=1}^d \lim_{n\to\infty} n\left(1-\exp\left(\frac{\mathbf{w}_k(\mathbf{v})}{n}\cdot\mathbf{z}\right)\right)} .$$

Finally, using L'Hôpital's rule, we have

$$\lim_{n\to\infty} n\left(1-\exp\left(\frac{\mathbf{w}_k(\mathbf{v})}{n}\cdot\mathbf{z}\right)\right) = -\mathbf{w}_k(\mathbf{v})\cdot\mathbf{z} ,$$

and the theorem follows. □

It turns out (Exercise 10.4) that for each vertex cone $\mathcal{K}_{\mathbf{v}}$,

$$\int_{\mathcal{K}_{\mathbf{v}}} \exp(\mathbf{x}\cdot\mathbf{z})\, d\mathbf{x} = (-1)^d \frac{\exp(\mathbf{v}\cdot\mathbf{z}) \left|\det\left(\mathbf{w}_1(\mathbf{v}),\dots,\mathbf{w}_d(\mathbf{v})\right)\right|}{\prod_{k=1}^d \left(\mathbf{w}_k(\mathbf{v})\cdot\mathbf{z}\right)} , \tag{10.7}$$

and Theorem 10.4 shows that the Fourier–Laplace transform of $\mathcal{P}$ equals the sum of the Fourier–Laplace transforms of the vertex cones. In other words,

$$\int_{\mathcal{P}} \exp(\mathbf{x}\cdot\mathbf{z})\, d\mathbf{x} = \sum_{\mathbf{v}\text{ a vertex of }\mathcal{P}} \int_{\mathcal{K}_{\mathbf{v}}} \exp(\mathbf{x}\cdot\mathbf{z})\, d\mathbf{x} .$$

We also remark that $\left|\det\left(\mathbf{w}_1(\mathbf{v}),\dots,\mathbf{w}_d(\mathbf{v})\right)\right|$ has a geometric meaning: it is the volume of the fundamental parallelepiped of the vertex cone $\mathcal{K}_{\mathbf{v}}$.

The curious reader might wonder what happens to the statement of Theorem 10.4 if we scale each of the generators $\mathbf{w}_k(\mathbf{v})$ by a different factor. It is immediate (Exercise 10.5) that the right-hand side of Theorem 10.4 remains invariant.

10.3 Polytopes Have Their Moments

The most common notion for moments of a set $\mathcal{P}\subset\mathbb{R}^d$ is

$$\mu_{\mathbf{a}} := \int_{\mathcal{P}} \mathbf{y}^{\mathbf{a}}\, d\mathbf{y} = \int_{\mathcal{P}} y_1^{a_1} y_2^{a_2}\cdots y_d^{a_d}\, d\mathbf{y}$$

for any fixed vector $\mathbf{a}=(a_1,a_2,\dots,a_d)\in\mathbb{C}^d$. For $\mathbf{a}=\mathbf{0}=(0,0,\dots,0)$, we get $\mu_{\mathbf{0}}=\operatorname{vol}\mathcal{P}$. As an application of moments, consider the problem of finding the center of mass of $\mathcal{P}$, which is defined by

$$\frac{1}{\operatorname{vol}\mathcal{P}}\left(\int_{\mathcal{P}} y_1\, d\mathbf{y}, \int_{\mathcal{P}} y_2\, d\mathbf{y}, \dots, \int_{\mathcal{P}} y_d\, d\mathbf{y}\right) .$$

This integral is equal to

$$\frac{1}{\mu_{\mathbf{0}}}\left(\mu_{(1,0,0,\dots,0)}, \mu_{(0,1,0,\dots,0)}, \dots, \mu_{(0,\dots,0,1)}\right).$$

Similarly, one can define the variance of $\mathcal{P}$ and other statistical data attached to $\mathcal{P}$ and use moments to compute them.

Our next task is to present the moments $\mu_{\mathbf{a}}$ in terms of Theorem 10.4. We make the change of variables $y_k = e^{x_k}$ in the defining integral of $\mu_{\mathbf{a}}$:

$$\mu_{\mathbf{a}} = \int_{\mathcal{P}} \mathbf{y}^{\mathbf{a}}\, d\mathbf{y} = \int_{\mathcal{P}} e^{x_1 a_1} e^{x_2 a_2} \cdots e^{x_d a_d}\, e^{x_1} e^{x_2} \cdots e^{x_d}\, d\mathbf{x} = \int_{\mathcal{P}} e^{\mathbf{x}\cdot(\mathbf{a}+\mathbf{1})}\, d\mathbf{x}.$$

Thus Theorem 10.4 gives us the following formulas for the moments of a simple rational d-polytope $\mathcal{P}$:

$$\mu_{\mathbf{a}} = (-1)^d \sum_{\mathbf{v} \text{ a vertex of } \mathcal{P}} \frac{\exp\left(\mathbf{v}\cdot(\mathbf{a}+\mathbf{1})\right) \left|\det\left(\mathbf{w}_1(\mathbf{v}), \dots, \mathbf{w}_d(\mathbf{v})\right)\right|}{\prod_{k=1}^{d} \left(\mathbf{w}_k(\mathbf{v})\cdot(\mathbf{a}+\mathbf{1})\right)},$$

for all $\mathbf{a}$ such that the denominators on the right-hand side do not vanish. Going a step further, we can use Theorem 10.4 to obtain information for a different set of moments. Along the way, we stumble on an amazing formula for the continuous volume of a polytope.

Theorem 10.5. *Suppose $\mathcal{P}$ is a simple rational convex d-polytope. For a vertex cone $\mathcal{K}_{\mathbf{v}}$ of $\mathcal{P}$, fix a set of generators $\mathbf{w}_1(\mathbf{v}), \mathbf{w}_2(\mathbf{v}), \dots, \mathbf{w}_d(\mathbf{v}) \in \mathbb{Z}^d$. Then*

$$\operatorname{vol}\mathcal{P} = \frac{(-1)^d}{d!} \sum_{\mathbf{v} \text{ a vertex of } \mathcal{P}} \frac{(\mathbf{v}\cdot\mathbf{z})^d \left|\det\left(\mathbf{w}_1(\mathbf{v}), \dots, \mathbf{w}_d(\mathbf{v})\right)\right|}{\prod_{k=1}^{d} \left(\mathbf{w}_k(\mathbf{v})\cdot\mathbf{z}\right)}$$

for all $\mathbf{z}$ such that the denominators on the right-hand side do not vanish. More generally, for any integer $j \geq 0$,

$$\int_{\mathcal{P}} (\mathbf{x}\cdot\mathbf{z})^j\, d\mathbf{x} = \frac{(-1)^d j!}{(j+d)!} \sum_{\mathbf{v} \text{ a vertex of } \mathcal{P}} \frac{(\mathbf{v}\cdot\mathbf{z})^{j+d} \left|\det\left(\mathbf{w}_1(\mathbf{v}), \dots, \mathbf{w}_d(\mathbf{v})\right)\right|}{\prod_{k=1}^{d} \left(\mathbf{w}_k(\mathbf{v})\cdot\mathbf{z}\right)}.$$

Proof. We replace the variable $\mathbf{z}$ in the identity of Theorem 10.4 by $s\mathbf{z}$, where s is a scalar:

$$\int_{\mathcal{P}} \exp\left(\mathbf{x}\cdot(s\mathbf{z})\right) d\mathbf{x} = (-1)^d \sum_{\mathbf{v} \text{ a vertex of } \mathcal{P}} \frac{\exp\left(\mathbf{v}\cdot(s\mathbf{z})\right) \left|\det\left(\mathbf{w}_1(\mathbf{v}), \dots, \mathbf{w}_d(\mathbf{v})\right)\right|}{\prod_{k=1}^{d} \left(\mathbf{w}_k(\mathbf{v})\cdot(s\mathbf{z})\right)},$$

which can be rewritten as

$$\int_{\mathcal{P}} \exp\left(s\,(\mathbf{x}\cdot\mathbf{z})\right) d\mathbf{x} = (-1)^d \sum_{\mathbf{v} \text{ a vertex of } \mathcal{P}} \frac{\exp\left(s\,(\mathbf{v}\cdot\mathbf{z})\right) \left|\det\left(\mathbf{w}_1(\mathbf{v}), \dots, \mathbf{w}_d(\mathbf{v})\right)\right|}{s^d \prod_{k=1}^{d} \left(\mathbf{w}_k(\mathbf{v})\cdot(\mathbf{z})\right)}.$$

The general statement of the corollary follows now by first expanding the exponential functions as Taylor series in s, and then comparing coefficients on both sides:

$$\begin{aligned}
&\sum_{j\geq 0} \int_{\mathcal{P}} (\mathbf{x}\cdot\mathbf{z})^j \, d\mathbf{x} \, \frac{s^j}{j!} \\
&= (-1)^d \sum_{\mathbf{v} \text{ a vertex of } \mathcal{P}} \sum_{j\geq 0} (\mathbf{v}\cdot\mathbf{z})^j \, \frac{s^{j-d}}{j!} \, \frac{|\det(\mathbf{w}_1(\mathbf{v}),\dots,\mathbf{w}_d(\mathbf{v}))|}{\prod_{k=1}^d (\mathbf{w}_k(\mathbf{v})\cdot(\mathbf{z}))} \\
&= \sum_{j\geq -d} (-1)^d \sum_{\mathbf{v} \text{ a vertex of } \mathcal{P}} \frac{(\mathbf{v}\cdot\mathbf{z})^{j+d} \, |\det(\mathbf{w}_1(\mathbf{v}),\dots,\mathbf{w}_d(\mathbf{v}))|}{\prod_{k=1}^d (\mathbf{w}_k(\mathbf{v})\cdot(\mathbf{z}))} \, \frac{s^j}{(j+d)!} \, . \qquad \square
\end{aligned}$$

The proof of this corollary reveals yet more identities between rational functions. Namely, the coefficients of the negative powers of s in the last line of the proof have to be zero. This immediately yields the following curious set of d identities for simple d-polytopes:

Corollary 10.6. *Suppose $\mathcal{P}$ is a simple rational convex d-polytope. For a vertex cone $\mathcal{K}_{\mathbf{v}}$ of $\mathcal{P}$, fix a set of generators $\mathbf{w}_1(\mathbf{v}), \mathbf{w}_2(\mathbf{v}), \dots, \mathbf{w}_d(\mathbf{v}) \in \mathbb{Z}^d$. Then for $0 \leq j \leq d-1$,*

$$\sum_{\mathbf{v} \text{ a vertex of } \mathcal{P}} \frac{(\mathbf{v}\cdot\mathbf{z})^j \, |\det(\mathbf{w}_1(\mathbf{v}),\dots,\mathbf{w}_d(\mathbf{v}))|}{\prod_{k=1}^d (\mathbf{w}_k(\mathbf{v})\cdot(\mathbf{z}))} = 0 \, . \qquad \square$$

10.4 From the Continuous to the Discrete Volume of a Polytope

In this section, we apply the Todd operator to a perturbation of the continuous volume. Namely, consider a simple full-dimensional polytope $\mathcal{P}$, which we may write as

$$\mathcal{P} = \left\{ \mathbf{x} \in \mathbb{R}^d : \mathbf{A}\,\mathbf{x} \leq \mathbf{b} \right\} .$$

Then we define the perturbed polytope

$$\mathcal{P}(\mathbf{h}) := \left\{ \mathbf{x} \in \mathbb{R}^d : \mathbf{A}\,\mathbf{x} \leq \mathbf{b} + \mathbf{h} \right\}$$

for a "small" vector $\mathbf{h} \in \mathbb{R}^m$. A famous theorem due to Askold Khovanskiĭ and Aleksandr Pukhlikov says that the integer-point count in $\mathcal{P}$ can be obtained from applying the Todd operator to $\operatorname{vol}(\mathcal{P}(\mathbf{h}))$. Here we prove the theorem for a certain class of polytopes, which we need to define first.

We call a rational pointed d-cone **unimodular** if its generators are a basis of $\mathbb{Z}^d$. An integral polytope is **unimodular** if each of its vertex cones is unimodular.[1]

[1] Unimodular polytopes go by two additional names, namely **smooth** and **Delzant**.

Theorem 10.7 (Khovanskiĭ–Pukhlikov theorem). *For a unimodular d-polytope $\mathcal{P}$,*

$$\#\left(\mathcal{P} \cap \mathbb{Z}^d\right) = \operatorname{Todd}_{\mathbf{h}} \operatorname{vol}\left(\mathcal{P}(\mathbf{h})\right)\big|_{\mathbf{h}=0} .$$

More generally,

$$\sigma_{\mathcal{P}}(\exp \mathbf{z}) = \operatorname{Todd}_{\mathbf{h}} \int_{\mathcal{P}(\mathbf{h})} \exp(\mathbf{x} \cdot \mathbf{z}) \, d\mathbf{x} \bigg|_{\mathbf{h}=0} .$$

Proof. We use Theorem 10.4, the continuous version of Brion's theorem; note that if $\mathcal{P}$ is unimodular, then $\mathcal{P}$ is automatically simple. For a vertex cone $\mathcal{K}_{\mathbf{v}}$ of $\mathcal{P}$, denote its generators by $\mathbf{w}_1(\mathbf{v}), \mathbf{w}_2(\mathbf{v}), \dots, \mathbf{w}_d(\mathbf{v}) \in \mathbb{Z}^d$. Then Theorem 10.4 states that

$$\begin{aligned} \int_{\mathcal{P}} \exp(\mathbf{x} \cdot \mathbf{z}) \, d\mathbf{x} &= (-1)^d \sum_{\mathbf{v} \text{ a vertex of } \mathcal{P}} \frac{\exp\left(\mathbf{v} \cdot \mathbf{z}\right) \left|\det\left(\mathbf{w}_1(\mathbf{v}), \dots, \mathbf{w}_d(\mathbf{v})\right)\right|}{\prod_{k=1}^{d} \left(\mathbf{w}_k(\mathbf{v}) \cdot \mathbf{z}\right)} \\ &= (-1)^d \sum_{\mathbf{v} \text{ a vertex of } \mathcal{P}} \frac{\exp\left(\mathbf{v} \cdot \mathbf{z}\right)}{\prod_{k=1}^{d} \left(\mathbf{w}_k(\mathbf{v}) \cdot \mathbf{z}\right)} , \end{aligned} \tag{10.8}$$

where the last identity follows from Exercise 10.3. A similar formula holds for $\mathcal{P}(\mathbf{h})$, except that we have to account for the shift of the vertices. The vector $\mathbf{h}$ shifts the facet-defining hyperplanes. This shift of the facets induces a shift of the vertices; let's say that the vertex $\mathbf{v}$ gets moved along each edge direction $\mathbf{w}_k$ (the vectors that generate the vertex cone $\mathcal{K}_{\mathbf{v}}$) by $h_k(\mathbf{v})$, so that $\mathcal{P}(\mathbf{h})$ has now the vertex $\mathbf{v} - \sum_{k=1}^{d} h_k(\mathbf{v}) \mathbf{w}_k(\mathbf{v})$. If $\mathbf{h}$ is small enough, $\mathcal{P}(\mathbf{h})$ will still be simple,[2] and we can apply Theorem 10.4 to $\mathcal{P}(\mathbf{h})$:

$$\begin{aligned} \int_{\mathcal{P}(\mathbf{h})} \exp(\mathbf{x} \cdot \mathbf{z}) \, d\mathbf{x} &= (-1)^d \sum_{\mathbf{v} \text{ a vertex of } \mathcal{P}} \frac{\exp\left(\left(\mathbf{v} - \sum_{k=1}^{d} h_k(\mathbf{v}) \mathbf{w}_k(\mathbf{v})\right) \cdot \mathbf{z}\right)}{\prod_{k=1}^{d} \left(\mathbf{w}_k(\mathbf{v}) \cdot \mathbf{z}\right)} \\ &= (-1)^d \sum_{\mathbf{v} \text{ a vertex of } \mathcal{P}} \frac{\exp\left(\mathbf{v} \cdot \mathbf{z} - \sum_{k=1}^{d} h_k(\mathbf{v}) \mathbf{w}_k(\mathbf{v}) \cdot \mathbf{z}\right)}{\prod_{k=1}^{d} \left(\mathbf{w}_k(\mathbf{v}) \cdot \mathbf{z}\right)} \\ &= (-1)^d \sum_{\mathbf{v} \text{ a vertex of } \mathcal{P}} \frac{\exp(\mathbf{v} \cdot \mathbf{z}) \prod_{k=1}^{d} \exp\left(-h_k(\mathbf{v}) \mathbf{w}_k(\mathbf{v}) \cdot \mathbf{z}\right)}{\prod_{k=1}^{d} \left(\mathbf{w}_k(\mathbf{v}) \cdot \mathbf{z}\right)} . \end{aligned}$$

Strictly speaking, this formula holds only for $\mathbf{h} \in \mathbb{Q}^m$, so that the vertices of $\mathcal{P}(\mathbf{h})$ are rational. Since we will eventually set $\mathbf{h} = 0$, this is a harmless restriction. Now we apply the Todd operator:

---

[2] The cautious reader may consult [193, p. 66] to confirm this fact.

$$
\begin{aligned}
&\mathrm{Todd}_{\mathbf{h}} \int_{\mathcal{P}(\mathbf{h})} \exp(\mathbf{x}\cdot\mathbf{z})\, d\mathbf{x}\Big|_{\mathbf{h}=0} \\
&\quad = (-1)^d \sum_{\mathbf{v} \text{ vertex of } \mathcal{P}} \mathrm{Todd}_{\mathbf{h}} \left. \frac{\exp(\mathbf{v}\cdot\mathbf{z}) \prod_{k=1}^d \exp\left(-h_k(\mathbf{v})\mathbf{w}_k(\mathbf{v})\cdot\mathbf{z}\right)}{\prod_{k=1}^d \left(\mathbf{w}_k(\mathbf{v})\cdot\mathbf{z}\right)} \right|_{\mathbf{h}=0} \\
&\quad = (-1)^d \sum_{\mathbf{v} \text{ vertex of } \mathcal{P}} \frac{\exp(\mathbf{v}\cdot\mathbf{z})}{\prod_{k=1}^d \left(\mathbf{w}_k(\mathbf{v})\cdot\mathbf{z}\right)} \\
&\qquad\qquad \times \prod_{k=1}^d \mathrm{Todd}_{h_k(\mathbf{v})} \exp\left(-h_k(\mathbf{v})\mathbf{w}_k(\mathbf{v})\cdot\mathbf{z}\right)\Big|_{h_k(\mathbf{v})=0} .
\end{aligned}
$$

By a multivariate version of Lemma 10.1,

$$
\begin{aligned}
&\mathrm{Todd}_{\mathbf{h}} \int_{\mathcal{P}(\mathbf{h})} \exp(\mathbf{x}\cdot\mathbf{z})\, d\mathbf{x}\Big|_{\mathbf{h}=0} \\
&\quad = (-1)^d \sum_{\mathbf{v} \text{ vertex of } \mathcal{P}} \frac{\exp\left(\mathbf{v}\cdot\mathbf{z}\right)}{\prod_{k=1}^d \left(\mathbf{w}_k(\mathbf{v})\cdot\mathbf{z}\right)} \prod_{k=1}^d \frac{-\mathbf{w}_k(\mathbf{v})\cdot\mathbf{z}}{1-\exp(\mathbf{w}_k(\mathbf{v})\cdot\mathbf{z})} \\
&\quad = \sum_{\mathbf{v} \text{ vertex of } \mathcal{P}} \exp\left(\mathbf{v}\cdot\mathbf{z}\right) \prod_{k=1}^d \frac{1}{1-\exp(\mathbf{w}_k(\mathbf{v})\cdot\mathbf{z})} .
\end{aligned}
$$

However, Brion's theorem (Theorem 9.7), together with the fact that $\mathcal{P}$ is unimodular, says that the right-hand side of this last formula is precisely the integer-point transform of $\mathcal{P}$ (see also (10.8)):

$$
\mathrm{Todd}_{\mathbf{h}} \int_{\mathcal{P}(\mathbf{h})} \exp(\mathbf{x}\cdot\mathbf{z})\, d\mathbf{x}\Big|_{\mathbf{h}=0} = \sigma_{\mathcal{P}}(\exp\mathbf{z}) .
$$

Finally, setting $\mathbf{z}=0$ gives

$$
\mathrm{Todd}_{\mathbf{h}} \int_{\mathcal{P}(\mathbf{h})} d\mathbf{x}\Big|_{\mathbf{h}=0} = \sum_{\mathbf{m}\in\mathcal{P}\cap\mathbb{Z}^d} 1 ,
$$

as claimed. □

We note that $\int_{\mathcal{P}(\mathbf{h})} \exp(\mathbf{x}\cdot\mathbf{z})\, d\mathbf{x}$ is, by definition, the continuous Fourier–Laplace transform of $\mathcal{P}(\mathbf{h})$. Upon being acted on by the discretizing operator $\mathrm{Todd}_{\mathbf{h}}$, $\int_{\mathcal{P}(\mathbf{h})} \exp(\mathbf{x}\cdot\mathbf{z})\, d\mathbf{x}$ gives us the discrete integer-point transform $\sigma_{\mathcal{P}}(\mathbf{z})$.

Notes

1. The classical Euler–Maclaurin formula states that

$$\sum_{k=1}^{n} f(k) = \int_0^n f(x)\,dx + \frac{f(0)+f(n)}{2} + \sum_{m=1}^{p} \frac{B_{2m}}{(2m)!}\left[f^{(2m-1)}(x)\right]_0^n + \frac{1}{(2p+1)!}\int_0^n B_{2p+1}\left(\{x\}\right) f^{(2p+1)}(x)\,dx\,,$$

where $B_k(x)$ denotes the kth Bernoulli polynomial. It was discovered independently by Leonhard Euler and Colin Maclaurin (1698–1746).[3] This formula provides an explicit error term, whereas Theorem 10.2 provides a summation formula with no error term.

2. The Todd operator was introduced by Friedrich Hirzebruch in the 1950s [100], following a more complicated definition by J. A. Todd [181, 182] some twenty years earlier. The Khovanskiĭ–Pukhlikov theorem (Theorem 10.7) can be interpreted as a combinatorial analogue of the algebro-geometric Hirzebruch–Riemann–Roch theorem, in which the Todd operator plays a prominent role.

3. Theorem 10.4, the continuous form of Brion's theorem, was generalized by Alexander Barvinok to *any* polytope [11]. In fact, [11] contains a certain extension of Brion's theorem to irrational polytopes as well. The decomposition formula for moments of a polytope in Theorem 10.5 is due to Michel Brion and Michèle Vergne [46].

4. Theorem 10.7 was first proved in 1992 by Askold Khovanskiĭ and Aleksandr Pukhlikov [108]. The proof we give here is essentially theirs. Their paper [108] also draws parallels between toric varieties and lattice polytopes. Subsequently, many attempts to provide formulas for Ehrhart quasipolynomials—some based on Theorem 10.7—have provided fertile ground for deeper connections and future work; a long but by no means complete list of references is [9, 33, 46, 55, 60, 61, 76, 91, 106, 107, 119, 125, 137, 146, 178].

Exercises

10.1. ♣ Show that $\operatorname{Todd}_h h = h + \frac{1}{2}$. More generally, prove that $\operatorname{Todd}_h h^k = B_k(h+1)$ for $k \ge 1$, where $B_k(x)$ denotes the kth Bernoulli polynomial.

10.2. ♣ Prove (10.5): Suppose $\mathcal{P}$ is a simple integral d-polytope. For a vertex cone $\mathcal{K}_{\mathbf{v}}$ of $\mathcal{P}$, denote its generators by $\mathbf{w}_1(\mathbf{v}), \mathbf{w}_2(\mathbf{v}), \dots, \mathbf{w}_d(\mathbf{v}) \in \mathbb{Z}^d$ and its fundamental parallelepiped by $\Pi_{\mathbf{v}}$. Then

$$\sum_{\mathbf{m}\in\mathcal{P}\cap\left(\frac{1}{n}\mathbb{Z}\right)^d} \exp(\mathbf{m}\cdot\mathbf{z}) = \sum_{\mathbf{v}\text{ a vertex of }\mathcal{P}} \frac{\exp(\mathbf{v}\cdot\mathbf{z})\sum_{\mathbf{m}\in\Pi_{\mathbf{v}}\cap\mathbb{Z}^d}\exp\left(\frac{\mathbf{m}}{n}\cdot\mathbf{z}\right)}{\prod_{k=1}^{d}\left(1-\exp\left(\frac{\mathbf{w}_k(\mathbf{v})}{n}\cdot\mathbf{z}\right)\right)}\,.$$

---

[3] For more information about Maclaurin, see
http://www-groups.dcs.st-and.ac.uk/~history/Biographies/Maclaurin.html.

10.3. ♣ Given a unimodular cone

$$\mathcal{K} = \{\mathbf{v} + \lambda_1 \mathbf{w}_1 + \lambda_2 \mathbf{w}_2 + \cdots + \lambda_d \mathbf{w}_d : \lambda_1, \lambda_2, \ldots, \lambda_d \ge 0\},$$

where $\mathbf{v}, \mathbf{w}_1, \mathbf{w}_2, \ldots, \mathbf{w}_d \in \mathbb{Z}^d$ such that $\mathbf{w}_1, \mathbf{w}_2, \ldots, \mathbf{w}_d$ are a basis for $\mathbb{Z}^d$, show that

$$\sigma_{\mathcal{K}}(\mathbf{z}) = \frac{\mathbf{z}^{\mathbf{v}}}{\prod_{k=1}^{d} \left(1 - \mathbf{z}^{\mathbf{w}_k}\right)}$$

and $|\det(\mathbf{w}_1, \ldots, \mathbf{w}_d)| = 1$.

10.4. ♣ Prove (10.7). That is, for the simple cone

$$\mathcal{K} = \left\{ \mathbf{v} + \sum_{k=1}^{d} \lambda_k \mathbf{w}_k : \lambda_k \ge 0 \right\}$$

with $\mathbf{v}, \mathbf{w}_1, \mathbf{w}_2, \ldots, \mathbf{w}_d \in \mathbb{Q}^d$, show that

$$\int_{\mathcal{K}} \exp(\mathbf{x} \cdot \mathbf{z}) \, d\mathbf{x} = (-1)^d \frac{\exp(\mathbf{v} \cdot \mathbf{z}) \, |\det(\mathbf{w}_1(\mathbf{v}), \ldots, \mathbf{w}_d(\mathbf{v}))|}{\prod_{k=1}^{d} (\mathbf{w}_k(\mathbf{v}) \cdot \mathbf{z})}.$$

10.5. Show that in the statement of Theorem 10.4, the expression

$$\frac{|\det(\mathbf{w}_1(\mathbf{v}), \ldots, \mathbf{w}_d(\mathbf{v}))|}{\prod_{k=1}^{d} (\mathbf{w}_k(\mathbf{v}) \cdot \mathbf{z})}$$

remains invariant upon scaling each $\mathbf{w}_k(\mathbf{v})$ by an independent positive integer.

Open Problems

10.6. Find all differentiable eigenfunctions of the Todd operator.

10.7. Classify all polytopes whose discrete and continuous volumes coincide, that is, $L_{\mathcal{P}}(1) = \operatorname{vol} \mathcal{P}$.

10.8. Which integer polytopes have a triangulation into d-simplices such that each of the simplices is unimodular?

11

Solid Angles

Everything you've learned in school as "obvious" becomes less and less obvious as you begin to study the universe. For example, there are no solids in the universe. There's not even a suggestion of a solid. There are no absolute continuums. There are no surfaces. There are no straight lines.

Buckminster Fuller (1895–1983)

The natural generalization of a two-dimensional angle to higher dimensions is called a *solid angle.* Given a pointed cone $\mathcal{K} \subset \mathbb{R}^d$, the solid angle at its apex is the proportion of space that the cone $\mathcal{K}$ occupies. In slightly different words, if we pick a point $\mathbf{x} \in \mathbb{R}^d$ "at random," then the probability that $\mathbf{x} \in \mathcal{K}$ is precisely the solid angle at the apex of $\mathcal{K}$. Yet another view of solid angles is that they are in fact volumes of spherical polytopes: the region of intersection of a cone with a sphere. There is a theory here that parallels the Ehrhart theory of Chapters 3 and 4, but which has some genuinely new ideas.

11.1 A New Discrete Volume Using Solid Angles

Suppose $\mathcal{P} \subset \mathbb{R}^d$ is a convex rational d-polyhedron. The **solid angle** $\omega_{\mathcal{P}}(\mathbf{x})$ of a point $\mathbf{x}$ (with respect to $\mathcal{P}$) is a real number equal to the proportion of a small ball centered at $\mathbf{x}$ that is contained in $\mathcal{P}$. That is, we let $B_\epsilon(\mathbf{x})$ denote the ball of radius ϵ centered at $\mathbf{x}$ and define

$$\omega_{\mathcal{P}}(\mathbf{x}) := \frac{\operatorname{vol}\left(B_\epsilon(\mathbf{x}) \cap \mathcal{P}\right)}{\operatorname{vol} B_\epsilon(\mathbf{x})}$$

for all positive ϵ sufficiently small. We note that when $\mathbf{x} \notin \mathcal{P}$, $\omega_{\mathcal{P}}(\mathbf{x}) = 0$; when $\mathbf{x} \in \mathcal{P}^\circ$, $\omega_{\mathcal{P}}(\mathbf{x}) = 1$; when $\mathbf{x} \in \partial\mathcal{P}$, $0 < \omega_{\mathcal{P}}(\mathbf{x}) < 1$. The **solid angle of a face** $\mathcal{F}$ of $\mathcal{P}$ is defined by picking any point $\mathbf{x}$ in the relative interior $\mathcal{F}^\circ$ and setting $\omega_{\mathcal{P}}(\mathcal{F}) = \omega_{\mathcal{P}}(\mathbf{x})$.

Example 11.1. We compute the solid angles of the faces belonging to the standard 3-simplex $\Delta = \operatorname{conv}\{(0,0,0),(1,0,0),(0,1,0),(0,0,1)\}$. As we just mentioned, a point interior to Δ has solid angle 1. Every facet has solid angle $\frac{1}{2}$ (and this remains true for any polytope).

The story gets interesting with the edges: here we are computing dihedral angles. The **dihedral angle** of any one-dimensional edge is defined by the angle between the outward-pointing normal to one of its defining facets and the inward-pointing normal to its other defining facet.

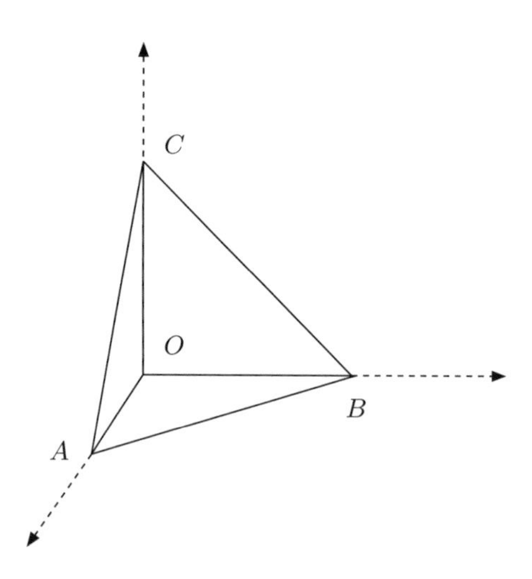

Each of the edges OA, OB, and OC in the above figure have the same solid angle $\frac{1}{4}$. Turning to the edge AB, we compute the angle between its defining facets as follows:

$$\cos^{-1}\left(\frac{1}{\sqrt{3}}(-1,-1,-1)\cdot(0,0,-1)\right) = \cos^{-1}\left(\frac{1}{\sqrt{3}}\right).$$

The edges AC and BC have the same solid angle by symmetry.

Finally, we compute the solid angle of the vertices: the origin has solid angle $\frac{1}{8}$, and the other three vertices all have the same solid angle ω. With Corollary 11.9 below (the Brianchon–Gram relation), we can compute this angle via

$$0 = \sum_{\mathcal{F}\subseteq\mathcal{P}} (-1)^{\dim \mathcal{F}} \omega_{\mathcal{P}}(\mathcal{F}) = -1 + 4\cdot\frac{1}{2} - 3\cdot\frac{1}{4} - 3\cdot\cos^{-1}\left(\frac{1}{\sqrt{3}}\right) + \frac{1}{8} + 3\cdot\omega\,,$$

which gives $\omega = \cos^{-1}\left(\frac{1}{\sqrt{3}}\right) - \frac{1}{8}$. □

We now introduce another measure of discrete volume; namely, we let

$$A_{\mathcal{P}}(t) := \sum_{\mathbf{m} \in t\mathcal{P} \cap \mathbb{Z}^d} \omega_{t\mathcal{P}}(\mathbf{m}),$$

the sum of the solid angles at all integer points in $t\mathcal{P}$; recalling that $\omega_{\mathcal{P}}(\mathbf{x}) = 0$ if $\mathbf{x} \notin \mathcal{P}$, we can also write

$$A_{\mathcal{P}}(t) = \sum_{\mathbf{m} \in \mathbb{Z}^d} \omega_{t\mathcal{P}}(\mathbf{m}).$$

This new discrete volume measure differs in a substantial way from the Ehrhart counting function $L_{\mathcal{P}}(t)$. Namely, suppose $\mathcal{P}$ is a d-polytope that can be written as the union of the polytopes $\mathcal{P}_1$ and $\mathcal{P}_2$ such that $\dim(\mathcal{P}_1 \cap \mathcal{P}_2) < d$, that is, $\mathcal{P}_1$ and $\mathcal{P}_2$ are glued along a lower-dimensional subset. Then at each lattice point $\mathbf{m} \in \mathbb{Z}^d$, $\omega_{\mathcal{P}_1}(\mathbf{m}) + \omega_{\mathcal{P}_2}(\mathbf{m}) = \omega_{\mathcal{P}}(\mathbf{m})$, and so the function $A_{\mathcal{P}}$ has an additive property:

$$A_{\mathcal{P}}(t) = A_{\mathcal{P}_1}(t) + A_{\mathcal{P}_2}(t). \tag{11.1}$$

In contrast, the Ehrhart counting functions satisfy

$$L_{\mathcal{P}}(t) = L_{\mathcal{P}_1}(t) + L_{\mathcal{P}_2}(t) - L_{\mathcal{P}_1 \cap \mathcal{P}_2}(t).$$

On the other hand, we can transfer computational effort from the Ehrhart counting functions to the solid-angle sum and vice versa, with the use of the following lemma.

Lemma 11.2. *Let $\mathcal{P}$ be a polytope. Then*

$$A_{\mathcal{P}}(t) = \sum_{\mathcal{F} \subseteq \mathcal{P}} \omega_{\mathcal{P}}(\mathcal{F})\, L_{\mathcal{F}^\circ}(t).$$

Proof. The dilated polytope $t\mathcal{P}$ is the disjoint union of its relative open faces $t\mathcal{F}^\circ$, so that we can write

$$A_{\mathcal{P}}(t) = \sum_{\mathbf{m} \in \mathbb{Z}^d} \omega_{t\mathcal{P}}(\mathbf{m}) = \sum_{\mathcal{F} \subseteq \mathcal{P}} \sum_{\mathbf{m} \in \mathbb{Z}^d} \omega_{t\mathcal{P}}(\mathbf{m})\, 1_{t\mathcal{F}^\circ}(\mathbf{m}).$$

But $\omega_{t\mathcal{P}}(\mathbf{m})$ is constant on each relatively open face $t\mathcal{F}^\circ$, and we called this constant $\omega_{\mathcal{P}}(\mathcal{F})$, whence

$$A_{\mathcal{P}}(t) = \sum_{\mathcal{F} \subseteq \mathcal{P}} \omega_{\mathcal{P}}(\mathcal{F}) \sum_{\mathbf{m} \in \mathbb{Z}^d} 1_{t\mathcal{F}^\circ}(\mathbf{m}) = \sum_{\mathcal{F} \subseteq \mathcal{P}} \omega_{\mathcal{P}}(\mathcal{F})\, L_{\mathcal{F}^\circ}(t). \qquad \square$$

Thus $A_{\mathcal{P}}(t)$ is a polynomial (respectively quasipolynomial) in t for an integral (respectively rational) polytope $\mathcal{P}$. We claim that Lemma 11.2 is in fact useful in practice. To drive the point home, we illustrate this identity by computing the solid-angle sum over all integer points of Δ in Example 11.1.

Example 11.3. We continue the solid-angle computation for the 3-simplex $\Delta = \operatorname{conv}\{(0,0,0),(1,0,0),(0,1,0),(0,0,1)\}$. We recall from Section 2.3 that $L_{\Delta^\circ} = \binom{t-1}{3}$. The facets of Δ are three standard triangles and one triangle that appeared in the context of the Frobenius problem. All four facets have the same interior Ehrhart polynomial $\binom{t-1}{2}$. A similar phenomenon holds for the edges of Δ: all six of them have the same interior Ehrhart polynomial $t-1$. These polynomials add up, by Lemma 11.2 and Example 11.1, to the solid-angle sum

$$\begin{aligned} A_\Delta(t) &= \binom{t-1}{3} + 4 \cdot \frac{1}{2}\binom{t-1}{2} + \left(3 \cdot \frac{1}{4} + 3 \cdot \cos^{-1}\left(\frac{1}{\sqrt{3}}\right)\right)(t-1) \\ &\quad + \frac{1}{8} + 3 \cdot \left(\cos^{-1}\left(\frac{1}{\sqrt{3}}\right) - \frac{1}{8}\right) \\ &= \frac{1}{6}t^3 + \left(3\cos^{-1}\left(\frac{1}{\sqrt{3}}\right) - \frac{5}{12}\right)t\,. \end{aligned}$$

The magic cancellation of the even terms of this polynomial is not a coincidence, as we will discover in Theorem 11.7. The curious reader may notice that the coefficient of t in this example is not a rational number, in stark contrast with Ehrhart polynomials. □

The analogue of Ehrhart's theorem (Theorem 3.23) in the world of solid angles is as follows.

Theorem 11.4 (Macdonald's theorem). *Suppose $\mathcal{P}$ is a rational convex d-polytope. Then $A_\mathcal{P}$ is a quasipolynomial of degree d whose leading coefficient is $\operatorname{vol}\mathcal{P}$ and whose period divides the denominator of $\mathcal{P}$.*

Proof. The denominator of a face $\mathcal{F} \subset \mathcal{P}$ divides the denominator of $\mathcal{P}$, and hence so does the period of $L_\mathcal{F}$, by Ehrhart's theorem (Theorem 3.23). By Lemma 11.2, $A_\mathcal{P}$ is a quasipolynomial with period dividing the denominator of $\mathcal{P}$. The leading term of $A_\mathcal{P}$ equals the leading term of $L_{\mathcal{P}^\circ}$, which is $\operatorname{vol}\mathcal{P}$, by Corollary 3.20 and its extension in Exercise 3.29. □

11.2 Solid-Angle Generating Functions and a Brion-Type Theorem

By analogy with the integer-point transform of a polyhedron $\mathcal{P} \subseteq \mathbb{R}^d$, which lists all lattice points in $\mathcal{P}$, we form the **solid-angle generating function**

$$\alpha_\mathcal{P}(\mathbf{z}) := \sum_{\mathbf{m} \in \mathcal{P} \cap \mathbb{Z}^d} \omega_\mathcal{P}(\mathbf{m})\, \mathbf{z}^\mathbf{m}.$$

Using the same reasoning as in (11.1) for $A_\mathcal{P}$, this function satisfies a nice additivity relation. Namely, if the d-polyhedron $\mathcal{P}$ equals $\mathcal{P}_1 \cup \mathcal{P}_2$, where $\dim(\mathcal{P}_1 \cap \mathcal{P}_2) < d$, then

$$\alpha_{\mathcal{P}}(\mathbf{z}) = \alpha_{\mathcal{P}_1}(\mathbf{z}) + \alpha_{\mathcal{P}_2}(\mathbf{z}) . \tag{11.2}$$

This generating function obeys the following reciprocity relation, which parallels both the statement and proof of Theorem 4.3:

Theorem 11.5. *Suppose $\mathcal{K}$ is a rational pointed d-cone $\mathcal{K}$ with the origin as apex, and $\mathbf{v} \in \mathbb{R}^d$. Then the solid-angle generating function $\alpha_{\mathbf{v}+\mathcal{K}}(\mathbf{z})$ of the pointed d-cone $\mathbf{v}+\mathcal{K}$ is a rational function that satisfies*

$$\alpha_{\mathbf{v}+\mathcal{K}}\left(\frac{1}{\mathbf{z}}\right) = (-1)^d \alpha_{-\mathbf{v}+\mathcal{K}}(\mathbf{z}) .$$

Proof. Because solid angles are additive by (11.2), it suffices to prove this theorem for simplicial cones. The proof for this case proceeds along the same lines as the proof of Theorem 4.2; the main geometric ingredient is Exercise 4.2. We invite the reader to finish the proof (Exercise 11.4). □

The analogue of Brion's theorem in terms of solid angles is as follows.

Theorem 11.6. *Suppose $\mathcal{P}$ is a rational convex polytope. Then we have the following identity of rational functions:*

$$\alpha_{\mathcal{P}}(\mathbf{z}) = \sum_{\mathbf{v} \text{ a vertex of } \mathcal{P}} \alpha_{\mathcal{K}_{\mathbf{v}}}(\mathbf{z}) .$$

Proof. As in the proof of Theorem 9.7, it suffices to prove Theorem 11.6 for simplices. So let Δ be a rational simplex. We write Δ as the disjoint union of its open faces and use Brion's theorem for open polytopes (Exercise 9.7) on each face. That is, if we denote the vertex cone of $\mathcal{F}$ at vertex $\mathbf{v}$ by $\mathcal{K}_{\mathbf{v}}(\mathcal{F})$, then by a monomial version of Lemma 11.2,

$$\begin{aligned}
\alpha_{\Delta}(\mathbf{z}) &= \sum_{\mathcal{F} \subseteq \Delta} \omega_{\Delta}(\mathcal{F})\, \sigma_{\mathcal{F}^\circ}(\mathbf{z}) \\
&= \sum_{\mathbf{v} \text{ a vertex of } \Delta} \omega_{\Delta}(\mathbf{v})\, \mathbf{z}^{\mathbf{v}} + \sum_{\substack{\mathcal{F} \subseteq \Delta \\ \dim \mathcal{F} > 0}} \omega_{\Delta}(\mathcal{F}) \sum_{\mathbf{v} \text{ a vertex of } \mathcal{F}} \sigma_{\mathcal{K}_{\mathbf{v}}(\mathcal{F})^\circ}(\mathbf{z}) ,
\end{aligned}$$

where we used Brion's theorem for open polytopes (Exercise 9.7) in the second step. By Exercise 11.5,

$$\sum_{\substack{\mathcal{F} \subseteq \Delta \\ \dim \mathcal{F} > 0}} \omega_{\Delta}(\mathcal{F}) \sum_{\mathbf{v} \text{ a vertex of } \mathcal{F}} \sigma_{\mathcal{K}_{\mathbf{v}}(\mathcal{F})^\circ}(\mathbf{z}) = \sum_{\mathbf{v} \text{ a vertex of } \Delta} \sum_{\substack{\mathcal{F} \subseteq \mathcal{K}_{\mathbf{v}} \\ \dim \mathcal{F} > 0}} \omega_{\mathcal{K}_{\mathbf{v}}}(\mathcal{F})\, \sigma_{\mathcal{F}^\circ}(\mathbf{z}) ,$$

and so

$$\begin{aligned}
\alpha_{\Delta}(\mathbf{z}) &= \sum_{\mathbf{v} \text{ a vertex of } \Delta} \omega_{\Delta}(\mathbf{v})\, \mathbf{z}^{\mathbf{v}} + \sum_{\mathbf{v} \text{ a vertex of } \Delta} \sum_{\substack{\mathcal{F} \subseteq \mathcal{K}_{\mathbf{v}} \\ \dim \mathcal{F} > 0}} \omega_{\mathcal{K}_{\mathbf{v}}}(\mathcal{F})\, \sigma_{\mathcal{F}^\circ}(\mathbf{z}) \\
&= \sum_{\mathbf{v} \text{ a vertex of } \Delta} \sum_{\mathcal{F} \subseteq \mathcal{K}_{\mathbf{v}}} \omega_{\mathcal{K}_{\mathbf{v}}}(\mathcal{F})\, \sigma_{\mathcal{F}^\circ}(\mathbf{z}) \\
&= \sum_{\mathbf{v} \text{ a vertex of } \Delta} \alpha_{\mathcal{K}_{\mathbf{v}}}(\mathbf{z}) .
\end{aligned}$$

□

11.3 Solid-Angle Reciprocity and the Brianchon–Gram Relations

With the help of Theorems 11.5 and 11.6, we can now prove the solid-angle analogue of Ehrhart–Macdonald reciprocity (Theorem 4.1):

Theorem 11.7 (Macdonald's reciprocity theorem). *Suppose $\mathcal{P}$ is a rational convex polytope. Then the quasipolynomial $A_{\mathcal{P}}$ satisfies*

$$A_{\mathcal{P}}(-t) = (-1)^{\dim \mathcal{P}} A_{\mathcal{P}}(t) .$$

Proof. We give the proof for an *integral* polytope $\mathcal{P}$ and invite the reader to generalize it to the rational case. The solid-angle counting function of $\mathcal{P}$ can be computed through the generating function:

$$A_{\mathcal{P}}(t) = \alpha_{t\mathcal{P}}(1, 1, \ldots, 1) = \lim_{\mathbf{z} \to \mathbf{1}} \alpha_{t\mathcal{P}}(\mathbf{z}) .$$

By Theorem 11.6,

$$A_{\mathcal{P}}(t) = \lim_{\mathbf{z} \to \mathbf{1}} \sum_{\mathbf{v} \text{ a vertex of } \mathcal{P}} \alpha_{t\mathcal{K}_{\mathbf{v}}}(\mathbf{z}) ,$$

where $\mathcal{K}_{\mathbf{v}}$ is the tangent cone of $\mathcal{P}$ at the vertex $\mathbf{v}$. We write $\mathcal{K}_{\mathbf{v}} = \mathbf{v} + \mathcal{K}(\mathbf{v})$, where $\mathcal{K}(\mathbf{v}) := \mathcal{K}_{\mathbf{v}} - \mathbf{v}$ is a rational cone with the origin as its apex. Then $t\mathcal{K}_{\mathbf{v}} = t\mathbf{v} + \mathcal{K}(\mathbf{v})$ because a cone whose apex is the origin does not change under dilation. Hence we obtain, with the help of Exercise 11.3,

$$A_{\mathcal{P}}(t) = \lim_{\mathbf{z} \to \mathbf{1}} \sum_{\mathbf{v} \text{ a vertex of } \mathcal{P}} \alpha_{t\mathbf{v}+\mathcal{K}(\mathbf{v})}(\mathbf{z}) = \lim_{\mathbf{z} \to \mathbf{1}} \sum_{\mathbf{v} \text{ a vertex of } \mathcal{P}} \mathbf{z}^{t\mathbf{v}} \alpha_{\mathcal{K}(\mathbf{v})}(\mathbf{z}) .$$

The rational functions $\alpha_{\mathcal{K}(\mathbf{v})}(\mathbf{z})$ on the right-hand side do not depend on t. If we think of the sum over all vertices as one big rational function, to which we apply L'Hôpital's rule to compute the limit as $\mathbf{z} \to \mathbf{1}$, this gives an alternative proof that $A_{\mathcal{P}}(t)$ is a polynomial, in line with our proof for the polynomiality of $L_{\mathcal{P}}(t)$ in Section 9.4. At the same time, this means we can view the identity

$$A_{\mathcal{P}}(t) = \lim_{\mathbf{z} \to \mathbf{1}} \sum_{\mathbf{v} \text{ a vertex of } \mathcal{P}} \mathbf{z}^{t\mathbf{v}} \alpha_{\mathcal{K}(\mathbf{v})}(\mathbf{z})$$

in a purely algebraic fashion: on the left-hand side we have a polynomial that makes sense for any complex t, and on the right-hand side we have a rational function of $\mathbf{z}$, whose limit we compute, for example, by L'Hôpital's rule. So the right-hand side, as a function of t, makes sense for any integer t. Hence we have the algebraic relation, for integral t,

$$A_{\mathcal{P}}(-t) = \lim_{\mathbf{z} \to \mathbf{1}} \sum_{\mathbf{v} \text{ a vertex of } \mathcal{P}} \mathbf{z}^{-t\mathbf{v}} \alpha_{\mathcal{K}(\mathbf{v})}(\mathbf{z}) .$$

But now by Theorem 11.5, $\alpha_{\mathcal{K}(\mathbf{v})}(\mathbf{z}) = (-1)^d \alpha_{\mathcal{K}(\mathbf{v})}\left(\frac{1}{\mathbf{z}}\right)$, and so

$$\begin{aligned}
A_{\mathcal{P}}(-t) &= \lim_{\mathbf{z}\to\mathbf{1}} \sum_{\mathbf{v} \text{ a vertex of } \mathcal{P}} \mathbf{z}^{-t\mathbf{v}}(-1)^d\, \alpha_{\mathcal{K}(\mathbf{v})}\left(\frac{1}{\mathbf{z}}\right) \\
&= (-1)^d \lim_{\mathbf{z}\to\mathbf{1}} \sum_{\mathbf{v} \text{ a vertex of } \mathcal{P}} \left(\frac{1}{\mathbf{z}}\right)^{t\mathbf{v}} \alpha_{\mathcal{K}(\mathbf{v})}\left(\frac{1}{\mathbf{z}}\right) \\
&= (-1)^d \lim_{\mathbf{z}\to\mathbf{1}} \sum_{\mathbf{v} \text{ a vertex of } \mathcal{P}} \alpha_{t\mathbf{v}+\mathcal{K}(\mathbf{v})}\left(\frac{1}{\mathbf{z}}\right) \\
&= (-1)^d \lim_{\mathbf{z}\to\mathbf{1}} \alpha_{t\mathcal{P}}\left(\frac{1}{\mathbf{z}}\right) \\
&= (-1)^d A_{\mathcal{P}}(t)\,.
\end{aligned}$$

In the third step we used Exercise 11.3 again.

This proves Theorem 11.7 for integral polytopes. The proof for *rational* polytopes follows along the same lines; one deals with rational vertices in the same manner as in our second proof of Ehrhart's theorem in Section 9.4. We invite the reader to finish the details in Exercise 11.6. □

We remark that throughout the proof, we cannot simply take the limit inside the finite sum over the vertices of $\mathcal{P}$, since $z = 1$ is a pole of each rational function $\alpha_{\mathcal{K}(\mathbf{v})}$. It is precisely the magic of Brion's theorem that makes these poles cancel each other, to yield $A_{\mathcal{P}}(t)$.

If $\mathcal{P}$ is an *integral* polytope, then $A_{\mathcal{P}}$ is a polynomial, and Theorem 11.7 tells us that $A_{\mathcal{P}}$ is always even or odd:

$$A_{\mathcal{P}}(t) = c_d\, t^d + c_{d-2} t^{d-2} + \cdots + c_0\,.$$

We can say more.

Theorem 11.8. *Suppose $\mathcal{P}$ is a rational convex polytope. Then $A_{\mathcal{P}}(0) = 0$.*

This is a meaningful zero. We note that the constant term of $A_{\mathcal{P}}$ is given by

$$A_{\mathcal{P}}(0) = \sum_{\mathcal{F}\subseteq\mathcal{P}} \omega_{\mathcal{P}}(\mathcal{F})\, L_{\mathcal{F}^\circ}(0) = \sum_{\mathcal{F}\subseteq\mathcal{P}} \omega_{\mathcal{P}}(\mathcal{F})\, (-1)^{\dim\mathcal{F}},$$

by Lemma 11.2 and Ehrhart–Macdonald reciprocity (Theorem 4.1). Hence Theorem 11.8 implies a classical and useful geometric identity:

Corollary 11.9 (Brianchon–Gram relation). *For a rational convex polytope $\mathcal{P}$,*

$$\sum_{\mathcal{F}\subseteq\mathcal{P}} (-1)^{\dim\mathcal{F}} \omega_{\mathcal{P}}(\mathcal{F}) = 0\,.$$

Example 11.10. Consider a triangle $\mathcal{T}$ in $\mathbb{R}^2$ with vertices $\mathbf{v}_1, \mathbf{v}_2, \mathbf{v}_3$ and edges E_1, E_2, E_3. The Brianchon–Gram relation tells us that for this triangle,

$$\omega_{\mathcal{T}}(\mathbf{v}_1) + \omega_{\mathcal{T}}(\mathbf{v}_2) + \omega_{\mathcal{T}}(\mathbf{v}_3) - (\omega_{\mathcal{T}}(E_1) + \omega_{\mathcal{T}}(E_2) + \omega_{\mathcal{T}}(E_3)) + \omega_{\mathcal{T}}(\mathcal{T}) = 0 .$$

Since the solid angles of the edges are all $\frac{1}{2}$ and $\omega_{\mathcal{T}}(\mathcal{T}) = 1$, we recover our friendly high-school identity "the sum of the angles in a triangle is 180 degrees":

$$\omega_{\mathcal{T}}(\mathbf{v}_1) + \omega_{\mathcal{T}}(\mathbf{v}_2) + \omega_{\mathcal{T}}(\mathbf{v}_3) = \frac{1}{2} .$$

Thus the Brianchon–Gram relation is the extension of this well-known fact to any dimension and any convex polytope. □

Proof of Theorem 11.8. It suffices to prove $A_\Delta(0) = 0$ for a rational simplex Δ, since solid angles of a triangulation simply add, by (11.1). Theorem 11.7 gives $A_\Delta(0) = 0$ if $\dim \Delta$ is odd.

So now suppose Δ is a rational d-simplex, where d is even, with vertices $\mathbf{v}_1, \mathbf{v}_2, \ldots, \mathbf{v}_{d+1}$. Let $\mathcal{P}(n)$ be the $(d+1)$-dimensional pyramid that we obtain by taking the convex hull of $(\mathbf{v}_1, 0), (\mathbf{v}_2, 0), \ldots, (\mathbf{v}_{d+1}, 0)$, and $(0, 0, \ldots, 0, n)$, where n is a positive integer (see Figure 11.1). Note that, since $d+1$ is odd,

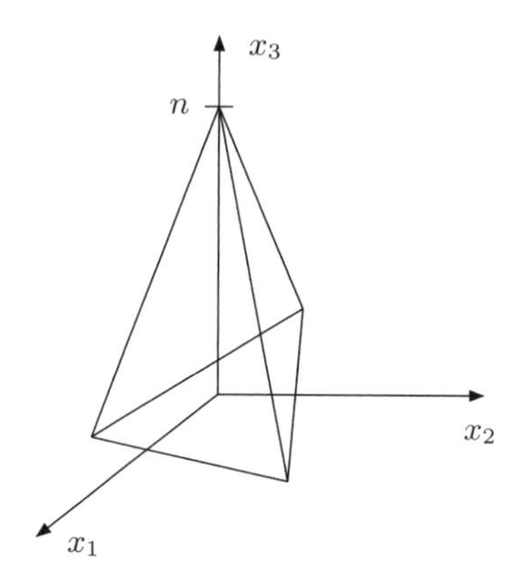

Fig. 11.1. The pyramid $\mathcal{P}(n)$ for a triangle Δ.

$$A_{\mathcal{P}(n)}(0) = \sum_{\mathcal{F}(n) \subseteq \mathcal{P}(n)} (-1)^{\dim \mathcal{F}(n)} \omega_{\mathcal{P}(n)}(\mathcal{F}(n)) = 0 .$$

We will conclude from this identity that $\sum_{\mathcal{F} \subseteq \Delta} (-1)^{\dim \mathcal{F}} \omega_\Delta(\mathcal{F}) = 0$, which implies that $A_\Delta(0) = 0$. To this end, we consider two types of faces of $\mathcal{P}(n)$:

(a) those that are also faces of Δ, and

(b) those that are not contained in Δ.

We start with the latter: Aside from the vertex $(0,0,\ldots,0,n)$, every face $\mathcal{F}(n)$ of $\mathcal{P}(n)$ that is not a face of Δ is the pyramid over a face $\mathcal{G}$ of Δ; let's denote this pyramid by $\operatorname{Pyr}(\mathcal{G},n)$. Further, as n grows, the solid angle of $\operatorname{Pyr}(\mathcal{G},n)$ (in $\mathcal{P}(n)$) approaches the solid angle of $\mathcal{G}$ (in Δ):

$$\lim_{n\to\infty} \omega_{\mathcal{P}(n)}\left(\operatorname{Pyr}(\mathcal{G},n)\right) = \omega_{\Delta}\left(\mathcal{G}\right),$$

since we're forming $\Delta\times[0,\infty)$ in the limit. On the other hand, a face $\mathcal{F}(n)=\mathcal{G}$ of $\mathcal{P}(n)$ that is also a face of Δ obeys the following limit behavior:

$$\lim_{n\to\infty} \omega_{\mathcal{P}(n)}\left(\mathcal{F}(n)\right) = \frac{1}{2}\,\omega_{\Delta}\left(\mathcal{G}\right).$$

The only face of $\mathcal{P}(n)$ that we still have to account for is the vertex $\mathbf{v} := (0,0,\ldots,0,n)$. Hence

$$\begin{aligned} 0 &= \sum_{\mathcal{F}(n)\subseteq\mathcal{P}(n)} (-1)^{\dim\mathcal{F}(n)}\omega_{\mathcal{P}(n)}(\mathcal{F}(n)) \\ &= \omega_{\mathcal{P}(n)}(\mathbf{v}) + \sum_{\mathcal{G}\subseteq\Delta}(-1)^{\dim\mathcal{G}+1}\omega_{\mathcal{P}(n)}\left(\operatorname{Pyr}(\mathcal{G},n)\right) \\ &\quad + \sum_{\mathcal{G}\subseteq\Delta}(-1)^{\dim\mathcal{G}}\omega_{\mathcal{P}(n)}\left(\mathcal{G}\right). \end{aligned}$$

Now we take the limit as $n\to\infty$ on both sides; note that $\lim_{n\to\infty}\omega_{\mathcal{P}(n)}(\mathbf{v}) = 0$, so that we obtain

$$\begin{aligned} 0 &= \sum_{\mathcal{G}\subseteq\Delta}(-1)^{\dim\mathcal{G}+1}\omega_{\Delta}\left(\mathcal{G}\right) + \sum_{\mathcal{G}\subseteq\Delta}(-1)^{\dim\mathcal{G}}\frac{1}{2}\,\omega_{\Delta}\left(\mathcal{G}\right) \\ &= \frac{1}{2}\sum_{\mathcal{G}\subseteq\Delta}(-1)^{\dim\mathcal{G}+1}\omega_{\Delta}\left(\mathcal{G}\right), \end{aligned}$$

and so

$$A_{\Delta}(0) = \sum_{\mathcal{G}\subseteq\Delta}(-1)^{\dim\mathcal{G}}\omega_{\Delta}(\mathcal{G}) = 0.$$

□

The combination of Theorems 11.7 and 11.8 implies that summing solid angles in a polygon is equivalent to computing its area:

Corollary 11.11. *Suppose $\mathcal{P}$ is a 2-dimensional integral polytope with area A. Then $A_{\mathcal{P}}(t) = A\,t^2$.*

11.4 The Generating Function of Macdonald's Solid-Angle Polynomials

We conclude this chapter with the study of the solid-angle analogue of Ehrhart series. Given an integral polytope $\mathcal{P}$, we define the **solid-angle series** of $\mathcal{P}$ as the generating function of the solid-angle polynomial, encoding the solid-angle sum over all dilates of $\mathcal{P}$ simultaneously:

$$\mathrm{Solid}_{\mathcal{P}}(z) := \sum_{t \geq 0} A_{\mathcal{P}}(t)\, z^t .$$

The following theorem is the solid-angle analogue to Theorems 3.12 and 4.4, with the added bonus that we get the palindromy of the numerator of $\mathrm{Solid}_{\mathcal{P}}$ for free.

Theorem 11.12. *Suppose $\mathcal{P}$ is an integral d-polytope. Then* $\mathrm{Solid}_{\mathcal{P}}$ *is a rational function of the form*

$$\mathrm{Solid}_{\mathcal{P}}(z) = \frac{a_d z^d + a_{d-1} z^{d-1} + \cdots + a_1 z}{(1-z)^{d+1}} .$$

Furthermore, we have the identity

$$\mathrm{Solid}_{\mathcal{P}}\left(\frac{1}{z}\right) = (-1)^{d+1}\, \mathrm{Solid}_{\mathcal{P}}(z)$$

or, equivalently, $a_k = a_{d+1-k}$ *for* $1 \leq k \leq \frac{d}{2}$.

Proof. The form of the rational function $\mathrm{Solid}_{\mathcal{P}}$ follows by Lemma 3.9 from the fact that $A_{\mathcal{P}}$ is a polynomial. The palindromy of $a_1, a_2, \dots, a_d$ is equivalent to the relation

$$\mathrm{Solid}_{\mathcal{P}}\left(\frac{1}{z}\right) = (-1)^{d+1}\, \mathrm{Solid}_{\mathcal{P}}(z) ,$$

which, in turn, follows from Theorem 11.7:

$$\mathrm{Solid}_{\mathcal{P}}(z) = \sum_{t \geq 0} A_{\mathcal{P}}(t)\, z^t = \sum_{t \geq 0} (-1)^d A_{\mathcal{P}}(-t)\, z^t = (-1)^d \sum_{t \leq 0} A_{\mathcal{P}}(t)\, z^{-t} .$$

Now we use Exercise 4.5:

$$(-1)^d \sum_{t \leq 0} A_{\mathcal{P}}(t)\, z^{-t} = (-1)^{d+1} \sum_{t \geq 1} A_{\mathcal{P}}(t)\, z^{-t} = (-1)^{d+1}\, \mathrm{Solid}_{\mathcal{P}}\left(\frac{1}{z}\right) .$$

In the last step we used the fact that $A_{\mathcal{P}}(0) = 0$ (Theorem 11.8). □

Notes

1. I. G. Macdonald inaugurated the systematic study of solid-angle sums in integral polytopes. The fundamental Theorems 11.4, 11.7, and 11.8 can be found in his 1971 paper [123]. The proof of Theorem 11.7 we give here follows [26].

2. The Brianchon–Gram relation (Corollary 11.9) is the solid-angle analogue of the Euler relation for face numbers (Theorem 5.2). The 2-dimensional case discussed in Example 11.10 is ancient; it was most certainly known to Euclid. The 3-dimensional case of Corollary 11.9 was discovered by Charles Julien Brianchon in 1837 and—as far as we know—was independently reproved by Jørgen Gram in 1874 [88]. It is not clear who first proved the general d-dimensional case of Corollary 11.9. The oldest proofs we could find were from the 1960s, by Branko Grünbaum [90], Micha A. Perles, and Geoffrey C. Shephard [142, 162].

3. Theorem 11.5 is a particular case of a reciprocity relation for *simple lattice-invariant valuations* due to Peter McMullen [128], who also proved a parallel extension of Ehrhart–Macdonald reciprocity to general lattice-invariant valuations. There is a current resurgence of activity on solid angles; see, for example, [52].

Exercises

11.1. Compute $A_{\mathcal{P}}(t)$, where $\mathcal{P}$ is the regular tetrahedron with vertices $(0,0,0)$, $(1,1,0)$, $(1,0,1)$, and $(0,1,1)$ (see Exercise 2.13).

11.2. Compute $A_{\mathcal{P}}(t)$, where $\mathcal{P}$ is the rational triangle with vertices $(0,0)$, $\left(\frac{1}{2},\frac{1}{2}\right)$, and $(1,0)$.

11.3. ♣ Let $\mathcal{K}$ be a rational d-cone and let $\mathbf{m} \in \mathbb{Z}^d$. By analogy with Exercise 3.5, show that $\alpha_{\mathbf{m}+\mathcal{K}}(\mathbf{z}) = \mathbf{z}^{\mathbf{m}} \alpha_{\mathcal{K}}(\mathbf{z})$.

11.4. ♣ Complete the proof of Theorem 11.5: For a rational pointed d-cone $\mathcal{K}$, $\alpha_{\mathcal{K}}(\mathbf{z})$ is a rational function that satisfies

$$\alpha_{\mathcal{K}}\left(\frac{1}{\mathbf{z}}\right) = (-1)^d \alpha_{\mathcal{K}}(\mathbf{z}) .$$

11.5. ♣ Suppose Δ is a rational simplex. Prove that

$$\sum_{\substack{\mathcal{F} \subseteq \Delta \\ \dim \mathcal{F} > 0}} \omega_{\Delta}(\mathcal{F}) \sum_{\mathbf{v} \text{ a vertex of } \mathcal{F}} \sigma_{\mathcal{K}_{\mathbf{v}}(\mathcal{F})^\circ}(\mathbf{z}) = \sum_{\mathbf{v} \text{ a vertex of } \Delta} \sum_{\substack{\mathcal{F} \subseteq \mathcal{K}_{\mathbf{v}} \\ \dim \mathcal{F} > 0}} \omega_{\mathcal{K}_{\mathbf{v}}}(\mathcal{F})\, \sigma_{\mathcal{F}^\circ}(\mathbf{z}) .$$

11.6. ♣ Provide the details of the proof of Theorem 11.7 for *rational* polytopes: Prove that if $\mathcal{P}$ is a rational convex polytope, then the quasipolynomial $A_{\mathcal{P}}$ satisfies

$$A_{\mathcal{P}}(-t) = (-1)^{\dim \mathcal{P}} A_{\mathcal{P}}(t) \, .$$

11.7. Recall from Exercise 3.1 that to any permutation $\pi \in S_d$ on d elements we can associate the simplex

$$\Delta_\pi := \operatorname{conv} \left\{ \mathbf{0}, \mathbf{e}_{\pi(1)}, \mathbf{e}_{\pi(1)} + \mathbf{e}_{\pi(2)}, \dots, \mathbf{e}_{\pi(1)} + \mathbf{e}_{\pi(2)} + \cdots + \mathbf{e}_{\pi(d)} \right\} .$$

Prove that for all $\pi \in S_n$, $A_{\Delta_\pi}(t) = \frac{1}{d!} \, t^d$.

11.8. Give a direct proof of Corollary 11.11, e.g., using Pick's theorem (Theorem 2.8).

11.9. State and prove the analogue of Theorem 11.12 for *rational* polytopes.

Open Problems

11.10. Study the roots of solid-angle polynomials.

11.11. Classify all polytopes that have only rational solid angles at their vertices.

11.12. Among all d-simplices Δ, does the regular d-simplex have the property that $\sum_{\mathbf{v} \text{ a vertex of } \Delta} \omega_\Delta(\mathbf{v})$ is a minimum?

11.13. Which integral polytopes $\mathcal{P}$ have solid-angle polynomials $A_{\mathcal{P}}(t) \in \mathbb{Q}[t]$? That is, for which integral polytopes $\mathcal{P}$ are all the coefficients of $A_{\mathcal{P}}(t)$ rational?

11.14. Are there solid-angle polynomials with negative coefficients?

11.15. Does the numerator polynomial of the rational generating function $\operatorname{Solid}_{\mathcal{P}}(z)$ always have nonnegative coefficients, by analogy with Theorem 3.12?

12

A Discrete Version of Green's Theorem Using Elliptic Functions

The shortest route between two truths in the real domain passes through the complex domain.

Jacques Salomon Hadamard (1865–1963)

We now allow ourselves the luxury of using basic complex analysis. In particular, we assume that the reader is familiar with contour integration and the residue theorem. We may view the residue theorem as yet another result that intimately connects the continuous and the discrete: it transforms a continuous integral into a discrete sum of residues.

Using the Weierstraß $\wp$ and ζ functions, we show here that Pick's theorem is a discrete version of Green's theorem in the plane. As a bonus, we also obtain an integral formula (Theorem 12.5 below) for the discrepancy between the area enclosed by a general curve C and the number of integer points contained in C.

12.1 The Residue Theorem

We begin this chapter by reviewing a few concepts from complex analysis. Suppose the complex-valued function f has an **isolated singularity** $w \in G$; that is, there is an open set $G \subset \mathbb{C}$ such that f is analytic on $G \setminus \{w\}$. Then f can be expressed locally by the **Laurent series**

$$f(z) = \sum_{n \in \mathbb{Z}} c_n (z - w)^n ,$$

valid for all $z \in G$; here $c_n \in \mathbb{C}$. The coefficient c_{-1} is called the **residue** of f at w; we will denote it by $\operatorname{Res}(z = w)$. The reason to give c_{-1} a special name can be found in the following theorem. We call a function **meromorphic** if it is analytic in $\mathbb{C}$ with the exception of isolated poles.

Theorem 12.1 (Residue theorem). *Suppose f is meromorphic and C is a positively oriented, piecewise differentiable, simple, closed curve that does not pass through any pole of f. Then*

$$\int_C f = 2\pi i \sum_w \operatorname{Res}(z=w),$$

where the sum is taken over all singularities w inside C. □

If f is a rational function, Theorem 12.1 gives the same result as the partial fraction expansion of f. We illustrate this philosophy by returning to the elementary beginnings of Chapter 1.

Example 12.2. Recall our constant-term identity for the restricted partition function for $A = \{a_1, a_2, \dots, a_d\}$ in Chapter 1:

$$p_A(n) = \operatorname{const}\left(\frac{1}{(1-z^{a_1})(1-z^{a_2})\cdots(1-z^{a_d})\,z^n}\right).$$

Computing the constant term of the Laurent series of $\frac{1}{(1-z^{a_1})\cdots(1-z^{a_d})z^n}$ expanded about $z=0$ is, naturally, equivalent to "shifting" this function by one exponent and computing the residue at $z=0$ of the function

$$f(z) := \frac{1}{(1-z^{a_1})(1-z^{a_2})\cdots(1-z^{a_d})\,z^{n+1}}.$$

Now let C_r be a positively oriented circle of radius $r>1$, centered at the origin. The residue $\operatorname{Res}(z=0) = p_A(n)$ is one of the residues that are picked up by the integral

$$\frac{1}{2\pi i}\int_{C_r} f = \operatorname{Res}(z=0) + \sum_w \operatorname{Res}(z=w),$$

where the sum is over all nonzero poles w of f that lie inside C_r. These poles are at the $a_1^{\text{th}}, a_2^{\text{th}}, \dots, a_d^{\text{th}}$ roots of unity. Moreover, with the help of Exercise 12.1 we can show that

$$\begin{aligned}
0 &= \lim_{r\to\infty}\frac{1}{2\pi i}\int_{C_r} f \\
&= \lim_{r\to\infty}\left(\operatorname{Res}(z=0) + \sum_w \operatorname{Res}(z=w)\right) \\
&= \operatorname{Res}(z=0) + \sum_w \operatorname{Res}(z=w),
\end{aligned}$$

where the sum extends over all $a_1^{\text{th}}, a_2^{\text{th}}, \dots, a_d^{\text{th}}$ roots of unity. In other words,

$$p_A(n) = \operatorname{Res}(z=0) = -\sum_w \operatorname{Res}(z=w).$$

To obtain the restricted partition function p_A, it remains to compute the residues at the roots of unity, and we invite the reader to realize that this computation is equivalent to the partial fraction expansion of Chapter 1 (Exercise 12.2). □

Analogous residue computations could replace any of the constant-term calculations that we performed in the earlier chapters.

12.2 The Weierstraß $\wp$ and ζ Functions

The main character in our play is the **Weierstraß ζ-function**, defined by

$$\zeta(z) = \frac{1}{z} + \sum_{(m,n)\in\mathbb{Z}^2\setminus(0,0)} \left(\frac{1}{z-(m+ni)} + \frac{1}{m+ni} + \frac{z}{(m+ni)^2} \right) . \tag{12.1}$$

This infinite sum converges absolutely for z belonging to compact subsets of the lattice-punctured plane $\mathbb{C} \setminus \mathbb{Z}^2$ (Exercise 12.4), and hence forms a meromorphic function of z.

The Weierstraß ζ-function possesses the following salient properties, which follow immediately from (12.1):

(1) ζ has a simple pole at every integer point $m+ni$ and is analytic elsewhere.
(2) The residue of ζ at each integer point $m+ni$ equals 1.

We can easily check (Exercise 12.5) that

$$\wp(z) := -\zeta'(z) = \frac{1}{z^2} + \sum_{(m,n)\in\mathbb{Z}^2\setminus(0,0)} \left(\frac{1}{(z-(m+ni))^2} - \frac{1}{(m+ni)^2} \right) , \tag{12.2}$$

the **Weierstraß $\wp$-function**. The $\wp$-function has a pole of order 2 at each integer point $m+ni$ and is analytic elsewhere, but has residue equal to zero at each integer point $m+ni$. However, $\wp$ possesses a very pleasant property that ζ does not: $\wp$ is **doubly periodic** on $\mathbb{C}$. We may state this more concretely:

Lemma 12.3. $\wp(z+1) = \wp(z+i) = \wp(z)$.

Proof. We first invite the reader to prove the following two properties of $\wp'$ (Exercises 12.6 and 12.7):

$$\wp'(z+1) = \wp'(z) , \tag{12.3}$$

$$\int_{z_0}^{z_1} \wp'(z)\, dz \quad \text{is path independent.} \tag{12.4}$$

By (12.3),

$$\frac{d}{dz} \left(\wp(z+1) - \wp(z) \right) = \wp'(z+1) - \wp'(z) = 0 ,$$

so $\wp(z+1) - \wp(z) = c$ for some constant c. On the other hand, $\wp$ is an even function (Exercise 12.8), and so $z = -\frac{1}{2}$ gives us

$$c = \wp\left(\tfrac{1}{2}\right) - \wp\left(-\tfrac{1}{2}\right) = 0\,.$$

This shows that $\wp(z+1) = \wp(z)$ for all $z \in \mathbb{C}\backslash\mathbb{Z}^2$. An analogous proof, which we invite the reader to construct in Exercise 12.9, shows that $\wp(z+i) = \wp(z)$. □

Lemma 12.3 implies that $\wp(z+m+ni) = \wp(z)$ for all $m, n \in \mathbb{Z}$. The following lemma shows that the Weierstraß ζ-function is only a conjugate-analytic term away from being doubly periodic.

Lemma 12.4. *There is a constant α such that the function $\zeta(z)+\alpha\bar{z}$ is doubly periodic with periods 1 and i.*

Proof. We begin with $w = m + ni$:

$$\zeta(z+m+ni) - \zeta(z) = -\int_{w=0}^{m+ni} \wp(z+w)\,dw\,, \tag{12.5}$$

by definition of $\wp(z) = -\zeta'(z)$. To make sure that (12.5) makes sense, we should also check that the definite integral in (12.5) is path independent (Exercise 12.10).

Due to the double periodicity of $\wp$,

$$\begin{aligned}\int_{w=0}^{m+ni} \wp(z+w)\,dw &= m\int_0^1 \wp(z+t)\,dt + ni\int_0^1 \wp(z+it)\,dt \\ &:= m\alpha(z) + ni\beta(z)\,,\end{aligned}$$

where

$$\alpha(z) := \int_0^1 \wp(z+t)\,dt \qquad \text{and} \qquad \beta(z) := \int_0^1 \wp(z+it)\,dt\,.$$

Now we observe that $\alpha(z+x_0) = \alpha(z)$ for any $x_0 \in \mathbb{R}$, so that $\alpha(x+iy)$ depends only on y. Similarly, $\beta(x+iy)$ depends only on x. But

$$\zeta(z+m+in) - \zeta(z) = -\left(m\alpha(y) + in\beta(x)\right)$$

must be analytic for all $z \in \mathbb{C}\setminus\mathbb{Z}^2$. If we now set $m = 0$, we conclude that $\beta(x)$ must be analytic in $\mathbb{C}\setminus\mathbb{Z}^2$, so that $\beta(x)$ must be a constant by the Cauchy–Riemann equations for analytic functions. Similarly, setting $n = 0$ implies that $\alpha(y)$ is constant. Thus

$$\zeta(z+m+in) - \zeta(z) = -\left(m\alpha + in\beta\right)$$

with constants α and β. Going back to the Weierstraß $\wp$-function, we can integrate the identity (Exercise 12.11)

$$\wp(iz) = -\wp(z) \tag{12.6}$$

to obtain the relationship $\beta = -\alpha$, since

$$\beta = \int_0^1 \wp(z + it)\, dt = \int_0^1 \wp(it)\, dt = -\int_0^1 \wp(t)\, dt = -\alpha\,.$$

To summarize, we have

$$\zeta(z + m + in) - \zeta(z) = -m\alpha + in\alpha = -\alpha\left(\overline{z + m + in} - \overline{z}\right),$$

so that $\zeta(z) + \alpha\overline{z}$ is doubly periodic. □

12.3 A Contour-Integral Extension of Pick's Theorem

For the remainder of this chapter, let C be any piecewise-differentiable, simple, closed curve in the plane, with a counterclockwise parametrization. We let D denote the region that C contains in its interior.

Theorem 12.5. *Let C avoid any integer point, that is, $C \cap \mathbb{Z}^2 = \varnothing$. Let I denote the number of integer points interior to C, and A the area of the region D enclosed by the curve C. Then*

$$\frac{1}{2\pi i}\int_C (\zeta(z) - \pi\overline{z})\, dz = I - A\,.$$

Proof. We have

$$\int_C (\zeta(z) + \alpha\overline{z})\, dz = \int_C \zeta(z)\, dz + \alpha\int_C (x - iy)\,(dx + idy)\,.$$

By Theorem 12.1, $\int_C \zeta(z)\, dz$ is equal to the sum of the residues of ζ at all of its interior poles. There are I such poles, and each pole of ζ has residue 1. Thus

$$\frac{1}{2\pi i}\int_C \zeta(z)\, dz = I\,. \tag{12.7}$$

On the other hand, Green's theorem tells us that

$$\begin{aligned}\int_C (x - iy)\,(dx + idy) &= \int_C (x - iy)\, dx + (y + ix)\, dy \\ &= \int_D \frac{\partial}{\partial x}(y + ix) - \frac{\partial}{\partial y}(x - iy) \\ &= \iint_D 2i \\ &= 2iA\,.\end{aligned}$$

Going back to (12.7), we get

$$\int_C (\zeta(z) + \alpha\bar{z})\, dz = 2\pi i I + \alpha\,(2iA)\,. \tag{12.8}$$

We only have to show that $\alpha = -\pi$. Consider the particular curve C that is a square path, centered at the origin, traversing the origin counterclockwise, and bounding a square of area 1. Thus $I = 1$ for this path. Since $\zeta(z) + \alpha\bar{z}$ is doubly periodic by Lemma 12.4, the integral in (12.8) vanishes. We can conclude that

$$0 = 2\pi i \cdot 1 + \alpha\,(2i \cdot 1)\,,$$

so that $\alpha = -\pi$. □

Notice that Theorem 12.5 has given us information about the Weierstraß ζ-function, namely that $\alpha = -\pi$.

This chapter offers a detour into an infinite landscape of discrete results that meet their continuous counterparts. Equipped with the modest tools offered in this book, we hope we have motivated the reader to explore this landscape further...

Notes

1. The Weierstraß $\wp$-function, named after Karl Theodor Wilhelm Weierstraß (1815–1897),[1] can be extended to any two-dimensional lattice $\mathcal{L} = \{kw_1 + jw_2 : k, j \in \mathbb{Z}\}$ for some $w_1, w_2 \in \mathbb{C}$ that are linearly independent over $\mathbb{R}$:

$$\wp_{\mathcal{L}}(z) = \frac{1}{z^2} + \sum_{m \in \mathcal{L}\setminus\{0\}} \left(\frac{1}{(z-m)^2} - \frac{1}{m^2} \right).$$

The Weierstraß $\wp_{\mathcal{L}}$-function and its derivative $\wp'_{\mathcal{L}}$ satisfy a polynomial relationship, namely, $\left(\wp'_{\mathcal{L}}\right)^2 = 4\left(\wp_{\mathcal{L}}\right)^3 - g_2\,\wp_{\mathcal{L}} - g_3$ for some constants g_2 and g_3 that depend on $\mathcal{L}$. This is the beginning of a wonderful friendship between complex analysis and elliptic curves.

2. Theorem 12.5 appeared in [75]. There it is also shown that one can retrieve Pick's theorem (Theorem 2.8) from Theorem 12.5.

Exercises

12.1. ♣ Show that for positive integers $a_1, a_d, \dots, a_d, n$,

---

[1] For more information about Weierstraß, see
`http://www-groups.dcs.st-and.ac.uk/~history/Biographies/Weierstrass.html`.

$$\lim_{r \to \infty} \int_{C_r} \frac{1}{(1-z^{a_1}) \cdots (1-z^{a_d})\, z^{n+1}} = 0\,.$$

This computation shows that the integrand above "has no pole at infinity."

12.2. ♣ Compute the residues at the nontrivial roots of unity of

$$f(z) = \frac{1}{(1-z^{a_1}) \cdots (1-z^{a_d})\, z^{n+1}}\,.$$

For simplicity, you may assume that $a_1, a_2, \dots, a_d$ are pairwise relatively prime.

12.3. Give an integral version of Theorem 2.13.

12.4. ♣ Show that

$$\zeta(z) \;=\; \frac{1}{z} + \sum_{(m,n) \in \mathbb{Z}^2 \setminus (0,0)} \left(\frac{1}{z-(m+ni)} + \frac{1}{m+ni} + \frac{z}{(m+ni)^2} \right)$$

converges absolutely for z belonging to compact subsets of $\mathbb{C} \setminus \mathbb{Z}^2$.

12.5. ♣ Prove (12.2), that is,

$$\zeta'(z) = -\frac{1}{z^2} - \sum_{(m,n) \in \mathbb{Z}^2 \setminus (0,0)} \left(\frac{1}{(z-(m+ni))^2} - \frac{1}{(m+ni)^2} \right).$$

12.6. ♣ Prove (12.3), that is, show that $\wp'(z+1) = \wp'(z)$.

12.7. ♣ Prove (12.4), that is, show that for any $z_0, z_1 \in \mathbb{C} \setminus \mathbb{Z}^2$, $\int_{z_0}^{z_1} \wp'(w)\, dw$ is path independent.

12.8. ♣ Show that $\wp$ is even, that is, $\wp(-z) = \wp(z)$.

12.9. ♣ Finish the proof of Lemma 12.3 by showing that $\wp(z+i) = \wp(z)$.

12.10. ♣ Prove that the integral in (12.5),

$$\zeta(z+m+ni) - \zeta(z) = -\int_{w=0}^{w=m+ni} \wp(z+w)\, dw\,,$$

is path independent.

12.11. ♣ Prove (12.6), that is, $\wp(iz) = -\wp(z)$.

Open Problems

12.12. Can we get even more information about the Weierstraß $\wp$ and ζ functions by using more detailed knowledge of the discrepancy between I and A for special curves C?

12.13. Find a complex-analytic extension of Theorem 12.5 to higher dimensions.

A

Vertex and Hyperplane Descriptions of Polytopes

Everything should be made as simple as possible, but not simpler.

Albert Einstein

In this appendix, we prove that every polytope has a vertex and a hyperplane description. This appendix owes everything to Günter Ziegler's beautiful exposition in [193]; in fact, these pages contain merely a few cherries picked from [193, Lecture 1].

As in Chapter 3, it is easier to move to the world of cones. To be as concrete as possible, let us call $\mathcal{K} \subseteq \mathbb{R}^d$ an **h-cone** if

$$\mathcal{K} = \left\{ \mathbf{x} \in \mathbb{R}^d : \mathbf{A}\,\mathbf{x} \leq \mathbf{0} \right\}$$

for some $\mathbf{A} \in \mathbb{R}^{m \times d}$; in this case $\mathcal{K}$ is given as the intersection of m halfspaces determined by the rows of $\mathbf{A}$. We use the notation $\mathcal{K} = \operatorname{hcone}(\mathbf{A})$.

On the other hand, we call $\mathcal{K} \subseteq \mathbb{R}^d$ a **v-cone** if

$$\mathcal{K} = \left\{ \mathbf{B}\,\mathbf{y} : \mathbf{y} \geq \mathbf{0} \right\}$$

for some $\mathbf{B} \in \mathbb{R}^{d \times n}$, that is, $\mathcal{K}$ is a pointed cone with the column vectors of $\mathbf{B}$ as generators. In this case we use the notation $\mathcal{K} = \operatorname{vcone}(\mathbf{B})$.

Note that, according to our definitions, any h- or v-cone contains the origin in its apex. We will prove that every h-cone is a v-cone and vice versa. More precisely:

Theorem A.1. *For every* $\mathbf{A} \in \mathbb{R}^{m \times d}$ *there exists* $\mathbf{B} \in \mathbb{R}^{d \times n}$ *(for some* n*) such that* $\operatorname{hcone}(\mathbf{A}) = \operatorname{vcone}(\mathbf{B})$*. Conversely, for every* $\mathbf{B} \in \mathbb{R}^{d \times n}$ *there exists* $\mathbf{A} \in \mathbb{R}^{m \times d}$ *(for some* m*) such that* $\operatorname{vcone}(\mathbf{B}) = \operatorname{hcone}(\mathbf{A})$*.*

We will prove the two halves of Theorem A.1 in Sections A.1 and A.2. For now, let us record that Theorem A.1 implies our goal, that is, the equivalence of the vertex and halfspace description of a polytope:

Corollary A.2. *If $\mathcal{P}$ is the convex hull of finitely many points in $\mathbb{R}^d$, then $\mathcal{P}$ is the intersection of finitely many half spaces in $\mathbb{R}^d$. Conversely, if $\mathcal{P}$ is given as the bounded intersection of finitely many half spaces in $\mathbb{R}^d$, then $\mathcal{P}$ is the convex hull of finitely many points in $\mathbb{R}^d$.*

Proof. If $\mathcal{P} = \operatorname{conv}\{\mathbf{v}_1, \mathbf{v}_2, \dots, \mathbf{v}_n\}$ for some $\mathbf{v}_1, \mathbf{v}_2, \dots, \mathbf{v}_n \in \mathbb{R}^d$, then coning over $\mathcal{P}$ (as defined in Chapter 3) gives

$$\operatorname{cone}(\mathcal{P}) = \operatorname{vcone}\begin{pmatrix} \mathbf{v}_1 & \mathbf{v}_2 & \cdots & \mathbf{v}_n \\ 1 & 1 & & 1 \end{pmatrix}.$$

By Theorem A.1 we can find a matrix $(\mathbf{A}, \mathbf{b}) \in \mathbb{R}^{m \times (d+1)}$ such that

$$\operatorname{cone}(\mathcal{P}) = \operatorname{hcone}(\mathbf{A}, \mathbf{b}) = \left\{ \mathbf{x} \in \mathbb{R}^{d+1} : (\mathbf{A}, \mathbf{b})\,\mathbf{x} \le \mathbf{0} \right\}.$$

We recover the polytope $\mathcal{P}$ upon setting $x_{d+1} = 1$, that is,

$$\mathcal{P} = \left\{ \mathbf{x} \in \mathbb{R}^d : \mathbf{A}\mathbf{x} \le -\mathbf{b} \right\},$$

which is a hyperplane description of $\mathcal{P}$.

These steps can be reversed: Suppose the polytope $\mathcal{P}$ is given as

$$\mathcal{P} = \left\{ \mathbf{x} \in \mathbb{R}^d : \mathbf{A}\mathbf{x} \le -\mathbf{b} \right\}$$

for some $\mathbf{A} \in \mathbb{R}^{m \times d}$ and $\mathbf{b} \in \mathbb{R}^m$. Then $\mathcal{P}$ can be obtained from

$$\operatorname{hcone}(\mathbf{A}, \mathbf{b}) = \left\{ \mathbf{x} \in \mathbb{R}^{d+1} : (\mathbf{A}, \mathbf{b})\,\mathbf{x} \le \mathbf{0} \right\}$$

by setting $x_{d+1} = 1$. By Theorem A.1 we can construct a matrix $\mathbf{B} \in \mathbb{R}^{(d+1) \times n}$ such that

$$\operatorname{hcone}(\mathbf{A}, \mathbf{b}) = \operatorname{vcone}(\mathbf{B}).$$

We may normalize the generators of $\operatorname{vcone}(\mathbf{B})$, that is, the columns of $\mathbf{B}$, such that they all have $(d+1)$st variable equal to one:

$$\mathbf{B} = \begin{pmatrix} \mathbf{v}_1 & \mathbf{v}_2 & \cdots & \mathbf{v}_n \\ 1 & 1 & & 1 \end{pmatrix}.$$

Since $\mathcal{P}$ can be recovered from $\operatorname{vcone}(\mathbf{B})$ by setting $x_{d+1} = 1$, we conclude that $\mathcal{P} = \operatorname{conv}\{\mathbf{v}_1, \mathbf{v}_2, \dots, \mathbf{v}_n\}$. □

A.1 Every h-cone is a v-cone

Suppose

$$\mathcal{K} = \operatorname{hcone}(\mathbf{A}) = \left\{ \mathbf{x} \in \mathbb{R}^d : \mathbf{A}\mathbf{x} \le \mathbf{0} \right\}$$

for some $\mathbf{A} \in \mathbb{R}^{m \times d}$. We introduce an auxiliary m-dimensional variable $\mathbf{y}$ and write

$$\mathcal{K} = \left\{ \begin{pmatrix} \mathbf{x} \\ \mathbf{y} \end{pmatrix} \in \mathbb{R}^{d+m} : \mathbf{A}\,\mathbf{x} \le \mathbf{y} \right\} \cap \left\{ \begin{pmatrix} \mathbf{x} \\ \mathbf{y} \end{pmatrix} \in \mathbb{R}^{d+m} : \mathbf{y} = 0 \right\}. \tag{A.1}$$

(Strictly speaking, this is $\mathcal{K}$ lifted into a d-dimensional subspace of $\mathbb{R}^{d+m}$.) Our goal in this section is to prove the following two lemmas.

Lemma A.3. *The h-cone* $\left\{ \binom{\mathbf{x}}{\mathbf{y}} \in \mathbb{R}^{d+m} : \mathbf{A}\,\mathbf{x} \le \mathbf{y} \right\}$ *is a v-cone.*

Lemma A.4. *If* $\mathcal{K} \subseteq \mathbb{R}^d$ *is a v-cone, then so is* $\mathcal{K} \cap \left\{ \mathbf{x} \in \mathbb{R}^d : x_k = 0 \right\}$, *for any* k.

The first half of Theorem A.1 follows with these two lemmas, as we can start with (A.1) and intersect with one hyperplane $y_k = 0$ at a time.

Proof of Lemma A.3. We start by noting that

$$\begin{aligned} \mathcal{K} &= \left\{ \begin{pmatrix} \mathbf{x} \\ \mathbf{y} \end{pmatrix} \in \mathbb{R}^{d+m} : \mathbf{A}\,\mathbf{x} \le \mathbf{y} \right\} \\ &= \left\{ \begin{pmatrix} \mathbf{x} \\ \mathbf{y} \end{pmatrix} \in \mathbb{R}^{d+m} : (\mathbf{A}, -\mathbf{I}) \begin{pmatrix} \mathbf{x} \\ \mathbf{y} \end{pmatrix} \le \mathbf{0} \right\} \end{aligned}$$

is an h-cone; here $\mathbf{I}$ represents an $m \times m$ identity matrix. Let us denote the kth unit vector by $\mathbf{e}_k$. Then we can decompose

$$\begin{aligned} \begin{pmatrix} \mathbf{x} \\ \mathbf{y} \end{pmatrix} &= \sum_{j=1}^{d} x_j \begin{pmatrix} \mathbf{e}_j \\ \mathbf{A}\,\mathbf{e}_j \end{pmatrix} + \sum_{k=1}^{m} (y_k - (\mathbf{A}\,\mathbf{x})_k) \begin{pmatrix} \mathbf{0} \\ \mathbf{e}_k \end{pmatrix} \\ &= \sum_{j=1}^{d} |x_j| \operatorname{sign}(x_j) \begin{pmatrix} \mathbf{e}_j \\ \mathbf{A}\,\mathbf{e}_j \end{pmatrix} + \sum_{k=1}^{m} (y_k - (\mathbf{A}\,\mathbf{x})_k) \begin{pmatrix} \mathbf{0} \\ \mathbf{e}_k \end{pmatrix}. \end{aligned}$$

Note that if $\binom{\mathbf{x}}{\mathbf{y}} \in \mathcal{K}$ then $y_k - (\mathbf{A}\,\mathbf{x})_k \ge 0$ for all k, and so $\binom{\mathbf{x}}{\mathbf{y}}$ can be written as a nonnegative linear combination of the vectors $\operatorname{sign}(x_j) \binom{\mathbf{e}_j}{\mathbf{A}\,\mathbf{e}_j}$, $1 \le j \le d$, and $\binom{\mathbf{0}}{\mathbf{e}_k}$, $1 \le k \le m$. But this means that $\mathcal{K}$ is a v-cone. □

Proof of Lemma A.4. Suppose $\mathcal{K} = \operatorname{vcone}(\mathbf{B})$, where $\mathbf{B}$ has the column vectors $\mathbf{b}_1, \mathbf{b}_2, \dots, \mathbf{b}_n \in \mathbb{R}^d$; that is, $\mathbf{b}_1, \mathbf{b}_2, \dots, \mathbf{b}_n$ are the generators of $\mathcal{K}$. Fix $k \le d$ and construct a new matrix $\mathbf{B}_k$ whose column vectors are all $\mathbf{b}_j$ for which $b_{jk} = 0$, and the combinations $b_{ik}\mathbf{b}_j - b_{jk}\mathbf{b}_i$ whenever $b_{ik} > 0$ and $b_{jk} < 0$. We claim that

$$\mathcal{K} \cap \left\{ \mathbf{x} \in \mathbb{R}^d : x_k = 0 \right\} = \operatorname{vcone}(\mathbf{B}_k).$$

Every $\mathbf{x} \in \operatorname{vcone}(\mathbf{B}_k)$ satisfies $x_k = 0$ by construction of $\mathbf{B}_k$, and so $\operatorname{vcone}(\mathbf{B}_k) \subseteq \mathcal{K} \cap \left\{ \mathbf{x} \in \mathbb{R}^d : x_k = 0 \right\}$ follows immediately. We need to do some more work to prove the reverse containment.

Suppose $\mathbf{x} \in \mathcal{K} \cap \left\{ \mathbf{x} \in \mathbb{R}^d : x_k = 0 \right\}$, that is, $\mathbf{x} = \lambda_1 \mathbf{b}_1 + \lambda_2 \mathbf{b}_2 + \cdots + \lambda_n \mathbf{b}_n$ for some $\lambda_1, \lambda_2, \ldots, \lambda_n \geq 0$ and $x_k = \lambda_1 b_{1k} + \lambda_2 b_{2k} + \cdots + \lambda_n b_{nk} = 0$. This allows us to define

$$\Lambda = \sum_{i:\, b_{ik} > 0} \lambda_i b_{ik} = - \sum_{j:\, b_{jk} < 0} \lambda_j b_{jk} .$$

Note that $\Lambda \geq 0$. Now consider the decomposition

$$\mathbf{x} = \sum_{j:\, b_{jk} = 0} \lambda_j \mathbf{b}_j + \sum_{i:\, b_{ik} > 0} \lambda_i \mathbf{b}_i + \sum_{j:\, b_{jk} < 0} \lambda_j \mathbf{b}_j . \tag{A.2}$$

If $\Lambda = 0$ then $\lambda_i b_{ik} = 0$ for all i such that $b_{ik} > 0$, and so $\lambda_i = 0$ for these i. Similarly, $\lambda_j = 0$ for all j such that $b_{jk} < 0$. Thus we conclude from $\Lambda = 0$ that

$$\mathbf{x} = \sum_{j:\, b_{jk} = 0} \lambda_j \mathbf{b}_j \in \operatorname{vcone}(\mathbf{B}_k) .$$

Now assume $\Lambda > 0$. Then we can expand the decomposition (A.2) into

$$\begin{aligned}
\mathbf{x} &= \sum_{j:\, b_{jk} = 0} \lambda_j \mathbf{b}_j + \frac{1}{\Lambda} \left(- \sum_{j:\, b_{jk} < 0} \lambda_j b_{jk} \right) \left(\sum_{i:\, b_{ik} > 0} \lambda_i \mathbf{b}_i \right) \\
&\qquad + \frac{1}{\Lambda} \left(\sum_{i:\, b_{ik} > 0} \lambda_i b_{ik} \right) \left(\sum_{j:\, b_{jk} < 0} \lambda_j \mathbf{b}_j \right) \\
&= \sum_{j:\, b_{jk} = 0} \lambda_j \mathbf{b}_j + \frac{1}{\Lambda} \sum_{\substack{i:\, b_{ik} > 0 \\ j:\, b_{jk} < 0}} \lambda_i \lambda_j \left(b_{ik} \mathbf{b}_j - b_{jk} \mathbf{b}_i \right),
\end{aligned}$$

which is by construction in $\operatorname{vcone}(\mathbf{B}_k)$. □

A.2 Every v-cone is an h-cone

Suppose

$$\mathcal{K} = \operatorname{vcone}(\mathbf{B}) = \{ \mathbf{B}\,\mathbf{y} : \mathbf{y} \geq \mathbf{0} \}$$

for some $\mathbf{B} \in \mathbb{R}^{d \times n}$. Then $\mathcal{K}$ is the projection of

$$\left\{ \begin{pmatrix} \mathbf{x} \\ \mathbf{y} \end{pmatrix} \in \mathbb{R}^{d+n} : \mathbf{y} \geq \mathbf{0},\, \mathbf{x} = \mathbf{B}\,\mathbf{y} \right\} \tag{A.3}$$

to the subspace $\left\{ \binom{\mathbf{x}}{\mathbf{y}} \in \mathbb{R}^{d+n} : \mathbf{y} = \mathbf{0} \right\}$. The constraints for (A.3) can be written as

$$\mathbf{y} \geq \mathbf{0} \qquad \text{and} \qquad (\mathbf{I}, -\mathbf{B}) \begin{pmatrix} \mathbf{x} \\ \mathbf{y} \end{pmatrix} = \mathbf{0} .$$

Thus the set (A.3) is an h-cone, for which we can project one component of $\mathbf{y}$ at a time to obtain $\mathcal{K}$. This means that it suffices to prove the following lemma to finish the second half of Theorem A.1.

Lemma A.5. *If $\mathcal{K}$ is an h-cone, then the projection $\{\mathbf{x} - x_k \mathbf{e}_k : \mathbf{x} \in \mathcal{K}\}$ is also an h-cone, for any k.*

Proof. Suppose $\mathcal{K} = \operatorname{hcone}(\mathbf{A})$ for some $\mathbf{A} \in \mathbb{R}^{m \times d}$. Fix k and consider

$$\mathcal{P}_k = \{\mathbf{x} + \lambda \mathbf{e}_k : \mathbf{x} \in \mathcal{K}, \lambda \in \mathbb{R}\}.$$

The projection we're after can be constructed from this set as

$$\{\mathbf{x} - x_k \mathbf{e}_k : \mathbf{x} \in \mathcal{K}\} = \mathcal{P}_k \cap \left\{\mathbf{x} \in \mathbb{R}^d : x_k = 0\right\},$$

so that it suffices to prove that $\mathcal{P}_k$ is an h-cone.

Suppose $\mathbf{a}_1, \mathbf{a}_2, \ldots, \mathbf{a}_m$ are the row vectors of $\mathbf{A}$. We construct a new matrix $\mathbf{A}_k$ whose row vectors are all $\mathbf{a}_j$ for which $a_{jk} = 0$, and the combinations $a_{ik}\mathbf{a}_j - a_{jk}\mathbf{a}_i$ whenever $a_{ik} > 0$ and $a_{jk} < 0$. We claim that $\mathcal{P}_k = \operatorname{hcone}(\mathbf{A}_k)$.

If $\mathbf{x} \in \mathcal{K}$ then $\mathbf{A}\,\mathbf{x} \leq \mathbf{0}$, which implies $\mathbf{A}_k\,\mathbf{x} \leq \mathbf{0}$ because each row of $\mathbf{A}_k$ is a nonnegative linear combination of rows of $\mathbf{A}$; that is, $\mathcal{K} \subseteq \operatorname{hcone}(\mathbf{A}_k)$. However, the kth component of $\mathbf{A}_k$ is zero by construction, and so $\mathcal{K} \subseteq \operatorname{hcone}(\mathbf{A}_k)$ implies $\mathcal{P}_k \subseteq \operatorname{hcone}(\mathbf{A}_k)$.

Conversely, suppose $\mathbf{x} \in \operatorname{hcone}(\mathbf{A}_k)$. We need to find a $\lambda \in \mathbb{R}$ such that $\mathbf{A}(\mathbf{x} - \lambda \mathbf{e}_k) \leq \mathbf{0}$, that is,

$$\begin{aligned} a_{11}x_1 + \cdots + a_{1k}(x_k - \lambda) + \cdots + a_{1d}x_d &\leq 0 \\ &\vdots \\ a_{m1}x_1 + \cdots + a_{mk}(x_k - \lambda) + \cdots + a_{md}x_d &\leq 0. \end{aligned}$$

The jth constraint is $\mathbf{a}_j \cdot \mathbf{x} - a_{jk}\lambda \leq 0$, that is, $\mathbf{a}_j \cdot \mathbf{x} \leq a_{jk}\lambda$. This gives the following conditions on λ:

$$\begin{aligned} \lambda &\geq \frac{\mathbf{a}_i \cdot \mathbf{x}}{a_{ik}} \qquad \text{if } a_{ik} > 0, \\ \lambda &\leq \frac{\mathbf{a}_j \cdot \mathbf{x}}{a_{jk}} \qquad \text{if } a_{jk} < 0. \end{aligned}$$

Such a λ exists because if $a_{ik} > 0$ and $a_{jk} < 0$ then (since $\mathbf{x} \in \operatorname{hcone}(\mathbf{A}_k)$)

$$(a_{ik}\mathbf{a}_j - a_{jk}\mathbf{a}_i) \cdot \mathbf{x} \leq 0,$$

which is equivalent to

$$\frac{\mathbf{a}_i \cdot \mathbf{x}}{a_{ik}} \leq \frac{\mathbf{a}_j \cdot \mathbf{x}}{a_{jk}}.$$

Thus we can find a λ that satisfies

$$\frac{\mathbf{a}_i \cdot \mathbf{x}}{a_{ik}} \leq \lambda \leq \frac{\mathbf{a}_j \cdot \mathbf{x}}{a_{jk}},$$

which proves $\operatorname{hcone}(\mathbf{A}_k) \subseteq \mathcal{P}_k$. □

B

Triangulations of Polytopes

Obvious is the most dangerous word in mathematics.

Eric Temple Bell

The goal of this appendix is to prove Theorem 3.1. Recall that a triangulation of a convex d-polytope $\mathcal{P}$ is a finite collection T of d-simplices with the following properties:

- $\mathcal{P} = \bigcup_{\Delta \in T} \Delta$.
- For any $\Delta_1, \Delta_2 \in T$, $\Delta_1 \cap \Delta_2$ is a face common to both Δ_1 and Δ_2.

Theorem 3.1 says that $\mathcal{P}$ can be triangulated using no new vertices, that is, there exists a triangulation T such that the vertices of any $\Delta \in T$ are vertices of $\mathcal{P}$. In preparation, we first show that a triangulation of a polytope induces a triangulation on any of its facets in a natural way.

Lemma B.1. *Suppose $T(\mathcal{P})$ is a triangulation of the d-polytope $\mathcal{P}$, and $\mathcal{F}$ is a facet of $\mathcal{P}$. Then*

$$T(\mathcal{F}) := \{\mathcal{S} \cap \mathcal{F} : \mathcal{S} \in T(\mathcal{P}), \dim(\mathcal{S} \cap \mathcal{F}) = d-1\}$$

is a triangulation of $\mathcal{F}$.

Proof. To avoid unnecessary notation, we write $\bigcup T(\mathcal{F})$ for $\bigcup_{\Delta \in T(\mathcal{F})} \Delta$. We have to show:

(i) $\mathcal{F} = \bigcup T(\mathcal{F})$.
(ii) For any $\Delta_1, \Delta_2 \in T(\mathcal{F})$, $\Delta_1 \cap \Delta_2$ is a face common to both Δ_1 and Δ_2.

(i) First, $\bigcup T(\mathcal{F}) \subseteq \mathcal{F}$ by definition of $T(\mathcal{F})$. Now we will show that $\mathcal{F} \setminus \bigcup T(\mathcal{F}) = \varnothing$, by means of contradiction. Let $x \in \mathcal{F} \setminus \bigcup T(\mathcal{F})$. If there is a neighborhood N of x in $\mathcal{F}$ that contains no points of $\bigcup T(\mathcal{F})$, then N consists only of points contained in some simplices from $T(\mathcal{P})$ that meet $\mathcal{F}$ in a set of dimension less than $d-1$, which is impossible since $\dim N = d-1$ and there are only finitely many simplices in $T(\mathcal{P})$. Hence any neighborhood of $\mathbf{x}$ in $\mathcal{F}$ contains points in some $\Delta \in T(\mathcal{F})$. However, $\bigcup T(\mathcal{F})$ is closed, and so such an $\mathbf{x}$ cannot exist. Hence $\mathcal{F} \setminus \bigcup T(\mathcal{F}) = \varnothing$.

(ii) Given $\Delta_1, \Delta_2 \in T(\mathcal{F})$, there are $\mathcal{S}_1, \mathcal{S}_2 \in T(\mathcal{P})$ such that

$$\Delta_1 = \mathcal{S}_1 \cap \mathcal{F} \qquad \text{and} \qquad \Delta_2 = \mathcal{S}_2 \cap \mathcal{F}$$

and the intersections of both $\mathcal{S}_1$ and S_2 with $\mathcal{F}$ are $(d-1)$-dimensional. Now $\Delta_1 \cap \Delta_2 = \mathcal{S}_1 \cap \mathcal{S}_2 \cap \mathcal{F}$, and since $\mathcal{S}_1, \mathcal{S}_2 \in T(\mathcal{P})$, $\mathcal{S}_1 \cap \mathcal{S}_2$ is a face of both $\mathcal{S}_1$ and $\mathcal{S}_2$. That is, there are hyperplanes H_1 and H_2 in $\mathbb{R}^d$ such that

$$\mathcal{S}_1 \cap \mathcal{S}_2 = \mathcal{S}_1 \cap H_1 \qquad \text{and} \qquad \mathcal{S}_1 \cap \mathcal{S}_2 = \mathcal{S}_2 \cap H_2 \,.$$

The $(d-1)$-hyperplanes H_1, H_2 in $\mathbb{R}^d$ induce the $(d-2)$-hyperplanes

$$h_1 := H_1 \cap \operatorname{span} \mathcal{F} \qquad \text{and} \qquad h_2 := H_2 \cap \operatorname{span} \mathcal{F}$$

in $\operatorname{span} \mathcal{F}$. We claim that $h_1 \cap \Delta_1 = \Delta_1 \cap \Delta_2 = h_2 \cap \Delta_2$, that is, $\Delta_1 \cap \Delta_2$ is a face of both Δ_1 and Δ_2. Indeed,

$$\begin{aligned} h_1 \cap \Delta_1 &= h_1 \cap (\mathcal{S}_1 \cap \mathcal{F}) \\ &= (H_1 \cap \operatorname{span} \mathcal{F}) \cap (\mathcal{S}_1 \cap \mathcal{F}) \\ &= (H_1 \cap \mathcal{S}_1) \cap (\mathcal{F} \cap \operatorname{span} \mathcal{F}) \\ &= (\mathcal{S}_1 \cap \mathcal{S}_2) \cap \mathcal{F} \\ &= \Delta_1 \cap \Delta_2 \,, \end{aligned}$$

and a practically identical calculation gives $h_2 \cap \Delta_2 = \Delta_1 \cap \Delta_2$. □

Proof of Theorem 3.1. We use induction on the number of vertices of the d-polytope $\mathcal{P}$. If $\mathcal{P}$ has $d+1$ vertices, then $\mathcal{P}$ is a simplex, and $\{\mathcal{P}\}$ is a triangulation.

For the induction step, suppose we are given a d-polytope $\mathcal{P}$ with at least $d+2$ vertices. Fix a vertex $\mathbf{v}$ of $\mathcal{P}$ such that $\mathcal{Q}$, the convex hull of the remaining vertices of $\mathcal{P}$, is still of dimension d. By the induction hypothesis, we can triangulate $\mathcal{Q}$.

We call a facet $\mathcal{F}$ of $\mathcal{Q}$ **visible** from $\mathbf{v}$ if for any $\mathbf{x} \in \mathcal{F}$, the half-open line segment $(\mathbf{x}, \mathbf{v}]$ is disjoint from $\mathcal{Q}$. By the Lemma, the triangulation $T(\mathcal{Q})$ of $\mathcal{Q}$ induces the triangulation

$$T(\mathcal{F}) = \{\Delta \cap \mathcal{F} : \, \Delta \in T(\mathcal{Q}), \, \dim(\Delta \cap \mathcal{F}) = d-1\}$$

of a facet $\mathcal{F}$ of $\mathcal{Q}$.

Let T consist of the convex hulls of $\mathbf{v}$ with each $(d-1)$-simplex in the triangulations of the visible facets. We claim that $T \cup T(\mathcal{Q})$ forms a triangulation of $\mathcal{P}$. To prove this, we have to show:

(i) $\mathcal{P} = \bigcup (T \cup T(\mathcal{Q}))$.
(ii) For any $\Delta_1, \Delta_2 \in T \cup T(\mathcal{Q})$, $\Delta_1 \cap \Delta_2$ is a face common to both Δ_1 and Δ_2.

(i) $\mathcal{P} \supseteq \bigcup (T \cup T(\mathcal{Q}))$ follows by the definitions of T and $T(\mathcal{Q})$. To prove $\mathcal{P} \subseteq \bigcup (T \cup T(\mathcal{Q}))$, assume $\mathbf{x} \in \mathcal{P}$ is given. If $\mathbf{x} \in \mathcal{Q}$ then $\mathbf{x} \in \bigcup T(\mathcal{Q})$. If $\mathbf{x} \in \mathcal{P} \setminus \mathcal{Q}$, consider the line through $\mathbf{v}$ and $\mathbf{x}$. This line meets $\mathcal{Q}$ (because $\mathcal{P}$ is convex), so let $\mathbf{y} \in \mathcal{Q}$ be the first point in $\mathcal{Q}$ that we meet when traveling along the line towards $\mathcal{Q}$. This point $\mathbf{y}$ is on a facet of $\mathcal{Q}$ that is, by construction, visible from $\mathbf{v}$, and hence $\mathbf{x} \in \Delta$ for some $\Delta \in T$.

(ii) Given $\Delta_1, \Delta_2 \in T \cup T(\mathcal{Q})$, there are three cases:

(a) $\Delta_1, \Delta_2 \in T(\mathcal{Q})$;
(b) $\Delta_1, \Delta_2 \in T$;
(c) $\Delta_1 \in T, \Delta_2 \in T(\mathcal{Q})$.

In each case we have to show that $\Delta_1 \cap \Delta_2$ is a face common to Δ_1 and Δ_2.

(a) Since $T(\mathcal{Q})$ is a triangulation, $\Delta_1 \cap \Delta_2$ is a face of both Δ_1 and Δ_2.

(b) Given $\Delta_1, \Delta_2 \in T$, there exist $S_1, S_2 \in T(\mathcal{F})$ such that $\Delta_1 = \operatorname{conv}\{\mathbf{v}, S_1\}$ and $\Delta_2 = \operatorname{conv}\{\mathbf{v}, S_2\}$. Since $T(\mathcal{F})$ is a triangulation, $S_1 \cap S_2$ is a face common to S_1 and S_2. By convexity, $\Delta_1 \cap \Delta_2 = \operatorname{conv}\{\mathbf{v}, S_1 \cap S_2\}$. Exercise 2.6 shows that $S_1 \cap S_2$ is a simplex, and that this simplex is the convex hull of some of the common vertices of S_1 and S_2. But then $\Delta_1 \cap \Delta_2 = \operatorname{conv}\{\mathbf{v}, S_1 \cap S_2\}$ is the convex hull of some of the common vertices of Δ_1 and Δ_2 and hence, again by Exercise 2.6, a face of both Δ_1 and Δ_2.

(c) Since $\Delta_1 \in T$, there exists $S \in T(\mathcal{F})$ such that $\Delta_1 = \operatorname{conv}\{\mathbf{v}, S\}$. By construction, $\Delta_1 \cap \mathcal{Q} = S$, and S is a face of some $\Delta \in T(\mathcal{Q})$. Since $T(\mathcal{Q})$ is a triangulation, $\Delta \cap \Delta_2$ is a face common to Δ and Δ_2. But then

$$\Delta_1 \cap \Delta_2 = S \cap \Delta_2 = (S \cap \Delta) \cap \Delta_2 = S \cap (\Delta \cap \Delta_2)$$

is an intersection of two faces of Δ and hence by Exercise 2.6 again a face of Δ and a simplex. The vertices of $\Delta_1 \cap \Delta_2 = S \cap (\Delta \cap \Delta_2)$ form a subset of the vertices common to S and Δ_2. Since S is a face of Δ_1, $\Delta_1 \cap \Delta_2$ is a face of both Δ_1 and Δ_2, by Exercise 2.6. □

Hints for ♣ Exercises

Well here's another clue for you all.

John Lennon & Paul McCartney ("Glass Onion," *The White Album*)

Chapter 1

1.1 Set up the partial fraction expansion as

$$\frac{z}{1-z-z^2} = \frac{A}{1-\frac{1+\sqrt{5}}{2}\,z} + \frac{B}{1-\frac{1-\sqrt{5}}{2}\,z}$$

and clear denominators to compute A and B; one can do so, for example, by specializing z.

1.2 Multiply out $(1-z)\left(1+z+z^2+\cdots+z^n\right)$. For the infinite sum, note that $\lim_{k\to\infty} z^k = 0$ if $|z|<1$.

1.3 Start with the observation that there are $\lfloor x \rfloor + 1$ lattice points in the interval $[0, x]$.

1.4 (i) & (j) Write $n = qm + r$ for some integers q, r such that $0 \le r < m$. Distinguish the cases $r = 0$ and $r > 0$.

1.9 Use the fact that if m and n are relatively prime, given any $a \in \mathbb{Z}$ there exists $b \in \mathbb{Z}$ (which is unique modulo n) such that $mb \equiv a \pmod{n}$. For the second equality of sets, think about the case $a = 0$.

1.12 First translate the line segment to the origin and explain why this translation leaves the integer-point enumeration invariant. For the case $(a, b) = (0, 0)$, first study the problem under the restriction that $\gcd(c, d) = 1$.

1.17 Given a triangle $\mathcal{T}$ with vertices on the integer lattice, consider the parallelogram $\mathcal{P}$ formed by two fixed edges of $\mathcal{T}$. Use integral translates of $\mathcal{P}$

to tile the plane $\mathbb{R}^2$. Conclude from this tiling that $\mathcal{P}$ contains only its vertices as lattice points if and only if the area of $\mathcal{P}$ is 1.

1.20 Given an integer b, the Euclidean algorithm asserts the existence of $m_1, m_2, \dots, m_d \in \mathbb{Z}$ such that b can be represented as $b = m_1 a_1 + m_2 a_2 + \cdots + m_d a_d$. Convince yourself that we can demand that in this representation $0 \le m_2, m_3, \dots, m_d < a_1$. Conclude that all integers beyond $(a_1 - 1)(a_2 + a_3 + \cdots + a_d)$ are representable in terms of $a_1, a_2, \dots, a_d$. (This argument can be refined to yield another proof of Theorem 1.2.)

1.21 Use the setup

$$f(z) = \frac{A_1}{z} + \frac{A_2}{z^2} + \cdots + \frac{A_n}{z^n} + \frac{B_1}{z-1} + \frac{B_2}{(z-1)^2} + \sum_{k=1}^{a-1} \frac{C_k}{z - \xi_a^k} + \sum_{j=1}^{b-1} \frac{D_j}{z - \xi_b^j} .$$

To compute C_k, multiply both sides by $\left(z - \xi_a^k\right)$ and calculate the limit as $z \to \xi_a^k$. The coefficients D_j can be computed in a similar fashion.

1.22 Use Exercise 1.9 (with $m = b^{-1}$) on the left-hand side of the equation.

1.24 Suppose $a > b$. The integer $a + b$ certainly has a representation in terms of a and b, namely, $1 \cdot a + 1 \cdot b$. Think about how the coefficient of b would change if we changed the coefficient of a.

1.29 Use the partial fraction setup (1.11), multiply both sides by $\left(z - \xi_{a_1}^k\right)$, and take the limit as $z \to \xi_{a_1}^k$.

1.31 Convince yourself of the generating-function setup

$$\sum_{n \ge 1} p_A^\circ(n)\, z^n = \left(\frac{z^{a_1}}{1 - z^{a_1}}\right)\left(\frac{z^{a_2}}{1 - z^{a_2}}\right) \cdots \left(\frac{z^{a_d}}{1 - z^{a_d}}\right) .$$

Now use the machinery of Section 1.5.

Chapter 2

2.1 Use Exercise 1.3 for the closed interval. For open intervals, you can use Exercise 1.4(j) or the $\lceil \dots \rceil$ notation of Exercise 1.4(e). To show the quasipolynomial character, rewrite the greatest-integer function in terms of the fractional-part function.

2.2 Write $\mathcal{R}$ as a direct product of two intervals and use Exercise 1.3.

2.6 Start by showing that the convex hull of a d-element subset W of V is a face of Δ. This allows you to prove the first statement by induction (using Exercise 2.5). For the converse statement, given a supporting hyperplane H that defines the face $\mathcal{F}$ of Δ, let $W \subseteq V$ consist of those vertices of Δ that are in H. Now prove that any point

$$\mathbf{x} = \lambda_1 \mathbf{v}_1 + \lambda_2 \mathbf{v}_2 + \cdots + \lambda_{d+1} \mathbf{v}_{d+1}$$

in $\mathcal{F}$ has to satisfy $\lambda_k = 0$ for all $\mathbf{v}_k \notin W$.

2.7 First show that the linear inequalities and equations describing a rational polytope can be chosen with rational coefficients, and then clear denominators.

2.9 Write $\frac{1}{(1-z)^{d+1}} = \left(\sum_{k_1 \geq 0} z^{k_1}\right) \left(\sum_{k_2 \geq 0} z^{k_2}\right) \cdots \left(\sum_{k_{d+1} \geq 0} z^{k_{d+1}}\right)$ and come up with a combinatorial enumeration scheme to compute the coefficients of this power series.

2.10 Write $\binom{t+k}{d} = \frac{(t+k)(t+k-1)\cdots(t+k-d+1)}{d!}$ and switch t to $-t$.

2.14 Think about the poles of the function $\frac{z}{e^z-1}$ and use a theorem from complex analysis.

2.15 Compute the generating function of $B_d(1-x)$ and rewrite it as $\frac{ze^{-xz}}{1-e^{-z}}$.

2.16 Show that $\frac{z}{e^z-1} + \frac{1}{2}z$ is an even function of $\mathbb{Z}$.

2.23 Follow the steps of the proof of Theorem 2.4.

2.24 Extend $\mathcal{T}$ to a rectangle whose diagonal is the hypotenuse of $\mathcal{T}$, and consider the lattice points on this diagonal separately.

2.25 For the area use elementary calculus. For the number of boundary points on $t\mathcal{P}$, extend Exercise 1.12 to a set of line segments whose union forms a simple closed curve.

2.31 Rewrite the inequality as $\left(\left\lceil \frac{ta}{d} \right\rceil - 1\right) e + \left(\left\lceil \frac{tb}{d} \right\rceil - 1\right) f \leq tr$ and compare this with the definition of $\mathcal{T}$.

2.32 To compute C_3, multiply both sides of (2.20) by $(z-3)^2$ and compute the limit as $z \to 1$. The coefficients A_j and B_l can be computed in a similar fashion. To compute C_2, first move $\frac{C_3}{(z-1)^3}$ in (2.20) to the left-hand side, then multiply by $(z-1)^2$ and take the limit as $z \to 1$. A similar, even more elaborate, computation gives C_1. (Alternatively, compute the Laurent series of the function in (2.20) at $z = 1$ with a computer algebra system such as `Maple` or `Mathematica`.)

2.34 Follow the proof of Theorem 2.10. Use Exercise 2.33 to compute the additional coefficients in the partial fraction expansion of the generating function corresponding to this lattice-point count.

2.36 Start with computing the constant term of

$$\frac{1}{(1-z_1z_2)(1-z_1^2z_2)(1-z_1)(1-z_2)\, z_1^{3t} z_2^{2t}}$$

with respect to z_2 by treating z_1 as a constant and setting up a partial fraction expansion of this function with respect to z_2.

Chapter 3

3.2 Write the simplicial cones as cones over simplices and use Exercise 2.6.

3.4 Write down a typical term of the product

$$\sigma_S\left(z_1, z_2, \ldots, z_m\right) \sigma_T\left(z_{m+1}, z_{m+2}, \ldots, z_{m+n}\right).$$

3.5 Multiply out $\mathbf{z}^{\mathbf{m}}\sigma_{\mathcal{K}}(\mathbf{z})$.

3.6 Write a typical term in $\sigma_S\left(\frac{1}{z_1}, \frac{1}{z_2}, \ldots, \frac{1}{z_d}\right) = \sigma_S\left(z_1^{-1}, z_2^{-1}, \ldots, z_d^{-1}\right)$.

3.8 Given the polynomial f, split up the generating function on the left-hand side according to the terms of f and use (2.2). Conversely, if the polynomial g is given, use (2.6).

3.13 Show that $H \cap \mathbb{Z}^d$ is a $\mathbb{Z}$-module. Therefore it has a basis; extend this basis to a basis of $\mathbb{Z}^d$.

3.14 Start by proving the result for a single hyperplane, for example, by referring to Exercise 3.13.

3.19 Given f, split up the generating function on the left-hand side according to the constituents of f; then use Exercise 3.8. Conversely, given g and h, multiply both by a polynomial to get the denominator on the right-hand side into the form $(1 - z^p)^{d+1}$; then use (2.6).

3.20 Start with the setup on page 75, and closely orient yourself along the proof of Theorem 3.8.

3.29 Use Lemma 3.19.

Chapter 4

4.1 Use Exercise 2.1.

4.2 Use the explicit description of Π given by (4.3).

4.3 Consider each simplicial cone $\mathcal{K}_j$ separately, and look at the arrangement of its bounding hyperplanes. For each hyperplane, use Exercise 3.13.

4.5 For (a), convince yourself that $Q(-t)$ is also a quasipolynomial. For (b), use (1.3). For (c), differentiate (1.3). For (d), think about one constituent of the quasipolynomial at a time.

4.6 In the generating function for $L_{\mathcal{P}}(t-k)$, make a change in the summation variable; then use Theorem 4.4.

4.11 Use the fact that $\mathbf{A}$ has only integral entries. For the second part, write down the explicit hyperplane descriptions of $(t+1)\mathcal{P}^\circ$ and $t\mathcal{P}$.

4.12 Assume that there exist $t \in \mathbb{Z}$ and a facet hyperplane H of $\mathcal{P}$ such that there is a lattice point between tH and $(t+1)H$. Translate this lattice point to a lattice point that violates (4.12).

Chapter 5

5.4 Consider an *interval* $[\mathcal{F}, \mathcal{P}]$ in the face lattice of $\mathcal{P}$: $[\mathcal{F}, \mathcal{P}]$ contains all faces $\mathcal{G}$ such that $\mathcal{F} \subseteq \mathcal{G} \subseteq \mathcal{P}$. Prove that if $\mathcal{P}$ is simple, any such interval is isomorphic to a Boolean lattice.

5.5 Use Exercise 2.6 to show that the face lattice of a simplex is isomorphic to a Boolean lattice.

Chapter 6

6.1 Think permutation matrices.

6.3 Show that the rank of (6.5) is $2n - 1$.

6.5 Start by showing that all permutation matrices are indeed vertices. Then use Exercise 6.4 to show that there are no other vertices.

6.6 Establish a bijection between semimagic squares with line sum $t - n$ and semimagic squares with *positive* entries and line sum t.

6.7 Think about the smallest possible line sum if the entries of the square are positive integers.

6.8 Follow the computation on page 112 that led to the formula for H_2.

6.9 Multiply both sides of (6.7) by $\left(w - \frac{1}{z_k}\right)$ and take the limit as $w \to \frac{1}{z_k}$.

6.10 Orient yourself along the computation in (6.10).

6.16 Compute the matrix equivalent to (6.5) for the polytope describing all magic squares of a given size. Show that this matrix has rank $2n + 1$.

6.18 Orient yourself along the computation on page 112.

Chapter 7

7.1 Show that both polynomials have the same roots and the same constant term.

7.2 Use Exercise 7.1.

7.5 Differentiate (1.3).

7.6 Use (1.3).

7.7 Write an arbitrary function on $\mathbb{Z}$ with period b in terms of $\delta_m(x)$, $1 \leq m \leq b$.

7.8 Use the definition (7.6) of the inner product and the properties $z\overline{z} = |z|^2$ and $\overline{(zw)} = \overline{z} \cdot \overline{w}$ for complex numbers z and w.

7.14 Use the definition (7.4) and simplify the fractional-part function in the sum on the right-hand side.

7.22 Use the definition of **F**.

Chapter 8

8.5 Use Exercise 1.9.

8.7 Use the methods outlined in the hints for Exercises 1.21 and 2.32 to compute the partial fraction coefficients for $z = 1$ in (8.3).

8.9 Multiply out all the terms on the left-hand side and make use of Exercises 1.9 and 7.14.

8.11 Use the methods outlined in the hints for Exercises 1.21 and 2.32 to compute a partial fraction expansion of (8.7).

Chapter 9

9.1 Show that $(\operatorname{span}\mathcal{F})^{\perp} \cap \mathcal{K}_{\mathcal{F}}$ is a cone. Then prove that if H is a defining hyperplane for $\mathcal{F}$, then $H \cap (\operatorname{span}\mathcal{F})^{\perp}$ is a hyperplane in the vector space $(\operatorname{span}\mathcal{F})^{\perp}$. Finally, show that this hyperplane $H \cap (\operatorname{span}\mathcal{F})^{\perp}$ defines the apex of $(\operatorname{span}\mathcal{F})^{\perp} \cap \mathcal{K}_{\mathcal{F}}$, and that this apex is a point.

9.2 Consider the hyperplanes $H_1, H_2, \dots, H_{d+1}$ that bound Δ. For each hyperplane H_k, denote by H_k^+ the closed half-space bounded by H_k that contains Δ, and by H_k^- the open half-space bounded by H_k that does not contain Δ. Show that every tangent cone of Δ is the intersection of some of the H_k^+'s, and conversely, that every intersection of some of the H_k^+'s, except for $\Delta = \bigcup_{k=1}^{d+1} H_k^+$, is a tangent cone of Δ. Since $H_k^+ \cup H_k^- = \mathbb{R}^d$ as a disjoint union, for each k, the point $\mathbf{x}$ is either in H_k^+ or H_k^-. Prove that the intersection of those H_k^+ that contain $\mathbf{x}$ is the sought-after tangent cone.

9.4 As in Exercise 5.5, show that the face lattice of a simplex is a Boolean lattice. Note that any sublattice of a Boolean lattice is again Boolean.

9.6 One approach to this problem is first to dilate $\mathcal{P}$ and the corresponding hyperplanes in H by a small factor. To avoid subtleties, first translate $\mathcal{P}$ by an integer vector, if necessary, to ensure that none of the hyperplanes in H contains the origin. Use Exercise 3.13.

9.7 Adjust the steps in Section 9.3 to open polytopes. Start by proving a Brianchon–Gram identity for open simplices, by analogy with Theorem 9.5. This implies a Brion-type identity for open simplices, as in Corollary 9.6. Finally, adjust the proof of Theorem 9.7 to open polytopes.

Chapter 10

10.1 Use (10.3), Exercise 2.18, and (2.11).

10.3 Use the definition of unimodularity to show that the only integer point in the fundamental parallelepiped of $\mathcal{K}$ is $\mathbf{v}$.

10.4 Orient yourself along the proof of Theorem 10.4; instead of a sum over vertex cones, just consider one simple cone $\mathcal{K}$.

Chapter 11

11.3 Multiply out $\mathbf{z}^{\mathbf{m}} \alpha_{\mathcal{K}}(\mathbf{z})$.

11.4 Orient yourself along the proof of Theorem 4.2. Note that for solid angles, we do not require the condition that the boundary of $\mathcal{K}$ contains no lattice point.

11.5 As a warm-up exercise, show that

$$\sum_{\substack{\mathcal{F} \subseteq \Delta \\ \dim \mathcal{F} > 0}} \sum_{\mathbf{v} \text{ a vertex of } \mathcal{F}} \sigma_{\mathcal{K}_{\mathbf{v}}(\mathcal{F})^\circ}(\mathbf{z}) = \sum_{\mathbf{v} \text{ a vertex of } \Delta} \sum_{\substack{\mathcal{F} \subseteq \mathcal{K}_{\mathbf{v}} \\ \dim \mathcal{F} > 0}} \sigma_{\mathcal{F}^\circ}(\mathbf{z}) .$$

11.6 Start with the setup of our second proof of Ehrhart's theorem in Section 9.4; that is, it suffices to prove that if p is the denominator of $\mathcal{P}$, then $A_{\mathcal{P}}(-r - pt) = (-1)^{\dim \mathcal{P}} A_{\mathcal{P}}(r + pt)$ for any integers r and t with $0 \le r < p$ and $t > 0$. (Think of r as fixed and t as variable.) Now orient yourself along the proof on page 184.

Chapter 12

12.1 Bound the integral from above, using the length of C_r and an upper bound for the absolute value of the integrand.

12.2 The nontrivial roots of unity are simple poles of f, for which the residue computation boils down to a simple limit.

12.4 Start by combining the terms $\frac{1}{z-(m+ni)}$ and $\frac{1}{m+ni}$ into one fraction.

12.5 Differentiate (12.1) term by term.

12.6 Compute $\wp'$ explicitly.

12.7 Use a famous theorem from complex analysis.

12.8 Compute $\wp(-z)$ and use the fact that $(-(m+in))^2 = (m+in)^2$.

12.9 Repeat the proof of Lemma 12.3, but now starting with the proof of $\wp'(z+i) = \wp'(z)$.

12.10 Use a famous theorem from complex analysis.

12.11 Use the definition of the Weierstraß $\wp$ function.

References

1. Maya Ahmed, Jesús A. De Loera, and Raymond Hemmecke. Polyhedral cones of magic cubes and squares. In *Discrete and Computational Geometry*, volume 25 of *Algorithms Combin.*, pages 25–41. Springer, Berlin, 2003. arXiv:math.CO/0201108.
2. Maya Mohsin Ahmed. How many squares are there, Mr. Franklin?: constructing and enumerating Franklin squares. *Amer. Math. Monthly*, 111(5):394–410, 2004.
3. Harsh Anand, Vishwa Chander Dumir, and Hansraj Gupta. A combinatorial distribution problem. *Duke Math. J.*, 33:757–769, 1966.
4. W. S. Andrews. *Magic Squares and Cubes*. Dover Publications Inc., New York, 1960.
5. Tom M. Apostol. Generalized Dedekind sums and transformation formulae of certain Lambert series. *Duke Math. J.*, 17:147–157, 1950.
6. Tom M. Apostol and Thiennu H. Vu. Identities for sums of Dedekind type. *J. Number Theory*, 14(3):391–396, 1982.
7. Vladimir I. Arnold. *Arnold's problems*. Springer-Verlag, Berlin, 2004. Translated and revised edition of the 2000 Russian original, With a preface by V. Philippov, A. Yakivchik and M. Peters.
8. Christos A. Athanasiadis. Ehrhart polynomials, simplicial polytopes, magic squares and a conjecture of Stanley. *J. Reine Angew. Math.*, 583:163–174, 2005.
9. Welleda Baldoni-Silva and Michèle Vergne. Residue formulae for volumes and Ehrhart polynomials of convex polytopes. Preprint (arXiv:math.CO/0103097), 2001.
10. Philippe Barkan. Sur les sommes de Dedekind et les fractions continues finies. *C. R. Acad. Sci. Paris Sér. A-B*, 284(16):A923–A926, 1977.
11. Alexander Barvinok. Exponential integrals and sums over convex polyhedra. *Funktsional. Anal. i Prilozhen.*, 26(2):64–66, 1992.
12. Alexander Barvinok. *A Course in Convexity*, volume 54 of *Graduate Studies in Mathematics*. American Mathematical Society, Providence, RI, 2002.
13. Alexander Barvinok and James E. Pommersheim. An algorithmic theory of lattice points in polyhedra. In *New Perspectives in Algebraic Combinatorics (Berkeley, CA, 1996–97)*, volume 38 of *Math. Sci. Res. Inst. Publ.*, pages 91–147. Cambridge Univ. Press, Cambridge, 1999.

14. Alexander Barvinok and Kevin Woods. Short rational generating functions for lattice point problems. *J. Amer. Math. Soc.*, 16(4):957–979 (electronic), 2003. arXiv:math.CO/0211146.
15. Alexander I. Barvinok. A polynomial time algorithm for counting integral points in polyhedra when the dimension is fixed. *Math. Oper. Res.*, 19(4):769–779, 1994.
16. Victor V. Batyrev. Dual polyhedra and mirror symmetry for Calabi-Yau hypersurfaces in toric varieties. *J. Algebraic Geom.*, 3(3):493–535, 1994. arXiv:alg-geom/9310003.
17. Victor V. Batyrev. Lattice polytopes with a given h^*-polynomial. Preprint (arXiv:math.CO/0602593), 2006.
18. Victor V. Batyrev and Dimitrios I. Dais. Strong McKay correspondence, string-theoretic Hodge numbers and mirror symmetry. *Topology*, 35(4):901–929, 1996.
19. Matthias Beck. Counting lattice points by means of the residue theorem. *Ramanujan J.*, 4(3):299–310, 2000. arXiv:math.CO/0306035.
20. Matthias Beck. Multidimensional Ehrhart reciprocity. *J. Combin. Theory Ser. A*, 97(1):187–194, 2002. arXiv:math.CO/0111331.
21. Matthias Beck. Dedekind cotangent sums. *Acta Arith.*, 109(2):109–130, 2003. arXiv:math.NT/0112077.
22. Matthias Beck, Beifang Chen, Lenny Fukshansky, Christian Haase, Allen Knutson, Bruce Reznick, Sinai Robins, and Achill Schürmann. Problems from the Cottonwood Room. In *Integer Points in Polyhedra—Geometry, Number Theory, Algebra, Optimization*, volume 374 of *Contemp. Math.*, pages 179–191. Amer. Math. Soc., Providence, RI, 2005.
23. Matthias Beck, Moshe Cohen, Jessica Cuomo, and Paul Gribelyuk. The number of "magic" squares, cubes, and hypercubes. *Amer. Math. Monthly*, 110(8):707–717, 2003. arXiv:math.CO/0201013.
24. Matthias Beck, Jesús A. De Loera, Mike Develin, Julian Pfeifle, and Richard P. Stanley. Coefficients and roots of Ehrhart polynomials. In *Integer Points in Polyhedra—Geometry, Number Theory, Algebra, Optimization*, volume 374 of *Contemp. Math.*, pages 15–36. Amer. Math. Soc., Providence, RI, 2005. arXiv:math.CO/0402148.
25. Matthias Beck, Ricardo Diaz, and Sinai Robins. The Frobenius problem, rational polytopes, and Fourier-Dedekind sums. *J. Number Theory*, 96(1):1–21, 2002. arXiv:math.NT/0204035.
26. Matthias Beck and Richard Ehrenborg. Ehrhart–Macdonald reciprocity extended. Preprint (arXiv:math.CO/0504230), 2006.
27. Matthias Beck, Christian Haase, and Frank Sottile. Theorems of Brion, Lawrence, and Varchenko on rational generating functions for cones. Preprint (arXiv:math.CO/0506466), 2006.
28. Matthias Beck and Dennis Pixton. The Ehrhart polynomial of the Birkhoff polytope. *Discrete Comput. Geom.*, 30(4):623–637, 2003. arXiv:math.CO/0202267.
29. Matthias Beck and Sinai Robins. Explicit and efficient formulas for the lattice point count in rational polygons using Dedekind–Rademacher sums. *Discrete Comput. Geom.*, 27(4):443–459, 2002. arXiv:math.CO/0111329.
30. Matthias Beck and Frank Sottile. Irrational proofs for two theorems of Stanley. Preprint (arXiv:math.CO/0506315), to appear in *European J. Combin.*, 2006.

31. Matthias Beck and Thomas Zaslavsky. An enumerative geometry for magic and magilatin labellings. Preprint (arXiv:math.CO/0506315), to appear in *Ann. Combin.*, 2006.
32. Dale Beihoffer, Jemimah Hendry, Albert Nijenhuis, and Stan Wagon. Faster algorithms for Frobenius numbers. *Electron. J. Combin.*, 12(1):Research Paper 27, 38 pp. (electronic), 2005.
33. Nicole Berline and Michèle Vergne. Local Euler–Maclaurin formula for polytopes. Preprint (arXiv:math.CO/0507256), 2006.
34. Bruce C. Berndt. Reciprocity theorems for Dedekind sums and generalizations. *Advances in Math.*, 23(3):285–316, 1977.
35. Bruce C. Berndt and Ulrich Dieter. Sums involving the greatest integer function and Riemann-Stieltjes integration. *J. Reine Angew. Math.*, 337:208–220, 1982.
36. Bruce C. Berndt and Boon Pin Yeap. Explicit evaluations and reciprocity theorems for finite trigonometric sums. *Adv. in Appl. Math.*, 29(3):358–385, 2002.
37. Christian Bey, Martin Henk, and Jörg M. Wills. Notes on the roots of Ehrhart polynomials. Preprint (arXiv:math.MG/0606089), 2006.
38. Louis J. Billera and A. Sarangarajan. The combinatorics of permutation polytopes. In *Formal Power Series and Algebraic Combinatorics (New Brunswick, NJ, 1994)*, pages 1–23. Amer. Math. Soc., Providence, RI, 1996.
39. Garrett Birkhoff. Tres observaciones sobre el álgebra lineal. *Revista Facultad de Ciencias Exactas, Puras y Aplicadas Universidad Nacional de Tucumán, Serie A (Matemáticas y Física Teóretica)*, 5:147–151, 1946.
40. G. R. Blakley. Combinatorial remarks on partitions of a multipartite number. *Duke Math. J.*, 31:335–340, 1964.
41. Alfred Brauer. On a problem of partitions. *Amer. J. Math.*, 64:299–312, 1942.
42. Benjamin Braun. Norm bounds for Ehrhart polynomial roots. Preprint (arXiv:math.CO/0602464), to appear in *Discrete Comput. Geom.*, 2006.
43. Henrik Bresinsky. Symmetric semigroups of integers generated by 4 elements. *Manuscripta Math.*, 17(3):205–219, 1975.
44. Charles J. Brianchon. Théorème nouveau sur les polyèdres. *J. Ecole (Royale) Polytechnique*, 15:317–319, 1837.
45. Michel Brion. Points entiers dans les polyèdres convexes. *Ann. Sci. École Norm. Sup. (4)*, 21(4):653–663, 1988.
46. Michel Brion and Michèle Vergne. Residue formulae, vector partition functions and lattice points in rational polytopes. *J. Amer. Math. Soc.*, 10(4):797–833, 1997.
47. Arne Brøndsted. *An Introduction to Convex Polytopes*, volume 90 of *Graduate Texts in Mathematics*. Springer-Verlag, New York, 1983.
48. Richard A. Brualdi and Peter M. Gibson. Convex polyhedra of doubly stochastic matrices. I. Applications of the permanent function. *J. Combinatorial Theory Ser. A*, 22(2):194–230, 1977.
49. Richard A. Brualdi and Peter M. Gibson. Convex polyhedra of doubly stochastic matrices. III. Affine and combinatorial properties of U_n. *J. Combinatorial Theory Ser. A*, 22(3):338–351, 1977.
50. Heinz Bruggesser and Peter Mani. Shellable decompositions of cells and spheres. *Math. Scand.*, 29:197–205 (1972), 1971.
51. Daniel Bump, Kwok-Kwong Choi, Pär Kurlberg, and Jeffrey Vaaler. A local Riemann hypothesis. I. *Math. Z.*, 233(1):1–19, 2000.

52. Kristin A. Camenga. Vector spaces spanned by the angle sums of polytopes. Preprint (arXiv:math.MG/0508629), 2005.
53. Schuyler Cammann. Old Chinese magic squares. *Sinologica*, 7:14–53, 1962.
54. Schuyler Cammann. Islamic and Indian magic squares, Parts I and II. *History of Religions*, 8:181–209; 271–299, 1969.
55. Sylvain E. Cappell and Julius L. Shaneson. Euler-Maclaurin expansions for lattices above dimension one. *C. R. Acad. Sci. Paris Sér. I Math.*, 321(7):885–890, 1995.
56. Leonard Carlitz. Some theorems on generalized Dedekind sums. *Pacific J. Math.*, 3:513–522, 1953.
57. John W. S. Cassels. *An Introduction to the Geometry of Numbers.* Classics in Mathematics. Springer-Verlag, Berlin, 1997. Corrected reprint of the 1971 edition.
58. Clara S. Chan, David P. Robbins, and David S. Yuen. On the volume of a certain polytope. *Experiment. Math.*, 9(1):91–99, 2000. arXiv:math.CO/9810154.
59. Beifang Chen. Weight functions, double reciprocity laws, and volume formulas for lattice polyhedra. *Proc. Natl. Acad. Sci. USA*, 95(16):9093–9098 (electronic), 1998.
60. Beifang Chen. Lattice points, Dedekind sums, and Ehrhart polynomials of lattice polyhedra. *Discrete Comput. Geom.*, 28(2):175–199, 2002.
61. Beifang Chen and Vladimir Turaev. Counting lattice points of rational polyhedra. *Adv. Math.*, 155(1):84–97, 2000.
62. Louis Comtet. *Advanced Combinatorics.* D. Reidel Publishing Co., Dordrecht, enlarged edition, 1974.
63. Wolfgang Dahmen and Charles A. Micchelli. The number of solutions to linear Diophantine equations and multivariate splines. *Trans. Amer. Math. Soc.*, 308(2):509–532, 1988.
64. Vladimir I. Danilov. The geometry of toric varieties. *Uspekhi Mat. Nauk*, 33(2(200)):85–134, 247, 1978.
65. J. Leslie Davison. On the linear Diophantine problem of Frobenius. *J. Number Theory*, 48(3):353–363, 1994.
66. Jesús A. De Loera, David Haws, Raymond Hemmecke, Peter Huggins, and Ruriko Yoshida. A user's guide for LattE v1.1, software package LattE. 2004. Electronically available at http://www.math.ucdavis.edu/∼latte/.
67. Jesús A. De Loera, Raymond Hemmecke, Jeremiah Tauzer, and Ruriko Yoshida. Effective lattice point counting in rational convex polytopes. *J. Symbolic Comput.*, 38(4):1273–1302, 2004.
68. Jesús A. De Loera and Shmuel Onn. The complexity of three-way statistical tables. *SIAM J. Comput.*, 33(4):819–836 (electronic), 2004. arXiv:math.CO/0207200.
69. Jesús A. De Loera, Jörg Rambau, and Francisco Santos. *Triangulations of Point Sets: Applications, Structures, Algorithms.* Springer (to appear), 2006.
70. Richard Dedekind. Erläuterungen zu den Fragmenten xxviii. In *Collected Works of Bernhard Riemann*, pages 466–478. Dover Publ., New York, 1953.
71. Max Dehn. Die Eulersche Formel im Zusammenhang mit dem Inhalt in der nicht-euklidischen Geometrie. *Math. Ann.*, 61:279–298, 1905.
72. József Dénes and Anthony D. Keedwell. *Latin Squares and Their Applications.* Academic Press, New York, 1974.
73. Graham Denham. Short generating functions for some semigroup algebras. *Electron. J. Combin.*, 10:Research Paper 36, 7 pp. (electronic), 2003.

74. Persi Diaconis and Anil Gangolli. Rectangular arrays with fixed margins. In *Discrete Probability and Algorithms (Minneapolis, MN, 1993)*, pages 15–41. Springer, New York, 1995.
75. Ricardo Diaz and Sinai Robins. Pick's formula via the Weierstrass $\wp$-function. *Amer. Math. Monthly*, 102(5):431–437, 1995.
76. Ricardo Diaz and Sinai Robins. The Ehrhart polynomial of a lattice polytope. *Ann. of Math. (2)*, 145(3):503–518, 1997.
77. Ulrich Dieter. Das Verhalten der Kleinschen Funktionen $\log \sigma_{g,h}(\omega_1, \omega_2)$ gegenüber Modultransformationen und verallgemeinerte Dedekindsche Summen. *J. Reine Angew. Math.*, 201:37–70, 1959.
78. Ulrich Dieter. Cotangent sums, a further generalization of Dedekind sums. *J. Number Theory*, 18(3):289–305, 1984.
79. Eugène Ehrhart. Sur les polyèdres rationnels homothétiques à n dimensions. *C. R. Acad. Sci. Paris*, 254:616–618, 1962.
80. Eugène Ehrhart. Sur les carrés magiques. *C. R. Acad. Sci. Paris Sér. A-B*, 277:A651–A654, 1973.
81. Eugène Ehrhart. *Polynômes arithmétiques et méthode des polyèdres en combinatoire*. Birkhäuser Verlag, Basel, 1977. International Series of Numerical Mathematics, Vol. 35.
82. Günter Ewald. *Combinatorial Convexity and Algebraic Geometry*, volume 168 of *Graduate Texts in Mathematics*. Springer-Verlag, New York, 1996.
83. Leonid G. Fel and Boris Y. Rubinstein. Restricted partition function as Bernoulli and Euler polynomials of higher order. *Ramanujan J.*, 11(3):331–348, 2006. `arXiv:math.NT/0304356`.
84. William Fulton. *Introduction to Toric Varieties*, volume 131 of *Annals of Mathematics Studies*. Princeton University Press, Princeton, NJ, 1993.
85. Stavros Garoufalidis and James E. Pommersheim. Values of zeta functions at negative integers, Dedekind sums and toric geometry. *J. Amer. Math. Soc.*, 14(1):1–23 (electronic), 2001.
86. Ewgenij Gawrilow and Michael Joswig. polymake: a framework for analyzing convex polytopes. In *Polytopes—combinatorics and computation (Oberwolfach, 1997)*, volume 29 of *DMV Sem.*, pages 43–73. Birkhäuser, Basel, 2000. Software `polymake` available at `www.math.tu-berlin.de/polymake/`.
87. Ira M. Gessel. Generating functions and generalized Dedekind sums. *Electron. J. Combin.*, 4(2):Research Paper 11, approx. 17 pp. (electronic), 1997.
88. Jorgen P. Gram. Om rumvinklerne i et polyeder. *Tidsskrift for Math. (Copenhagen)*, 4(3):161–163, 1874.
89. Harold Greenberg. An algorithm for a linear Diophantine equation and a problem of Frobenius. *Numer. Math.*, 34(4):349–352, 1980.
90. Branko Grünbaum. *Convex Polytopes*, volume 221 of *Graduate Texts in Mathematics*. Springer-Verlag, New York, second edition, 2003. Prepared and with a preface by Volker Kaibel, Victor Klee, and Günter M. Ziegler.
91. Victor Guillemin. Riemann-Roch for toric orbifolds. *J. Differential Geom.*, 45(1):53–73, 1997.
92. Ulrich Halbritter. Some new reciprocity formulas for generalized Dedekind sums. *Results Math.*, 8(1):21–46, 1985.
93. Richard R. Hall, J. C. Wilson, and Don Zagier. Reciprocity formulae for general Dedekind–Rademacher sums. *Acta Arith.*, 73(4):389–396, 1995.

94. Martin Henk, Achill Schürmann, and Jörg M. Wills. Ehrhart polynomial and successive minima. Preprint (`arXiv:math.MG/0507528`), to appear in *Mathematika*, 2005.
95. Jürgen Herzog. Generators and relations of abelian semigroups and semigroup rings. *Manuscripta Math.*, 3:175–193, 1970.
96. Takayuki Hibi. *Algebraic Combinatorics on Convex Polytopes.* Carslaw, 1992.
97. Takayuki Hibi. Dual polytopes of rational convex polytopes. *Combinatorica*, 12(2):237–240, 1992.
98. Takayuki Hibi. A lower bound theorem for Ehrhart polynomials of convex polytopes. *Adv. Math.*, 105(2):162–165, 1994.
99. Dean Hickerson. Continued fractions and density results for Dedekind sums. *J. Reine Angew. Math.*, 290:113–116, 1977.
100. Friedrich Hirzebruch. *Neue topologische Methoden in der algebraischen Geometrie.* Ergebnisse der Mathematik und ihrer Grenzgebiete (N.F.), Heft 9. Springer-Verlag, Berlin, 1956.
101. Friedrich Hirzebruch and Don Zagier. *The Atiyah-Singer Theorem and Elementary Number Theory.* Publish or Perish Inc., Boston, Mass., 1974.
102. Jeffrey Hood and David Perkinson. Some facets of the polytope of even permutation matrices. *Linear Algebra Appl.*, 381:237–244, 2004.
103. Masa-Nori Ishida. Polyhedral Laurent series and Brion's equalities. *Internat. J. Math.*, 1(3):251–265, 1990.
104. Maruf Israilovich Israilov. Determination of the number of solutions of linear Diophantine equations and their applications in the theory of invariant cubature formulas. *Sibirsk. Mat. Zh.*, 22(2):121–136, 237, 1981.
105. Ravi Kannan. Lattice translates of a polytope and the Frobenius problem. *Combinatorica*, 12(2):161–177, 1992.
106. Jean-Michel Kantor and Askold G. Khovanskiĭ. Une application du théorème de Riemann-Roch combinatoire au polynôme d'Ehrhart des polytopes entiers de $\mathbf{R}^d$. *C. R. Acad. Sci. Paris Sér. I Math.*, 317(5):501–507, 1993.
107. Yael Karshon, Shlomo Sternberg, and Jonathan Weitsman. The Euler-Maclaurin formula for simple integral polytopes. *Proc. Natl. Acad. Sci. USA*, 100(2):426–433 (electronic), 2003.
108. Askold G. Khovanskiĭ and Aleksandr V. Pukhlikov. The Riemann-Roch theorem for integrals and sums of quasipolynomials on virtual polytopes. *Algebra i Analiz*, 4(4):188–216, 1992.
109. Peter Kirschenhofer, Attila Pethő, and Robert F. Tichy. On analytical and Diophantine properties of a family of counting polynomials. *Acta Sci. Math. (Szeged)*, 65(1-2):47–59, 1999.
110. Daniel A. Klain and Gian-Carlo Rota. *Introduction to Geometric Probability.* Lezioni Lincee. Cambridge University Press, Cambridge, 1997.
111. Victor Klee. A combinatorial analogue of Poincaré's duality theorem. *Canad. J. Math.*, 16:517–531, 1964.
112. Donald E. Knuth. Permutations, matrices, and generalized Young tableaux. *Pacific J. Math.*, 34:709–727, 1970.
113. Donald E. Knuth. Notes on generalized Dedekind sums. *Acta Arith.*, 33(4):297–325, 1977.
114. Donald E. Knuth. *The Art of Computer Programming. Vol. 2.* Addison-Wesley Publishing Co., Reading, Mass., second edition, 1981.

115. Matthias Köppe. A primal Barvinok algorithm based on irrational decompositions. Preprint (arXiv:math.CO/0603308), 2006. Software LattE macchiato available at http://www.math.uni-magdeburg.de/~mkoeppe/latte/.
116. Thomas W. Körner. *Fourier Analysis.* Cambridge University Press, Cambridge, 1988.
117. Maximilian Kreuzer and Harald Skarke. Classification of reflexive polyhedra in three dimensions. *Adv. Theor. Math. Phys.*, 2(4):853–871, 1998. arXiv:hep-th/9805190.
118. Maximilian Kreuzer and Harald Skarke. Complete classification of reflexive polyhedra in four dimensions. *Adv. Theor. Math. Phys.*, 4(6):1209–1230, 2000. arXiv:hep-th/0002240.
119. Jean B. Lasserre and Eduardo S. Zeron. On counting integral points in a convex rational polytope. *Math. Oper. Res.*, 28(4):853–870, 2003.
120. Jim Lawrence. A short proof of Euler's relation for convex polytopes. *Canad. Math. Bull.*, 40(4):471–474, 1997.
121. Arjen K. Lenstra, Hendrik W. Lenstra, Jr., and László Lovász. Factoring polynomials with rational coefficients. *Math. Ann.*, 261(4):515–534, 1982.
122. László Lovász. *Combinatorial Problems and Exercises.* North-Holland Publishing Co., Amsterdam, second edition, 1993.
123. Ian G. Macdonald. Polynomials associated with finite cell-complexes. *J. London Math. Soc. (2)*, 4:181–192, 1971.
124. Percy A. MacMahon. *Combinatory Analysis.* Chelsea Publishing Co., New York, 1960.
125. Evgeny N. Materov. The Bott formula for toric varieties. *Mosc. Math. J.*, 2(1):161–182, 2002. arXiv:math.AG/9904110.
126. Tyrrell B. McAllister and Kevin M. Woods. The minimum period of the Ehrhart quasi-polynomial of a rational polytope. *J. Combin. Theory Ser. A*, 109(2):345–352, 2005. arXiv:math.CO/0310255.
127. Peter McMullen. Valuations and Euler-type relations on certain classes of convex polytopes. *Proc. London Math. Soc. (3)*, 35(1):113–135, 1977.
128. Peter McMullen. Lattice invariant valuations on rational polytopes. *Arch. Math. (Basel)*, 31(5):509–516, 1978/79.
129. Curt Meyer. Über einige Anwendungen Dedekindscher Summen. *J. Reine Angew. Math.*, 198:143–203, 1957.
130. Jeffrey L. Meyer. Character analogues of Dedekind sums and transformations of analytic Eisenstein series. *Pacific J. Math.*, 194(1):137–164, 2000.
131. Werner Meyer and Robert Sczech. Über eine topologische und zahlentheoretische Anwendung von Hirzebruchs Spitzenauflösung. *Math. Ann.*, 240(1):69–96, 1979.
132. Ezra Miller and Bernd Sturmfels. *Combinatorial Commutative Algebra*, volume 227 of *Graduate Texts in Mathematics.* Springer-Verlag, New York, 2005.
133. Hermann Minkowski. *Geometrie der Zahlen.* Bibliotheca Mathematica Teubneriana, Band 40. Johnson Reprint Corp., New York, 1968.
134. Marcel Morales. Syzygies of monomial curves and a linear Diophantine problem of Frobenius. Preprint, Max-Planck-Institut für Mathematik, Bonn, 1986.
135. Marcel Morales. Noetherian symbolic blow-ups. *J. Algebra*, 140(1):12–25, 1991.
136. Louis J. Mordell. Lattice points in a tetrahedron and generalized Dedekind sums. *J. Indian Math. Soc. (N.S.)*, 15:41–46, 1951.
137. Robert Morelli. Pick's theorem and the Todd class of a toric variety. *Adv. Math.*, 100(2):183–231, 1993.

138. Gerald Myerson. On semiregular finite continued fractions. *Arch. Math. (Basel)*, 48(5):420–425, 1987.
139. Walter Nef. Zur Einführung der Eulerschen Charakteristik. *Monatsh. Math.*, 92(1):41–46, 1981.
140. C. D. Olds, Anneli Lax, and Giuliana P. Davidoff. *The Geometry of Numbers*, volume 41 of *Anneli Lax New Mathematical Library*. Mathematical Association of America, Washington, DC, 2000.
141. Paul C. Pasles. The lost squares of Dr. Franklin: Ben Franklin's missing squares and the secret of the magic circle. *Amer. Math. Monthly*, 108(6):489–511, 2001.
142. Micha A. Perles and Geoffrey C. Shephard. Angle sums of convex polytopes. *Math. Scand.*, 21:199–218 (1969), 1967.
143. Georg Alexander Pick. Geometrisches zur Zahlenlehre. *Sitzenber. Lotos (Prague)*, 19:311–319, 1899.
144. Clifford A. Pickover. *The Zen of Magic Squares, Circles, and Stars*. Princeton University Press, Princeton, NJ, 2002.
145. Christopher Polis. Pick's theorem extended and generalized. *Math. Teacher*, pages 399–401, 1991.
146. James E. Pommersheim. Toric varieties, lattice points and Dedekind sums. *Math. Ann.*, 295(1):1–24, 1993.
147. Bjorn Poonen and Fernando Rodriguez-Villegas. Lattice polygons and the number 12. *Amer. Math. Monthly*, 107(3):238–250, 2000.
148. Tiberiu Popoviciu. Asupra unei probleme de patitie a numerelor. *Acad. Republicii Populare Romane, Filiala Cluj, Studii si cercetari stiintifice*, 4:7–58, 1953.
149. Hans Rademacher. Generalization of the reciprocity formula for Dedekind sums. *Duke Math. J.*, 21:391–397, 1954.
150. Hans Rademacher. Some remarks on certain generalized Dedekind sums. *Acta Arith.*, 9:97–105, 1964.
151. Hans Rademacher and Emil Grosswald. *Dedekind Sums*. The Mathematical Association of America, Washington, D.C., 1972.
152. Jorge L. Ramírez-Alfonsín. Complexity of the Frobenius problem. *Combinatorica*, 16(1):143–147, 1996.
153. Jorge L. Ramírez-Alfonsín. *The Diophantine Frobenius Problem*. Oxford University Press, 2006.
154. J. E. Reeve. On the volume of lattice polyhedra. *Proc. london Math. Soc. (3)*, 7:378–395, 1957.
155. Les Reid and Leslie G. Roberts. Monomial subrings in arbitrary dimension. *J. Algebra*, 236(2):703–730, 2001.
156. Jürgen Richter-Gebert. *Realization Spaces of Polytopes*, volume 1643 of *Lecture Notes in Mathematics*. Springer-Verlag, Berlin Heidelberg, 1996.
157. Boris Y. Rubinstein. An explicit formula for restricted partition function through Bernoulli polynomials. Preprint, to appear in *Ramanujan Journal*, 2005.
158. Ludwig Schläfli. Theorie der vielfachen Kontinuität. In *Ludwig Schläfli, 1814–1895, Gesammelte Mathematische Abhandlungen, Vol. I*, pages 167–387. Birkhäuser, Basel, 1950.
159. Alexander Schrijver. *Combinatorial Optimization. Polyhedra and Efficiency. Vol. A–C*, volume 24 of *Algorithms and Combinatorics*. Springer-Verlag, Berlin, 2003.

160. Paul R. Scott. On convex lattice polygons. *Bull. Austral. Math. Soc.*, 15(3):395–399, 1976.
161. Sinan Sertöz. On the number of solutions of the Diophantine equation of Frobenius. *Diskret. Mat.*, 10(2):62–71, 1998.
162. Geoffrey C. Shephard. An elementary proof of Gram's theorem for convex polytopes. *Canad. J. Math.*, 19:1214–1217, 1967.
163. Carl Ludwig Siegel. *Lectures on the Geometry of Numbers.* Springer-Verlag, Berlin, 1989. Notes by B. Friedman, rewritten by Komaravolu Chandrasekharan with the assistance of Rudolf Suter, with a preface by Chandrasekharan.
164. R. Jamie Simpson and Robert Tijdeman. Multi-dimensional versions of a theorem of Fine and Wilf and a formula of Sylvester. *Proc. Amer. Math. Soc.*, 131(6):1661–1671 (electronic), 2003.
165. Neil J. A. Sloane. On-line encyclopedia of integer sequences. `http://www.research.att.com/~njas/sequences/index.html`.
166. David Solomon. Algebraic properties of Shintani's generating functions: Dedekind sums and cocycles on $\mathrm{PGL}_2(\mathbf{Q})$. *Compositio Math.*, 112(3):333–362, 1998.
167. Duncan M. Y. Sommerville. The relation connecting the angle-sums and volume of a polytope in space of n dimensions. *Proc. Roy. Soc. London, Ser. A*, 115:103–119, 1927.
168. Richard P. Stanley. Linear homogeneous Diophantine equations and magic labelings of graphs. *Duke Math. J.*, 40:607–632, 1973.
169. Richard P. Stanley. Combinatorial reciprocity theorems. *Advances in Math.*, 14:194–253, 1974.
170. Richard P. Stanley. Decompositions of rational convex polytopes. *Ann. Discrete Math.*, 6:333–342, 1980.
171. Richard P. Stanley. *Combinatorics and Commutative Algebra*, volume 41 of *Progress in Mathematics.* Birkhäuser Boston Inc., Boston, MA, second edition, 1996.
172. Richard P. Stanley. *Enumerative Combinatorics. Vol. 1*, volume 49 of *Cambridge Studies in Advanced Mathematics.* Cambridge University Press, Cambridge, 1997. With a foreword by Gian-Carlo Rota, Corrected reprint of the 1986 original.
173. Bernd Sturmfels. On vector partition functions. *J. Combin. Theory Ser. A*, 72(2):302–309, 1995.
174. Bernd Sturmfels. *Gröbner Bases and Convex Polytopes*, volume 8 of *University Lecture Series.* American Mathematical Society, Providence, RI, 1996.
175. James J. Sylvester. On the partition of numbers. *Quaterly J. Math.*, 1:141–152, 1857.
176. James J. Sylvester. On subinvariants, i.e. semi-invariants to binary quanties of an unlimited order. *Amer. J. Math.*, 5:119–136, 1882.
177. James J. Sylvester. Mathematical questions with their solutions. *Educational Times*, 41:171–178, 1884.
178. András Szenes and Michèle Vergne. Residue formulae for vector partitions and Euler-MacLaurin sums. *Adv. in Appl. Math.*, 30(1-2):295–342, 2003. `arXiv:math.CO/0202253`.
179. Lajos Takács. On generalized Dedekind sums. *J. Number Theory*, 11(2):264–272, 1979.

180. Audrey Terras. *Fourier Analysis on Finite Groups and Applications*, volume 43 of *London Mathematical Society Student Texts*. Cambridge University Press, Cambridge, 1999.
181. John A. Todd. The geometrical invariants of algebraic loci. *Proc. London Math. Soc.*, 43:127–141, 1937.
182. John A. Todd. The geometrical invariants of algebraic loci (second paper). *Proc. London Math. Soc.*, 45:410–424, 1939.
183. Amitabha Tripathi. The number of solutions to $ax+by=n$. *Fibonacci Quart.*, 38(4):290–293, 2000.
184. Helge Tverberg. How to cut a convex polytope into simplices. *Geometriae Dedicata*, 3:239–240, 1974.
185. Sven Verdoolaege. Software package `barvinok`. 2004. Electronically available at `http://freshmeat.net/projects/barvinok/`.
186. John von Neumann. A certain zero-sum two-person game equivalent to the optimal assignment problem. In *Contributions to the Theory of Games, vol. 2*, Annals of Mathematics Studies, no. 28, pages 5–12. Princeton University Press, Princeton, N. J., 1953.
187. Herbert S. Wilf. *generatingfunctionology*. Academic Press Inc., Boston, MA, second edition, 1994. Electronically available at `http://www.cis.upenn.edu/~wilf/`.
188. Kevin Woods. Computing the period of an Ehrhart quasi-polynomial. *Electron. J. Combin.*, 12(1):Research Paper 34, 12 pp. (electronic), 2005.
189. Guoce Xin. Constructing all magic squares of order three. Preprint (`arXiv: math.CO/0409468`), 2005.
190. Don Zagier. Higher dimensional Dedekind sums. *Math. Ann.*, 202:149–172, 1973.
191. Claudia Zaslavsky. *Africa Counts*. Prindle, Weber & Schmidt, Inc., Boston, Mass., 1973.
192. Doron Zeilberger. Proof of a conjecture of Chan, Robbins, and Yuen. *Electron. Trans. Numer. Anal.*, 9:147–148 (electronic), 1999.
193. Günter M. Ziegler. *Lectures on polytopes*. Springer-Verlag, New York, 1995. Revised edition, 1998; "Updates, corrections, and more" at `www.math.tu-berlin.de/~ziegler`.

List of Symbols

The following table contains a list of symbols that are frequently used throughout the book. The page numbers refer to the first appearance/definition of each symbol.

Symbol	Meaning	Page
$\hat{a}(m)$	Fourier coefficient of $a(n)$	126
$A(d,k)$	Eulerian number	28
$\mathcal{A}^{\perp}$	orthogonal complement of $\mathcal{A}$	159
$A_{\mathcal{P}}(t)$	solid-angle sum of $\mathcal{P}$	180
$\alpha_{\mathcal{P}}(\mathbf{z})$	solid-angle generating function	182
$B_k(x)$	Bernoulli polynomial	31
B_k	Bernoulli number	32
$\mathbf{B}_n$	Birkhoff polytope	108
$\operatorname{BiPyr}(\mathcal{P})$	bipyramid over $\mathcal{P}$	36
$\operatorname{cone}\mathcal{P}$	cone over $\mathcal{P}$	58
$\operatorname{const} f$	constant term of the generating function f	13
$\operatorname{conv} S$	convex hull of S	25
d-cone	d-dimensional cone	58
d-polytope	d-dimensional polytope	26
$\dim\mathcal{P}$	dimension of $\mathcal{P}$	26
$\delta_m(x)$	delta function	129
$\operatorname{Ehr}_{\mathcal{P}}(z)$	Ehrhart series of $\mathcal{P}$	28
$\operatorname{Ehr}_{\mathcal{P}^\circ}(z)$	Ehrhart series of the interior of $\mathcal{P}$	87
$\mathbf{e}_a(x)$	root-of-unity function $e^{2\pi i a x/b}$	129
f_k	face number	95
$F_k(t)$	lattice-point enumerator of the k-skeleton	96
$\mathbf{F}(f)$	Fourier transform of f	129
$g(a_1, a_2, \dots, a_d)$	Frobenius number	6
$H_n(t)$	number of semimagic $n \times n$ squares with line sum t	107
$\mathcal{K}_{\mathcal{F}}$	tangent cone of $\mathcal{F} \subseteq \mathcal{P}$	159

Symbol	Meaning	Page
$L_{\mathcal{P}}(t)$	lattice-point enumerator of $\mathcal{P}$	27
$L_{\mathcal{P}^\circ}(t)$	lattice-point enumerator of the interior of $\mathcal{P}$	28
$M_n(t)$	number of magic $n \times n$ squares with line sum t	107
$\omega_{\mathcal{P}}(\mathbf{x})$	solid angle of $\mathbf{x}$ (with respect to $\mathcal{P}$)	179
$p_A(n)$	restricted partition function	6
$\operatorname{poly}_A(n)$	polynomial part of $p_A(n)$	141
$\mathcal{P}$	a closed polytope	25
$\mathcal{P}^\circ$	interior of the polytope $\mathcal{P}$	28
$\mathcal{P}(\mathbf{h})$	perturbed polytope	174
$\operatorname{Pyr}(\mathcal{P})$	pyramid over $\mathcal{P}$	34
$\wp(z)$	Weierstraß $\wp$-function	193
Π	fundamental parallelepiped of a cone	62
$r_n(a,b)$	Dedekind–Rademacher sum	145
$s(a,b)$	Dedekind sum	128
$s_n(a_1, a_2, \dots, a_m; b)$	Fourier–Dedekind sum	14
$\operatorname{Solid}_{\mathcal{P}}(x)$	solid-angle series	188
$\operatorname{span} \mathcal{P}$	affine space spanned by $\mathcal{P}$	26
$\sigma_S(\mathbf{z})$	integer-point transform of S	60
$t\mathcal{P}$	t^{th} dilation of $\mathcal{P}$	27
Todd_h	Todd operator	168
$\operatorname{vol} \mathcal{P}$	(continuous) volume of $\mathcal{P}$	71
V_G	vector space of all complex-valued functions on $G = \{0, 1, 2, \dots, b-1\}$	129
ξ_a	root of unity $e^{2\pi i/a}$	8
$\zeta(z)$	Weierstraß ζ-function	193
$\lfloor x \rfloor$	greatest integer function	10
$\{x\}$	fraction-part function	10
$((x))$	sawtooth function	127
$\binom{m}{n}$	binomial coefficient	27
$\langle f, g \rangle$	inner product of f and g	130
$(f * g)(t)$	convolution of f and g	133
$1_S(\mathbf{x})$	characteristic function of S	160
$\#S$	number of elements in S	6
♣	an exercise that is used in the text	VIII

Index

Undergraduate Texts in Mathematics *(continued from p.ii)*

Printed in the United States of America